This volume provides a comprehensive overview of the use of ultraviolet laser radiation in the processing of materials. It is the first book to cover all the major up-to-date applications of UV lasers in a unified approach.

The development of lasers operating at ultraviolet wavelengths has provided materials scientists and engineers with a new set of tools. These combine the ability to vaporize the most refractory of materials with the precision to ablate micrometer-sized holes in polymers and the ability to remove thin layers from the cornea for correction of refractive errors in the human eye. This book outlines these applications and explores the use of UV laser radiation for the ablation and deposition of metals, insulating solids, polymers, semiconductors and superconductors. Emphasis has been placed on understanding the physical mechanisms accompanying these processes and the conversion of intense UV radiation to photo-thermal and photochemical energy in irradiated materials. An extensive bibliography has been included.

UV Lasers: effects and applications in materials science will be an invaluable source of current information in the rapidly developing field of laser applications for engineers, scientists, researchers and students in universities, government laboratories and the private sector. This book will also be valuable as a supplementary text for graduate courses in materials science.

UV Lasers: effects and applications in materials science

UV Lasers:
effects and applications in
materials science

W. W. DULEY
University of Waterloo, Ontario

CAMBRIDGE UNIVERSITY PRESS
Cambridge, New York, Melbourne, Madrid, Cape Town, Singapore, São Paulo

Cambridge University Press
The Edinburgh Building, Cambridge CB2 2RU, UK

Published in the United States of America by Cambridge University Press, New York

www.cambridge.org
Information on this title: www.cambridge.org/9780521464987

First published 1996
This digitally printed first paperback version 2005

A catalogue record for this publication is available from the British Library

Library of Congress Cataloguing in Publication data
Duley, W. W.
UV lasers: effects and applications in materials science / W.W. Duley.
p.   cm.
Includes bibliographical references.
ISBN 0 521 46498 6
1. Gas lasers – Industrial applications.   2. Ultraviolet radiation –
Industrial applications.   3. Materials – Effects of radiation on.
I. Title.
TA1695.D845   1996
620.1´1228 – dc20   96-13553 CIP

ISBN-13 978-0-521-46498-7 hardback
ISBN-10 0-521-46498-6 hardback

ISBN-13 978-0-521-02006-0 paperback
ISBN-10 0-521-02006-9 paperback

Contents

Preface

The development of lasers operating at ultraviolet wavelengths has provided mankind with a new set of unique tools. With characteristics which combine the precision to remove micrometer-thick layers of corneal tissue for the correction of refractive errors in the human eye and the ability to vaporize even the most refractory of materials, UV lasers have immediately developed into indispensable tools in many areas of materials science. The remarkable ability of high power pulsed excimer laser radiation to vaporize complex materials such as high temperature superconductors, while maintaining stoichiometry in thin films deposited from this vaporized material, offers many exciting opportunities in the creation of superconducting thin films and thin film devices. Similar unique capabilities are available in the deposition, doping and modification of semiconductors using UV laser radiation.

As a result of these and other applications, many of which can be immediately adopted by industry, UV lasers have a secure future in the field of materials science. Their implementation is limited only by our creativity in finding new applications and ways to use these new tools.

A fascinating aspect of the development of these applications involves the many fundamental questions that arise concerning the manner in which intense UV laser radiation interacts with matter. This is an area of great scientific interest and is truly interdisciplinary in nature so that answers to these questions will only come from both theoretical and experimental studies extending over a diverse range of disciplines.

In writing this book I have adopted the thesis that new applications, together with the refinement of present applications, of UV lasers will come only as we gain a deeper understanding of the mechanisms involved in the interaction of intense UV laser radiation with matter. My approach has therefore been to focus on these mechanisms and their relation to applications of UV laser radiation in materials science. Since interaction mechanisms tend to be most closely related in materials with similar properties, my discussion is then organized in this way with chapters on UV laser processing of metals, insulators, semiconductors and superconductors. The important field of laser deposition is discussed in a separate chapter. Sources of UV laser radiation together with ancillary components and systems are discussed in the first chapter. Chapter 2 provides a summary of the properties of materials at UV wavelengths, whereas Chapter 3 discusses the limitations of laser processing in a more general way. A comprehensive, but by no means exhaustive, bibliography is also provided to assist the reader in gaining access to the burgeoning literature on this subject.

Acknowledgments

My thanks go to Teresa Glaves who assisted with the production of the manuscript and who cheerfully met all the deadlines. Thanks also to Dr Hamid Jahani for preparation of a number of the figures and to Monica Kinsman for typing parts of the manuscript. My thanks also go to students, past and present, who contributed their ideas and insight, who took the courses, and who have been the source of constant inspiration. Special thanks in this regard to Drs K. Dunphy, G. Kinsman, S. Mihailov and M. Ogmen and to Y-L. Mao. I am also indebted to all those who freely granted permission to quote and reproduce various parts of their work and to Professor W. Steen who provided the opportunity to visit the University of Liverpool, where a major part of this book was written. Last, but not least, I thank my wife, Irmgardt, for constant support and for the challenging comment: 'What will you do with all your spare time now that the book is finished?'

Walter W. Duley
Waterloo

CHAPTER 1

Short wavelength lasers

With the demonstration of laser emission in ruby at 694.3 nm by
Maiman (1960) attention rapidly shifted to obtaining stimulated emission
at other wavelengths. Many other materials were found to exhibit laser
emission at visible, near infrared and infrared wavelengths but extension
of laser emission to the ultraviolet (UV) and vacuum ultraviolet (VUV)
remained elusive. Potential difficulties in achieving the conditions for
laser emission in the UV and VUV had been pointed out at a much
earlier date by Schawlow and Townes (1958) in their seminal work on
optical masers. The origin of these problems can be seen through an
analysis of the simple optically pumped three-level system shown in
Figure 1.1. If the rate of optical pumping of level 3 is R and all atoms
excited to level 3 are assumed to relax to level 2, then the rate equations
for the populations N_1 and N_2 in levels 1 and 2, respectively, are

$$\frac{dN_1}{dt} = -RN_1 + \gamma_{21}N_2 + \sigma F_{21}(N_2 - N_1) \tag{1}$$

$$\frac{dN_2}{dt} = RN_1 - \gamma_{21}N_2 - \sigma F_{21}(N_2 - N_1) \tag{2}$$

with

$$\frac{dN_1}{dt} = -\frac{dN_2}{dt} \tag{3}$$

where γ_{21} is the total relaxation rate from level 2 to level 1, σF_{21} is the
rate of stimulated transitions between these states, σ is a cross-section
for stimulated emission, and F_{21} is the photon flux (number of photons
$m^{-2} s^{-1}$) at frequency ν_{21}.

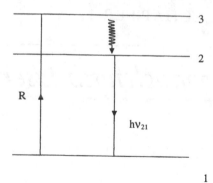

Figure 1.1. A three-level laser system with the lasing transition
occurring between levels 2 and 1.

In the steady state approximation and with F_{21} small (i.e. for the system near the threshold for stimulated emission) $dN_1/dt = dN_2/dt = 0$ and

$$\frac{N_2}{N_1} = \frac{R}{\gamma_{21}} \tag{4}$$

with $N_1 + N_2 = N$, the total number of atoms m^{-3}. We also have

$$N_2 - N_1 = \frac{R - \gamma_{21}}{R + \gamma_{21}} N \tag{5}$$

$$N_1 = \frac{\gamma_{21}}{R + \gamma_{21}} N \tag{6}$$

so that $R > \gamma_{21}$ is a requirement for obtaining a population inversion between levels 1 and 2. The rate of energy input to the system is

$$P = h\nu_{13} R N_1 \tag{7}$$

$$= \frac{h\nu_{13} R \gamma_{21} N}{R + \gamma_{21}} \ (\text{W m}^{-3}) \tag{8}$$

which reduces to

$$P \simeq \frac{h\nu_{13}\gamma_{21}}{2} N \tag{9}$$

for $R \sim \gamma_{21}$, the threshold for gain. γ_{21} would in general include several terms but, in the simplest case, is just A_{21}, the rate of spontaneous radiative decay. Since $A_{21} \propto \nu_{21}^3$

$$P \propto \nu_{13}\nu_{21}^3 \tag{10}$$

which shows that the pump power needed to achieve optical gain between two states separated by a UV transition will be orders of magnitude higher than that necessary to obtain gain at visible or IR frequencies.

Laser emission in the near UV was first reported by Heard (1963) and was obtained from N_2 gas excited with a high voltage, sub-microsecond pulse. Emission was observed at a number of wavelengths between 300 and 400 nm, but the strongest line was at 337.1 nm. The laser pulse had a duration of 20 ns. The N_2 laser has since developed into a simple and reliable source of pulsed near UV laser radiation, used primarily as an excitation source in luminescence studies and as a low cost pump laser for tunable dye lasers.

The demonstration of laser emission from N_2 at 337.1 nm led to theoretical (Bazhulin *et al.* 1965, Ali and Kolb 1965) and experimental (Hodgson 1970, 1971, Waynant *et al.* 1970) work on emission from other gases at shorter wavelengths. These studies led to the observation of stimulated emission near 160 nm in H_2 gas (Hodgson 1970, Waynant *et al.* 1970) and at 180–190 nm in CO (Hodgson 1971). In subsequent studies, stimulated emission in H_2 was extended to lines of the Werner system at 116.1 and 123 nm (Waynant 1972, Hodgson and Dreyfus 1972) although with low efficiency. Pulsed coherent radiation in the wavelength range $\lambda < 160$ nm is now routinely available using frequency mixing and harmonic generation in non-linear media (Reintjes 1980) or by Raman shifting techniques (Loree *et al.* 1977, Wallmeier and Zacharias 1988).

Pulsed laser emission at various near UV wavelengths in ion lasers was reported as early as 1965 (Bridges and Chester 1965, Cheo and Cooper 1965). Extension to VUV wavelengths was later obtained by Marling and Lang (1977) whereas CW operation of certain of these lasers to yield laser emission at near UV wavelengths was reported in 1971 (Bridges *et al.* 1971).

The formation of bound excited states in diatomic molecules with unstable or weakly bound ground states was soon recognized to be an efficient way to obtain population inversions and stimulated emission at short wavelengths (Basov *et al.* 1968, Jortner *et al.* 1965). Laser emission at 172.2 nm was soon reported from Xe_2 molecules in a high pressure

Xe gas excited with a relativistic electron beam by Koehler *et al.* (1972), Hoff *et al.* (1973), Gerardo and Johnson (1973) and Ault *et al.* (1973). Laser emission was also obtained from other inert gas dimers (Hughes *et al.* 1974, Koehler *et al.* 1975). However, the high operating pressures of these devices (up to 50 atm), together with the requirement for electron beam pumping, have limited the development of this type of laser.

Spectroscopic measurements of the emission from excited states of XeBr and XeCl molecules, formed by reaction of metastable Xe atoms and halogens, showed that excimer emission from molecules of this type could lead to stimulated emission at UV wavelengths (Velazco and Setser 1975). Laser emission at 281.8 nm was soon thereafter reported from XeBr (Searles and Hart 1975) and at 255 nm from XeI (Ewing and Brau 1975a) in electron beam pumped systems. Subsequent experiments showed that efficient excimer laser emission could be obtained at 193 nm from ArF (Hoffman *et al.* 1976), at 248 nm from KrF (Ewing and Brau 1975b), at 308 nm from XeCl (Ewing and Brau 1975b) and at 351 nm from XeF (Brau and Ewing 1975). Emission from F_2 at 157.5 nm was also reported by Rice *et al.* (1977). Electron beam pumping was used as an excitation source in these first experiments, but has now been replaced by electric discharge pumping (Burnham *et al.* 1976a,b). Today, the family of discharge lasers based on rare gas halide excimers is the dominant source for laser materials processing applications at UV wavelengths.

1.1 ULTRAVIOLET LASERS

Ultraviolet laser emission can be excited from numerous atoms, ions and molecules in pulsed and CW gaseous discharges. A wealth of emission wavelengths is available from laboratory devices (see Beck *et al.* (1978), and Waynant and Ediger (1993) for a list of wavelengths) but high power emission suitable for materials processing is available only from a limited subset of these systems. A list of commercial UV laser types and their general characteristics is given in Table 1.1.

High power CW UV laser radiation is available only from argon and krypton ion lasers and from harmonics of the 1.06 μm Nd:YAG laser. CW dye laser radiation may be doubled to wavelengths in the 250–350 nm range although conversion efficiencies and resulting output powers are small. The free electron laser (FEL) which is, in principle,

Table 1.1. *Commercial UV lasers (compiled in part using data from Laser Focus World Buyers' Guide, 1994 Edition).*

Types of CW lasers	Wavelength (nm)	Power (W)	Linewidth	Note
Ar^+	229–264	0.1		TEM_{00}
	275–306	1.5		Multimode
	300–336	2.4		Multimode
	333–364	7		TEM_{00}
	351–364	0.1		$M^2 \leq 11.1$
Dye	220–390	0.01	1 MHz	Doubled
He–Cd	325	0.10		Multimode
		0.05		TEM_{00}
Kr^+	338–356	0.5		TEM_{00}
Nd:YAG	266, 355	2		TEM_{00}
				3 kHz, Q-switched, quadrupled, tripled
Ti:sapphire	365–410	0.03		Doubled

Types of pulsed lasers	Wavelength (nm)	Energy/pulse (J)	Pulse length (ns)	Repetition rate (Hz)	Notes
Alexandrite	200–360	0.005–0.1	10–50	30	Flashlamp pumped
ArF	193	≤ 0.6	10–50	Up to 500	
Cu	255, 271, 289	2×10^{-4}	20	5000	
Dye	200–400	0.01–0.1	3.50	$\leq 10\,000$	Excimer or YAG pumped
	260–400	1	4000	0.1	Flashlamp pumped
F_2	157	0.05	10	100	
KrCl	222	≤ 0.2	10–50	Up to 500	
KrF	248	≤ 1.5	10–50	Up to 500	
N_2	337	0.01	3–10	50	
Nd:glass	263, 266, 351, 355	0.5–3	20–25	0.01–0.33	Quadrupled/tripled
Nd:YAG	266	0.05	5–20	10–50	Quadrupled/multimode
	355	0.05–0.5	5–20	10–50	Tripled multimode/TEM_{00}
Ti:sapphire	210–225	0.002	10	10	Tunable
	240–300	0.005	10	10	Tunable
	345–450	0.025	10	10	Tunable
	360–450		0.2–2 ps	92 MHz	Ar^+ pumped
XeCl	308	≤ 2	10–50	≤ 500	
XeF	351, 353	≤ 0.8	10–50	≤ 500	

tunable from soft X-ray to visible wavelengths may, in the future, be an alternative source of CW laser radiation for materials processing. However, at present, these devices are not available as commercial units. The projected performance of FELs in the UV and soft X-ray region has been discussed by Newman (1988).

At this time excimer lasers are the primary source for high peak and average power at UV wavelengths and are available as reliable industrially rated products from a variety of suppliers. Proven industrial applications have been developed for these lasers which make use of the unique combination of short wavelength, high intensity and large output beam area available from these devices.

High peak power and high repetition rates are available from harmonics of the Nd:YAG laser at 266 and 355 nm. However, the output of these devices is in the form of a relatively low order multimode or TEM_{00} mode beam, which is not as suitable for applications involving irradiation over extended areas or imaging of masks as are the low coherence beams from excimer lasers.

Short wavelength tunable laser radiation is also available from harmonics of the $Ti:Al_2O_3$ laser. However, pulse energies are small (Table 1.1) and the output occurs in a low order spatial mode, making these devices uncompetitive with excimer and YAG lasers.

Perhaps the greatest potential for the development of a solid-state replacement for the excimer laser lies in harmonic generation from high output semi-diode arrays based on GaAlAs emitters in the 780–820 nm range. Quasi-CW operation at power levels of 10^3 W at these wavelengths has now been reported (Groussin *et al.* 1993) with the primary limiting factor being diode lifetime. As this technology evolves, phased arrays of short wavelength harmonics of these devices may yield large area steerable beams at UV wavelengths, which will provide compact, portable sources for materials processing applications. Other laser diode materials such as AlGaInP have outputs at even shorter wavelengths (630–645 nm) (Serreze and Chen 1993).

1.2 RARE GAS HALIDE LASERS

The family of lasers based on exciplexes between excited rare gas atoms and halogen atoms provides a wide range of output wavelengths extending from 193 to 351 nm. Exciplexes are created in collisions between positive ions of the rare gas and a negative ion of the halogen or

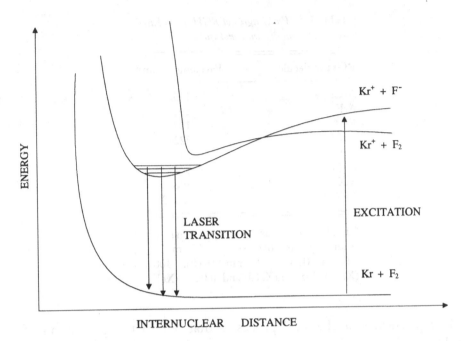

Figure 1.2. Electronic states in excimer laser emission.

between metastable rare gas atoms and halogen molecules. The resulting bound electronic state (Figure 1.2) has a deep minimum and a dissociation energy that may be as large as 4–5 eV. The lifetime of this electronic state with respect to UV radiative emission, which involves a transition to an unbound ground state of the molecule, is typically 10–30 ns. Dissociation of the ground state occurs in one vibrational period ($\lesssim 1$ ps). Thus the ground state of the emitter has essentially zero population. Studies of the radiative relaxation of rare gas halide (RGH) molecules by Velazco and Setser (1975) and by Golde and Thrush (1974) confirmed this model and led to the suggestion that RGH molecules might be the source of efficient laser emission at UV wavelengths.

Subsequent experiments using electron beam pumping led to the observation of laser emission from XeF (Brau and Ewing 1975), XeCl and KrF (Ewing and Brau 1975b), XeBr (Searles and Hart 1975) and ArF (Hoffman *et al.* 1976). Laser emission from XeF, KrF and ArF excited in an electric discharge was subsequently reported by Burnham *et al.* (1976a,b), Burnham and Djeu (1976) and Wang *et al.* (1976). These experiments formed the basis for the development of commercial RGH lasers based on a variety of molecules (Table 1.2).

Table 1.2. *Wavelengths of RGH lasers based on diatomic molecules.*

RGH molecule	Wavelength (nm)[a]
ArCl	175
ArF	193
F_2	157
KrCl	222
KrF	248
XeBr	282
XeCl	308
XeF	351, 353

[a] Highest gain transition. Lower gain transitions are also observed at 275 nm (KrF), 240 nm (KrCl), 275 nm (KrF), 300 nm (XeBr), 345 nm (XeCl) and 460 nm (XeF).

The physical and chemical processes that led to the generation of RGH excimers are complex and have been discussed in some detail in several excellent reviews (Shaw 1979, Brau 1984, Hutchinson 1980, Kannari *et al.* 1985, Obara and Kannari 1991, Smiley 1990). To illustrate this complexity in XeCl, the basic processes in a discharge pumped system are the following.

Secondary electron production

$$e + Xe \rightarrow Xe^+ + e + e \tag{11}$$

$$e + Ne \rightarrow Ne^+ + e + e \tag{12}$$

Production of metastable species (Xe^*, Ne^*)

$$e + Xe \rightarrow Xe^* + e \tag{13}$$

$$e + Ne \rightarrow Ne^* + e \tag{14}$$

Halogen ion production

$$e + HCl \rightarrow H + Cl^- \tag{15}$$

$$e + Cl_2 \rightarrow Cl^- + Cl \tag{16}$$

XeCl* production

$$Xe^* + HCl(v) \rightarrow XeCl^* + H \qquad (17)$$

$$Xe^+ + Cl^- \rightarrow XeCl^* \qquad (18)$$

$$Xe_2^+ + Cl^- \rightarrow XeCl^* + Xe \qquad (19)$$

$$NeXe^* + Cl \rightarrow XeCl^* + Ne \qquad (20)$$

Stimulated/spontaneous emission

$$XeCl^* + h\nu \rightarrow Xe + Cl + 2h\nu \qquad (21)$$

$$XeCl^* \rightarrow Xe + Cl + h\nu \qquad (22)$$

Collision quenching (M = Ne, Xe)

$$XeCl^* + M \rightarrow Xe + Cl + M \qquad (23)$$

$$XeCl^* + 2Xe \rightarrow Xe_2Cl^* + Xe \qquad (24)$$

$$XeCl^* + Xe + Ne \rightarrow Xe_2Cl^* + Ne \qquad (25)$$

Absorption of laser photons

$$Cl^- + h\nu \rightarrow Cl + e \qquad (26)$$

$$Xe^* + h\nu \rightarrow Xe^+ + e \qquad (27)$$

$$Xe_2^* + h\nu \rightarrow Xe_2^+ + e \qquad (28)$$

$$NeXe^* + h\nu \rightarrow Xe^+ + Ne \qquad (29)$$

$$Xe_2Cl^* + h\nu \rightarrow products \qquad (30)$$

Rate constants have been calculated or measured for many of these reactions so that comprehensive models of the kinetics of RGH discharges are available (Kannari *et al.* 1990, Johnson *et al.* 1990). The calculations of Johnson *et al.* (1990) for XeCl systems incorporate 203 chemical processes involving 42 chemical species.

Because the lower state in the laser transition is repulsive, or at best only weakly bound, emission occurs over a wider spectral bandwidth than in other gas lasers. RGH laser output is therefore not highly monochromatic. This, however, facilitates the generation of sub-picosecond pulses with very high brightness. Advantages of RGH lasers also include:

(a) low density in the active medium, which can support high intensities without self-focusing effects,

(b) that they are scalable to high active volume,

(c) operation at UV wavelengths,

(d) low spatial and temporal coherence minimizing laser speckle and fringe formation on imaging,

(e) operation at high repetition rates (up to 1 kHz) and

(f) high pulse energies available from large volume devices (10^4 J).

Problems and disadvantages with RGH lasers include:

(a) low spatial coherence prevents tight focusing,

(b) high divergence (2–10 mrad),

(c) short pulse length ($\lesssim 0.5\,\mu s$) is a particular problem when using fiber optic waveguides for beam delivery,

(d) a corrosive and hazardous gas mixture,

(e) limited gas lifetime and expensive mixture components (rare gases such as Xe, Kr and Ne) and

(f) limited lifetime of transmissive and reflective optical components, particularly those directly exposed to the lasing medium.

Limitations associated with these problems have been minimized in commercial excimer lasers, which have been engineered to yield reliable industrial products. Electrical components typically have lifetimes exceeding 10^9 pulses whereas optical components need service or replacement after 10^7–10^8 pulses. At a repetition rate of 100 Hz, 10^8 pulses accumulate after 280 h of operation. This emission can be excited from mixtures of the halogen in the form of HCl or F_2 with Ar, Kr and Xe. He plus Ne is also used as a buffer gas. A small quantity of H_2 can also be added to stabilize the chemistry of the active medium. The overall gas pressure is typically 3–4 atm, with He or Ne as the primary component (80–90%). The cost of Ne, Kr and Xe is a major factor in determining the operating cost of excimer lasers (Table 1.3).

Table 1.3. *Approximate cost per fill (USA dollars) using Ne buffer gas.*

Emitter	Cost ($)
ArF	35–45
KrF	45–55
XeCl	60–90

Table 1.4. *Maintenance schedule for a 35 W KrF laser operating at 200 Hz with a pulse energy of 125 mJ with halogen gas injection via microprocessor control (Austin et al. 1988).*

Maintenance	Accumulated pulses	Hours to maintenance	Downtime (min)
Halogen injection	1.5×10^6	2.1	
Partial gas replacement	6.0×10^6	8.3	
Total gas replacement	7.0×10^7	97	15
Window cleaning	1.5×10^8	208	20
Electrode exchange	6.0×10^8	833	240

Gas mixtures degrade over 10^6–10^7 pulses in fluoride systems (ArF and KrF) and 10^7–10^8 pulses when using chlorine (XeCl). The use of a gas regenerator in closed cycle operation will extend these lifetimes while yielding better pulse-to-pulse reproducibility. These regenerators condense chemical contaminants cryogenically and also filter out particulate matter created by corrosion of internal surfaces exposed to the laser discharge and its chemical products. A typical maintenance schedule is given in Table 1.4. Resonator optics will degrade after 10^7–10^8 pulses and must be cleaned or replaced. Commercial lasers usually have gate valves between the cavity optic and the main discharge volume to permit optics to be changed without the necessity of exposing the entire system to oxygen. This maintains the passivation of internal components and reduces the need for gas replacement after changing optics.

Passivation occurs when reactive surfaces accumulate a non-reactive layer which no longer combines rapidly with halogens. This layer is a metallic fluoride or chloride. Static and active passivation procedures are effective for chloride mixtures. In the former technique, a static mixture of HCl, H_2, He and Ne is allowed to react with internal laser surfaces over a period of 10–15 h. In active passivation, a low excitation discharge is operated in the laser until the color of the discharge

Table 1.5. *Output characteristics of commercial excimer lasers. Note that not all output characteristics are attainable in the same device or simultaneously in a single device, i.e. maximum power and maximum repetition rate.*

Characteristic	F_2	ArF	KrCl	KrF	XeCl	XeF
Wavelength (nm)	157	193	222	248	308	351, 353
Energy/pulse (J)	0.05	0.6	0.2	1.5	2.0	0.8
Average power (W)	3	60	10	160	180	70
Repetition rate			To 500 Hz			
Gas pressure			3–4 atm			
Pulse duration			Typically 10–50 ns			
Beam divergence			2–3 mrad			
Pulse-to-pulse stability			±3–6%			
Beam dimensions			10 × 30 nm			

Table 1.6. *Output characteristics of low power, high repetition rate RGH waveguide lasers (data courtesy of Potomac Photonics Inc.).*

	RGH excimer	
Characteristic	XeCl	KrF
Wavelength (nm)	308	248
Energy/pulse (µJ)	10	10
Repetition rate (kHz)	≤2	≤2
Pulse duration (ns)	125	50
Average power (mW)	15	18
Peak power (W)	80	200
Beam diameter (mm)	0.5	0.5

changes indicating that halogen species have been reduced in concentration or are no longer present.

A wide range of excimer lasers is available as commercial products extending from compact portable units to large industrial devices with average powers of several hundreds of watts. Speciality lasers which provide high repetition rates, spectrally narrowed emission or high pulse energy are also available. Output characteristics typical of many commercial excimer lasers are summarized in Table 1.5. Characteristics of lasers developed for special applications such as microlithography, high repetition rate, high brightness, high pulse energy, high average power and compactness are summarized in Tables 1.6 to 1.10, respectively. Not all of these devices have been commercialized to date.

Table 1.7. *Output characteristics of high energy, long pulse KrF lasers (data from the Institute of Fluid Mechanics, Laser–Matter Interaction Group, Marseille, France).*

Pumping	electron beam, double sided
Active volume (cm^3)	$20 \times 20 \times 100$
Output energy (J)	200
Pulse duration (ns)	400
Gas mixture (%)	$0.2 \, F_2$
	$4 \, Kr$
	$95.8 \, Ar$
Pressure (Pa)	2×10^5
Efficiency (%)	0.36

Table 1.8. *Output characteristics of high repetition rate, high average power RGH lasers (XeCl).*

Pumping	Discharge, UV/X-ray pre-ionization
Gas flow	Transverse, $\gtrsim 50 \, \mathrm{m \, s^{-1}}$
Active volume	Several liters
Repetition rate (kHz)	1–2
Energy/pulse (J)	1–10
Average power (kW)	$\gtrsim 1$
Wavelength (nm)	308

Table 1.9. *A summary of high pulse energy, low repetition rate RGH lasers (Smiley 1990).*

Laser	Wavelength (nm)	Pulse energy (kJ)	Pulse duration (µs)
XeF	353	5	2
XeCl	308	4	2
		6.5	0.65
KrF	248	5.5	2
		10–20	0.65

The vast majority of RGH excimer lasers operate in the mid-range of these characteristics, i.e. with pulse energies of 0.1–0.3 J and at repetition rates of about 200 Hz. Commercial excimer lasers operating under these conditions are highly reliable devices and offer 'turn-key' operation. Representative data showing the dependence of pulse energy on repetition rate for several RGH lasers are shown in Figure 1.3. Typically, the

Table 1.10. *Output characteristics of spectrally narrowed KrF lasers*
(Ishihara et al. 1990, Sengupta 1993).

Average power (W)	6
Repetition rate (Hz)	500
Energy/pulse (mJ)	12
Spectral bandwidth (pm)	≤ 1.3
Wavelength stability (pm)	$\leq \pm 0.25$ (active)
	$< \pm 6$ (inactive)
Wavelength tuning range (nm)	$2 + 8.2 – 248.5$
Pulse-to-pulse energy fluctuation	$<2.5\%$ (1σ)
Beam divergence (mrad)	<4 (H, V)
Beam size (H × V) (mm)	5.0×18.0 $(\pm 10\%)$
Polarization ratio (%)	>90 horizontal

pulse energy is independent of repetition rate at low repetition rate ($\lesssim 200\,\mathrm{Hz}$) and then declines fairly abruptly at higher frequencies. The average power emitted reaches a maximum at a repetition frequency that is slightly higher than the rollover frequency.

The output beam profile emitted by RGH excimer lasers is generally (but not exclusively) rectangular in cross section (Figure 1.4). The aspect ratio is (2–3):1 with the intensity approximating a Gaussian distribution in the smaller dimension. The intensity distribution along the broader dimension has quasi-Gaussian edges but is constant to within ±10–20% over most of the width of the beam.

Pulse-to-pulse instability can be a problem in RGH excimer lasers and arises from fluctuations in excitation and variations in gain. Internal sampling of the laser beam is commonly used to generate a signal that is utilized to change laser voltage or to replenish the gas mixture. Pulse-to-pulse energy fluctuations are usually ±3–6% for operation with KrF, XeCl and XeF but can be several times larger than this when operation occurs with F_2 or ArF.

The output of RGH lasers is highly multimode when using a plane parallel stable resonator. This has the positive effect of reducing interference effects and laser speckle, but makes it difficult to achieve a high quality tightly focused beam. The beam divergence in this configuration is typically 2–3 mrad. A reduction in the number of output modes and a decrease in beam divergence to about 0.2 mrad can be obtained with an unstable resonator as the optical cavity (Figure 1.5).

Microlithographical applications of excimer lasers (Sengupta 1993) require that the laser output be spectrally narrowed to 1–3 pm and that

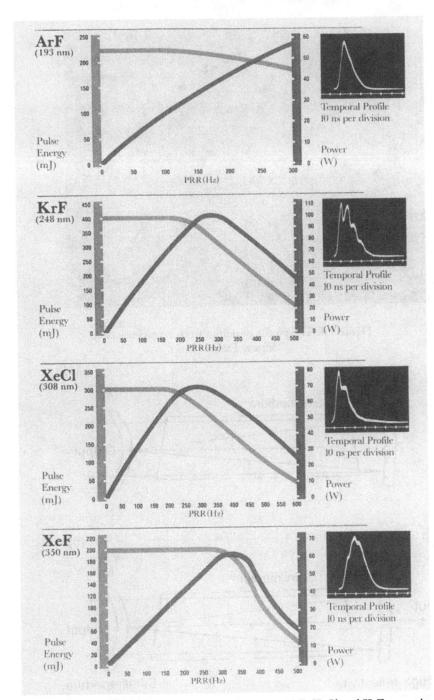

Figure 1.3. Pulse energy versus repetition rate for ArF, KrF, XeCl and XeF operation of a commercial RGH excimer laser. Representative temporal profiles are also shown (data courtesy of Lumonics Inc.).

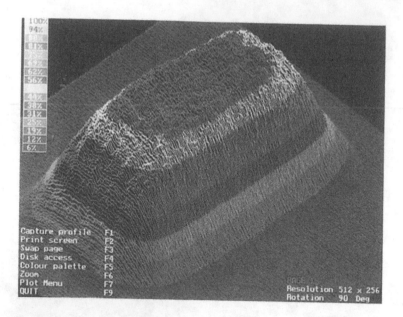

Figure 1.4. The profile of output from an excimer laser.
(Courtesy Exitech Ltd.)

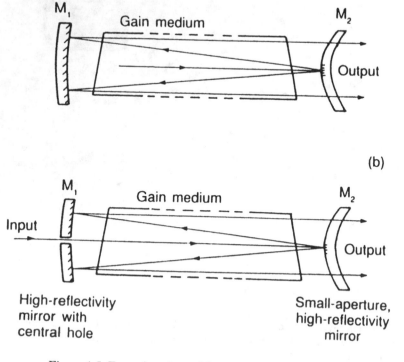

Figure 1.5. Examples of unstable resonator optics for RGH lasers.
(Courtesy Exitech Ltd.)

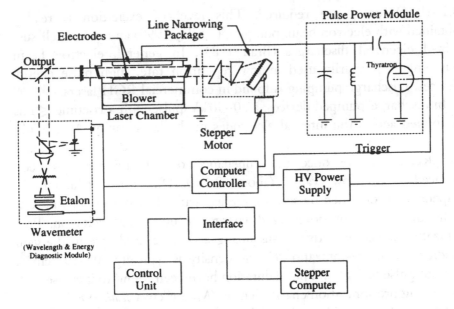

Figure 1.6. A laser lithography schematic block diagram showing major modules of the laser. The pulse power module and the chamber are very closely coupled in a real system (Sengupta 1993).

pulse-to-pulse energy variations be minimized. Wavelength narrowing can be obtained using intracavity dispersing elements such as gratings, prisms and etalons. Incorporation of one or more etalons will yield narrowed spectral emission of high brightness but these elements are susceptible to damage at the high intensities within the cavity. Prisms do not yield the required dispersion.

A straightforward solution to this problem is obtained using a prism–grating combination in the Lithrow configuration (Ishihara et al. 1992). This arrangement, coupled with a wavemeter to measure output wavelength (Figure 1.6) and a feedback control system, stabilizes both output wavelength and pulse-to-pulse energy to acceptable levels in a KrP based stepper microlithography system (Sengupta 1993). The spectral bandwidth can be varied by changing the widths of the two apertures at each end of the gain medium, whereas wavelength tuning is obtained by rotating the prisms and changing the angle of incidence for light on the grating.

The technology of energy deposition into RGH media is now well developed (Kompa 1986, Znotins and Norris 1988, Smiley 1990, McIntyre and Rhodes 1991, Obara and Kannari 1991). To achieve efficient laser emission in RGH systems, energy depositions of

$0.1-10\,\mathrm{MW\,cm^{-3}}$ are required. This level of excitation is readily obtained with electron beam pumping (Smiley 1990) and in fast self-sustained electrical discharges (Brau 1984). In general, electron beam pumping is primarily used in low repetition rate, high pulse energy devices. Discharge pumping is found in commercial RGH lasers.

In discharge-pumped devices, a $20-50\,\mathrm{kV}$ pulse with a risetime $\lesssim 1\,\mathrm{ns}$ is applied across the internal electrodes in the laser cavity. As a result, a current density of about $1\,\mathrm{kA\,cm^{-2}}$ flows through the gas containing the RGH mixture once the impedance of this mixture has been reduced to a low value (about $0.1-0.2\,\Omega$). This reduction in impedance is obtained by seeding the gas with electrons created by UV radiation from a pre-ionizing discharge or by X-ray excitation. Pre-ionization is also effective in stabilizing a volume discharge in the gain medium. The pre-ionization charge density is typically $10^6-10^{12}\,\mathrm{cm^{-3}}$. Output pulse energy and pulse duration have been found to increase with increasing pre-ionization charge density (Midorikawa *et al.* 1984).

Ionic and metastable states of the rare gas are involved in generating population in the excited states of RGH molecules. These states are created by electron impact; for example, in the KrF system

$$e + Kr \rightarrow Kr^+ + 2e \tag{31}$$

$$e + Kr \rightarrow Kr^* + e \tag{32}$$

where Kr^* is a metastable state. Since the ionization potential of Kr is about $14\,\mathrm{eV}$ and Kr^* is about $10\,\mathrm{eV}$ above the ground state in Kr, the limiting efficiency in the KrF system is $n \lesssim (h\nu/10)$, where $h\nu$ is the energy of an output photon. Since $h\nu = 5\,\mathrm{eV}$; $n \lesssim 0.5$.

In practice, the effective photon extraction efficiency, defined as the ratio of extracted laser energy to the energy deposited in the discharge volume, is typically $0.05-0.12$ in KrF and XeCl and about 0.05 in ArF (Rice *et al.* 1980). The primary loss mechanism for laser photons within the laser cavity is due to absorption in a variety of atoms and molecules which may be either transient species appearing only during excitation or more stable molecules such as O_2, CF_2 and HF that are the products of gas phase chemistry within the discharge. In the latter case, the concentration of these molecules may be reduced by circulation of the gas mixture through a cryogenic regenerator. The efficiency of conversion of electrical power to laser output power (the wall-plug efficiency) in RGH lasers is typically $0.5-2\%$.

RGH discharges are high gain media. The small signal gain co-efficient in XeCl and KrF amplifiers is $g_0 \simeq 0.1$–$0.2\,\mathrm{cm}^{-1}$ (Corkum and Taylor 1982, Szatmari and Schäfer 1987, Taylor *et al.* 1988a, b) which is sufficiently large that saturation readily occurs at low fluence. The saturation energy flux, E_s, in these systems is dependent on pulse length and the relation of pulse length to the gain recovery time (McIntyre and Rhodes 1991). For 2 ns pulses propagating in KrF, $E_s \simeq 3\,\mathrm{mJ\,cm}^{-2}$ (Banic *et al.* 1980). Amplification to high pulse energies then requires large aperture multi-stage devices.

1.3 OPTICAL COMPONENTS

There are only a few choices for highly transmissive optical materials at wavelengths shorter than 250 nm. In general, these are wide band gap oxides and fluorides. A summary of the most important of these materials together with their limiting physical and chemical characteristics is given in Table 1.11. The primary requirements for UV laser optics are

(1)   low absorption and high transmission at the laser wavelength,
(2)   radiation resistance, i.e. a high threshold for the formation of color centers,
(3)   that they be non-hygroscopic under laboratory conditions,
(4)   corrosion resistance when in contact with lasing gases,
(5)   suitable substrate for coating with metallic or dielectric over-layers for reflectivity modification,
(6)   low quantum efficiency for luminescent emission under experi-mental conditions and
(7)   low cost.

Materials that satisfy many of these criteria are fused SiO_2, MgF_2 and CaF_2, although the optimum choice will depend on wavelength. Fused SiO_2 is commonly used as a focusing optic material for KrF, XeCl and XeF laser radiation and may also be used at 193 nm. It is relatively inexpensive but is susceptible to the formation of color centers which induce an absorption band at 215–220 nm (Tsai and Griscom 1991). This degradation is highly sensitive to the presence of impurities, particularly hydroxyl (OH) radicals (Leclerc *et al.* 1991). Susceptibility to the formation of color centers increases with photon energy and is largest at 193 nm (Arai *et al.* 1988). Color center formation is also

Short wavelength lasers

Table 1.11. *Properties of crystalline solids used for UV laser transmissive optics (selected data from Weber 1982).*

Material	Transmission limit (nm)	Hardness knoop (kg mm^{-2})	Solubility H$_2$O (g/100 g)	Thermal conductivity (W m^{-1} K^{-1})	Radiation resistance
Al$_2$O$_3$	150	1000–1370	9.8×10^{-5}	35	Poor
BaF$_2$	140	80–500	0.12	12	Good
CaF$_2$	130	120	1.6×10^{-3}	10	Good
KCl	200	9–200	34.7	6.7	Poor
LiF	110	110–600	0.27	11.3	Poor
MgF$_2$	110	415	$<2 \times 10^{-4}$	21	Good
MgO	160	600–690	6.2×10^{-4}	59	Poor
SiO$_2$	160	500–740	$<10^{-3}$	6–10	Good

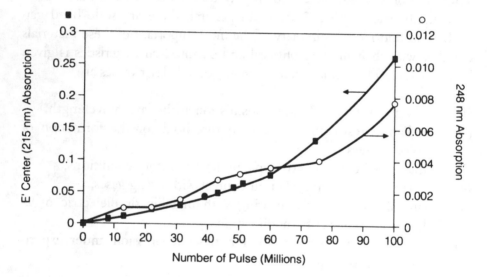

Figure 1.7. The dependence of induced absorption at 248 nm versus integrated exposure to 248 nm excimer laser pulses (after Schermerhorn 1992).

accompanied by lattice compaction and an increase in real refractive index (Fiori and Devine 1986). Figure 1.7 shows the induced absorption at 248 nm as a function of integrated fluence for a fused silica window irradiated with 248 nm excimer laser pulses (Schermerhorn 1992).

Radiation-induced compaction (Leung *et al.* 1991) has the effect of introducing an optical path difference (OPD) in fused SiO$_2$ in the region of exposure to the laser beam. The spatial dependence of the OPD then mimics that of the laser beam. Figure 1.8 shows the spatial

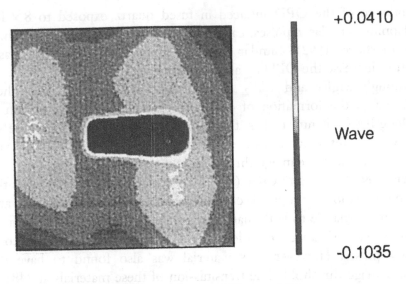

+0.0410

Wave

-0.1035

Phase Plot of Zygo XP

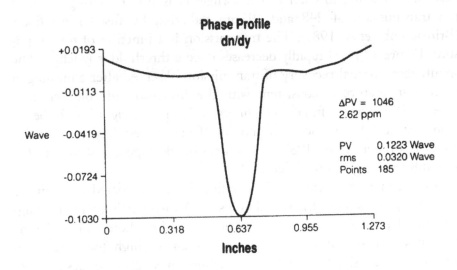

**Phase Profile
dn/dy**

+0.0193

-0.0113

Wave   -0.0419

-0.0724

-0.1030

0   0.318   0.637   0.955   1.273

Inches

ΔPV = 1046
2.62 ppm

PV   0.1223 Wave
rms   0.0320 Wave
Points 185

Phase Profile - Beam Height

Figure 1.8. The spatial dependence of OPD in fused quartz (Schermerhorn 1992).

dependence of the OPD induced in fused quartz exposed to 8×10^6 overlapping KrF laser pulses, each pulse having a fluence of $0.4 \, J \, cm^{-2}$ (Schermerhorn 1992). Similar exposure conditions using ArF laser radiation increase the OPD by about one order of magnitude.

Although MgF_2 and CaF_2 are both hydroscopic materials, their resistance to the formation of color centers and high transmission at wavelengths $<250 \, nm$ makes these materials well suited for use as windows and refractive optical components for KrF, ArF and F_2 laser radiation. A study of damage thresholds and radiation resistance in these materials by Toepke and Cope (1992) concluded that CaF_2 provides the best combination of optical characteristics, radiation resistance and price. An alternative to both materials is BaF_2, which was found to have the greatest resistance to color center formation for femtosecond excimer pulses. However, this material was also found to have the lowest damage threshold. The transmission of these materials at 248 nm can be a strong function of laser intensity (Taylor *et al.* 1988a, b, Simon *et al.* 1989), particularly at intensities in the range $I \simeq 10^8 \, W \, cm^{-2}$, for which two-photon absorption can become significant. Two-photon absorption leading to color center formation is also a limiting factor in the transmission of 248 and 193 nm radiation by fused silica fibers (Brimacombe *et al.* 1989). The transmission is a function of total exposure (Figure 1.9) and rapidly decreases once a threshold is reached. The spontaneous partial recovery of transmission observed after annealing at room temperature is consistent with the formation of color clusters. The transmission of fused silica fibers at 308 nm is significantly better than that at 248 nm since two-photon effects are less important at this wavelength (Pini *et al.* 1987). Some data on damage thresholds at 248 and 308 nm are given in Table 1.12.

For optimum performance, UV optics should be coated to minimize reflectivity and system transmission losses. Anti-reflection (AR) coatings can be deposited as dielectric layers, but the choice of dielectric materials is again limited by the requirement of high transmission at wavelengths as short as 193 nm. Coatings must also be designed to minimize stress and electrical field strength within the surface layer; a difficult constraint at intensities of up to $10^8 \, W \, cm^{-2}$. Coatings must also be adherent, free of defects and scattering centers, and abrasion and corrosion resistant. Common coating materials are oxides and fluorides such as AlF_3, Al_2O_3, HfO_2, LaF_3, MgF_2 and SiO_2. These are typically deposited in quarter-wave and half-wave layers using electron beam evaporation. Some data on changes in transmittance of representative

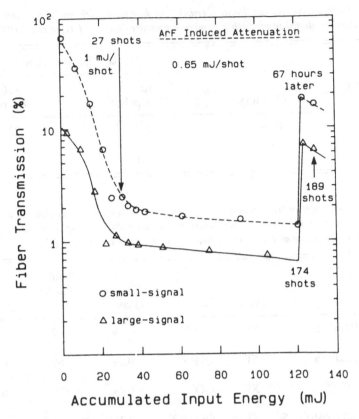

Figure 1.9. Transmission versus accumulated ArF input energy for $l = 50$ cm, Diaguide (ST-U) fiber. The dashed curve represents the best line through the small signal transmission data and the solid curve represents the best line through the high signal data. The fiber was run-in for 27 shots at 1 mJ per shot then at 0.65 mJ per shot for 162 shots (from Brimacombe *et al.* 1989).

AR coatings on single crystal CaF_2 windows reported by Krajnovich *et al.* (1992) are summarized in Table 1.13. The irradiation source was a KrF laser operating at a repetition rate of 200–300 Hz. The laser output was homogenized and focused onto a 7 mm square area on the optics to yield a fluence of 0.23–0.55 J cm^{-2}. All XF (extra-fine finish) samples had initial transmittances in the range $98.4 \leq T \leq 99.9\%$ whereas F (fine finish) samples had $95.0 \leq T \leq 99.4\%$. The best results in this study were obtained for oxide coatings on XF and F CaF_2 substrates. The coatings showed no degradation after exposure to 66×10^6 overlapping pulses of 248 nm radiation, each at a fluence of 0.25 J cm^{-2}. The most substantial changes in transmission were exhibited by coatings containing HfO_2.

Table 1.12. *A comparison between fluence thresholds for damage at KrF and XeCl wavelengths in silica fibers irradiated in air with a numerical aperture NA = 0.1 (Pini et al. 1987).*

Core diameter (μm)	Cladding	Damage fluence ($J\,cm^{-2}$)	
		KrF (15 ns pulse)	XeCl (20 ns pulse)
200	Silica	2.2	22.3
200	Plastic	3.2	15.9
600	Plastic	2.1	9.9

Table 1.13. *The change in transmission $\Delta T\%$ in AR-coated single CaF_2 windows at 248 nm (Krajnovich et al. 1992). Q and H denote quarterwave and halfwave, respectively. XF and F denote extra-fine and fine surface finish, respectively. S is substrate (CaF_2).*

AR coating	Finish	Design	Fluence ($J\,cm^{-2}$)	Number of pulses (10^6)	$\Delta T\%$
$MgF_2/LaF_3/MgF_2/S$	XF	QQ'Q	0.24	69	+0.3
	F				+0.6
$MgF_2/HfO_2/LaF_3/S$	XF	QHQ'	0.24	69	−0.8
	F				+0.8
$MgF_2/Al_2O_3/S$	XF	QQ'	0.55	75	−0.35
	F				−0.15
$SiO_2/Al_2O_3/SiO_2/Al_2O_3/S$	XF	QQ'QQ'	0.25	66	0.0
	F				0.0
$MgF_2/LaF_3/AlF_3/S$	XF	QQ'Q''	0.25	66	+0.2
	F				+0.2

The damage threshold for coated optics depends significantly on laser parameters such as pulse duration, intensity and average power as well as on the nature and composition of the coating and the roughness of the substrate. An experimental study of damage thresholds for AR coatings of various compositions on fused quartz and CaF_2 substrates (Itoh *et al.* 1989) at 248 and 193 nm (20 ns pulse duration) found that damage occurred at fluences in the range $0.92-1.65\,J\,cm^{-2}$ at 248 nm and at $0.4\,J\,cm^{-2}$ at 193 nm. This damage took the form of blistering over the area exposed to laser radiation.

Mixed oxide, oxide–fluoride and mixed fluoride coatings can also be used to produce high reflectivity surfaces. Multiple layers (10–20) are required to achieve optimum reflectivity on substrates such as SiO_2, sapphire and glass. These can be optimized for various angles of incidence. Reflectivity and transmission curves for dielectric mirrors with selected responses at 158 and 193 nm and 45° angle of incidence are shown in Figure 1.10. These curves are representative of high quality commercial optical reflectors (data courtesy of Acton Research Corporation).

Damage thresholds for dielectric coated mirrors at 248 nm (20 ns pulse) have been found to lie in the range $0.8–6.5\,J\,cm^{-2}$ (Itoh et al. 1989). Similar damage thresholds have been found for high reflectivity Sc_2O_3/SiO_2 coatings on fused SiO_2 Bk-7 glass and sapphire irradiated with 0.4 ns, 355 nm pulses from a tripled Q-switched Nd:YAG laser (Tamura et al. 1993).

Itoh et al. (1989) found that the initiation of damage in high reflectivity coatings irradiated with 248 nm KrF laser pulses occurs at the center of the beam, where the temperature rise due to absorption is largest. Overcoating with SiO_2 was found to raise the damage threshold, possibly by suppressing evaporation at the top of the dielectric stack.

1.4 OPTICS FOR MICROSTRUCTURE GENERATION

Many important industrial applications of excimer lasers involve the projection of a demagnified image of a mask or aperture onto a work-piece at a fluence sufficiently high to produce etching. An additional requirement is that the size of the irradiated field can be as large as several tens of mm^2. Over this area, the beam must be uniform to within about ±2% and the exposure reproducible to within about ±1% on a pulse-to-pulse basis. For highest spatial resolution and enhanced photochemical etching, laser wavelengths of 248 and 193 nm are preferred.

The design and performance of such systems have been discussed by Jain (1990), Rumsby and Gower (1991), Kahlert et al. (1992), Sercel (1992), Tabat et al. (1992), Smith et al. (1993) and Wittekoek et al. (1993). Figure 1.11 shows the basic elements of an excimer laser based projector system. Typical performance parameters for this system using a KrF laser generating 0.6 J per pulse are summarized in Table 1.14. These parameters can be improved by using a line-narrowed laser and by optimization of the imaging optics. Resolutions down to 0.25 μm have been reported using similar systems (Wittekoek et al. 1993, Smith et al. 1993).

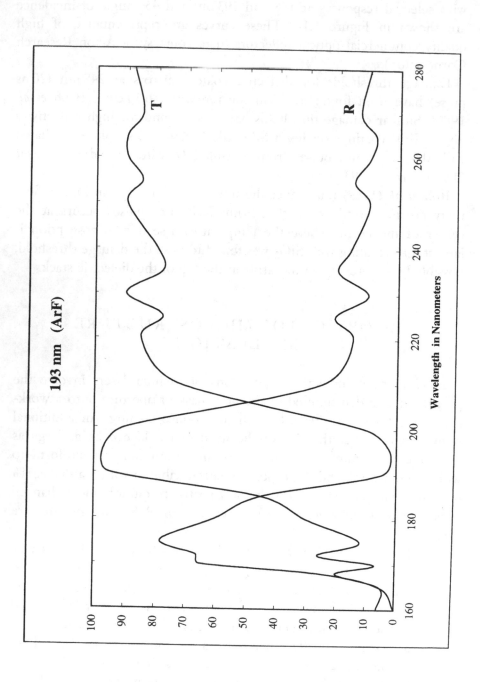

193 nm (ArF)

T

R

Wavelength in Nanometers

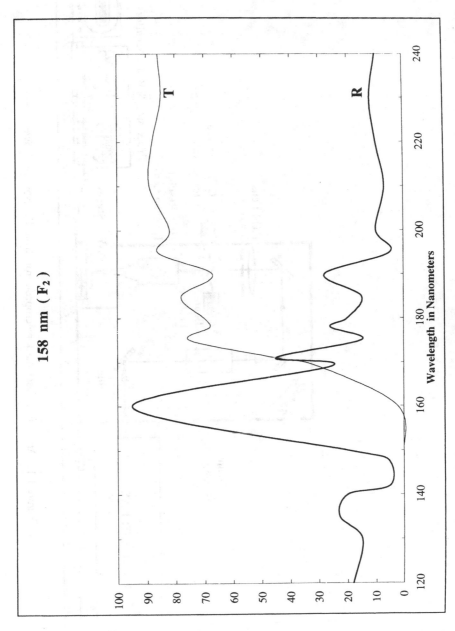

Figure 1.10. Wavelength dependences of reflectivity R and transmission T for multilayer dielectric mirrors optimized at 158 and 193 nm at 45° angle of incidence (data supplied by Acton Research Corporation).

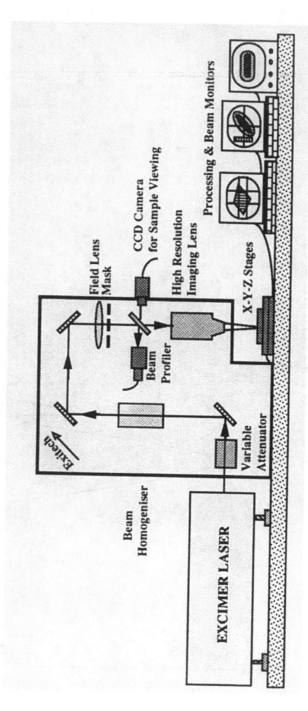

Figure 1.11. An excimer laser mask macroprojector (from Rumsby and Gower 1991).

Table 1.14. *The performance of the basic 248 nm excimer laser projection system (from Rumsby and Gower 1991).*

Magnification	Numerical aperture	Maximum imaging field (mm)	Minimum image size (μm)	Maximum image fluence (J cm^{-2})
16	0.65	1	5	40
8	0.65	2	10	10
4	0.50	4	10	2.5
2	0.40	8	20	1.0
1	0.28	16	20	0.25

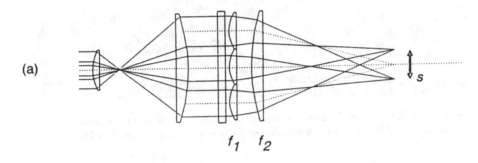

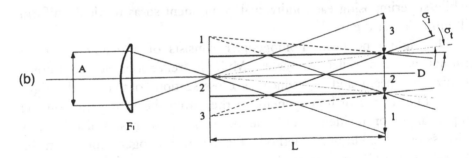

Figure 1.12. (a) An anamorphic fly's eye beam homogenizer.
(b) A light pipe homogenizer.

A primary requirement in these systems is that the excimer laser beam be homogenized. This can be accomplished either with an anamorphic 'fly's eye' homogenizer or by using a light tunnel or light pipe. These configurations are shown schematically in Figure 1.12. The anamorphic homogenizer utilized by Kahlert *et al.* (1992) consisted of a 7×7 matrix lens array. The rôle of this array is to expand each

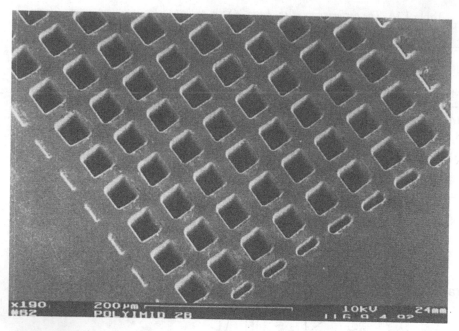

Figure 1.13. A scanning electron micrograph of a polyimide structure obtained with a homogenized beam and mask imaging at 248 nm (courtesy of Lambda Physik).

component in the incident laser beam to illuminate the entire mask while superimposing each individual component so as to yield uniform illumination at the mask.

The light pipe beam homogenizer consists of a quartz bar or a tunnel between reflective plates. Multiple reflections act to homogenize the beam and to reduce its cross-section to that of the light pipe. The degree of homogeneity in the output beam depends on the input angle of incidence and the number of internal reflections. A discussion of additional methods for beam homogenization can be found in Jain (1990). Figure 1.13 shows an example of the etching of polyimide using a mask imaging technique and 248 nm excimer laser radiation.

The output of the homogenizer can be imaged with a field lens through a mask that contains the pattern of interest or may simply be a circular or rectangular aperture that is imaged on the workpiece. In both cases a 'step and repeat' system can be used to repetitively etch structures. An alternative approach (Sercel 1992) is to scan the excimer beam over a fixed mask (Figure 1.14) while projecting a demagnified image onto the workpiece. This system was reported to yield a target

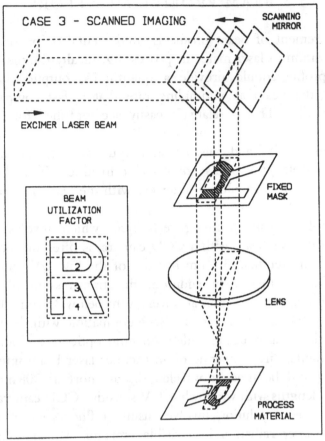

Figure 1.14. A scanned imaging micro-machining system (after Sercel 1992).

fluence of $20\,\mathrm{J\,cm^{-2}}$ with $3\,\mu\mathrm{m}$ resolution over a $2\,\mathrm{mm}$ diameter area (Sercel 1992). The design of a multiple lens system for use at UV wavelengths has been discussed by Hung and Su (1992).

The effect of prolonged exposure to $193\,\mathrm{nm}$ excimer laser radiation on pellicles for use in photolithographic systems has been discussed by Rothschild and Sedlacek (1993). They find that this exposure leads to thinning and eventually to rupture. The threshold for rupture was found to occur at $(25{-}50) \times 10^6$ pulses at a fluence of $5\,\mathrm{mJ\,cm^{-2}}$ per pulse and occurred after the thickness had been reduced by 20–50%. A reduction in fluence to $0.125\,\mathrm{mJ\,cm^{-2}}$ per pulse increased the threshold for rupture to $10^{10}{-}10^{11}$ pulses. No observable changes were seen in pellicles exposed to 10^7 pulses of $248\,\mathrm{nm}$ radiation at a fluence of $1\,\mathrm{mJ\,cm^{-2}}$.

1.5 BEAM PROFILE MONITORS

The measurement of beam profile is fundamental to process control and optimization in laser materials processing. Ideally the measurement of beam profile should provide a quantitative output of intensity throughout the beam at a sampling rate that is fast compared with process variables. This is relatively easily accomplished for CW laser outputs at visible, near IR and near UV wavelengths. However, the measurement of CW and pulsed laser output intensity distributions at deep UV wavelengths offers additional difficulties. In addition, the measurement system should not interfere with the beam in a significant way.

The simplest beam profiling technique, which involves looking directly at the laser beam with a CCD camera, is often not practical at wavelengths shorter than 300 nm because of the low UV sensitivity of standard CCD cameras. In addition, the transmission of standard camera optics rapidly decreases at wavelengths below 350 nm. The small active area of the CCD detector is also incompatible with the large size of most excimer laser beams unless imaging optics are used. Despite these constraints, direct imaging of an excimer laser beam into a CCD camera may still be used at wavelengths as short as 308 nm and at shorter wavelengths with windowless UV sensitive CCD cameras.

A more general solution involves using a fluorescent material to convert UV laser radiation into visible emission, which then can be detected with a conventional CCD camera. Nominally transparent solids such as MgO and sapphire are ideal as fluorescent converters when doped with a low level (ppm) of transition metal ions such as Cr^{3+} and Ni^{2+}. Since these materials are transparent at UV wavelengths, they can be placed in the UV beam without producing significant attenuation. A simple experimental configuration is shown schematically in Figure 1.15. Systems consisting of a fluorescent crystal suitable for use at 248 and 193 nm, a CCD camera for monitoring this fluorescence together with a video frame grabber and image analysis software are commercially available from a number of suppliers. The output from such a system is in the form of a false color contour plot which permits analysis of the two-dimensional intensity profile on a real-time basis. The fluorescent crystal can exhibit a linear intensity response of up to six orders of magnitude.

Alternate methods are available to directly image 190 and 258 nm excimer laser radiation. For example, Mann *et al.* (1992) have discussed

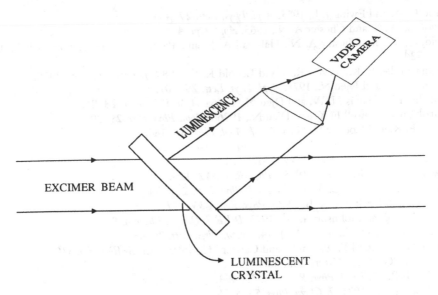

Figure 1.15. A beam profiler using a fluorescent crystal to convert UV
to visible radiation.

the use of a UV sensitive phosphor as an intermediary in converting
UV to visible radiation for detection with a standard CCD camera. The
use of a UV sensitive film with a resolution exceeding 10^3 lines mm^{-1}
to image the beam from excimer lasers has been discussed by Davis
(1990). A useful exposure range of 0.1–200 mJ cm^{-2} was reported.
Optimized results were obtained only after image processing techniques
had been applied to the recorded images.

1.6 REFERENCES

Ali A. W. and Kolb A. C., 1965. *Appl. Phys. Lett.* **13**, 259.
Arai K., Imai H., Hosono H., Abe Y. and Imagawa, H., 1988. *Appl. Phys. Lett.* **53**, 1891.
Ault E. R., Bhaumik M. L., Hughes W. M., Jensen R. J., Robinson C. P., Kolb A. C.
   and Shannon J., 1973. *IEEE J. Quant. Electron.* **9**, 1031.
Austin L., Basting D., Kahlert H. J., Muckenheim W. and Rebhan U., 1988. *Proc. SPIE*
   894, 2.
Banic J., Efthimopoulos T. and Stoicheff B. P., 1980. *Appl. Phys. Lett.* **37**, 686.
Basov N. G., Bogdankevich G. V., Danilychev V. A., Devyatkov A. G., Kashnikov G. N.
   and Lantsov N. P., 1968. *JETP Lett.* **7**, 317.
Bazhulin P. A., Knyazev I. N. and Petrash G. G., 1965. *Sov. Phys. JETP* **21**, 649.
Beck R., English W. and Guss K., 1978. *Table of Laser Lines in Gases and Vapors*, 2nd
   Edition, Springer-Verlag, Berlin.
Brau C. A., 1984 in *Excimer Lasers* ed. C. K. Rhodes, Spring, New York, p. 87.

Brau C. A. and Ewing J. J., 1975. *Appl. Phys. Lett.* **27**, 435.

Bridges, W. B. and Chester A. N., 1965. *Appl. Opt.* **4**, 573.

Bridges, W. B., Chester A. N., Halstead A. S. and Parker J. V., 1971. *Proc. IEEE* **59**, 724.

Brimacombe, R. K., Taylor R. S. and Lepold K. E., 1989. *J. Appl. Phys.* **66**, 4035.

Burnham R. and Djeu N., 1976. *Appl. Phys. Lett.* **29**, 707.

Burnham R., Harris N. W. and Djeu N., 1976a. *Appl. Phys. Lett.* **28**, 86.

Burnham R., Powell F. X. and Djeu N., 1976b. *Appl. Phys. Lett.* **29**, 30.

Cheo P. K. and Cooper H. G., 1965. *J. Appl. Phys.* **36**, 1862.

Corkum P. B. and Taylor R. S., 1982. *IEEE J. Quant. Electron.* **18**, 1962.

Davis G., 1990. *Laser Focus World* Nov., p. 117.

Ewing J. J. and Brau C. A., 1975a. *Phys. Rev.* **A12**, 129.

Ewing J. J. and Brau C. A., 1975b. *Appl. Phys. Lett.* **27**, 350.

Fiori C. and Devine R. A. B., 1986. *Phys. Rev.* **B33**, 2972.

Gerardo J. B. and Johnson A. W., 1973. *IEEE J. Quant. Electron.* **9**, 748.

Golde M. G. and Thrush B. A., 1974. *Chem. Phys. Lett.* **29**, 486.

Groussin B., Pitard F., Parent A. and Carriere C., 1993. *Proc. SPIE* **1850**, 330.

Heard, H. G., 1963. *Nature* **200**, 667.

Hodgson R. T., 1970. *Phys. Rev. Lett.* **25**, 494.

Hodgson R. T., 1971. *J. Chem. Phys.* **55**, 5378.

Hodgson R. T. and Dreyfus R. W., 1972. *Phys. Rev. Lett.* **28**, 536.

Hoff P. W., Swingle J. C. and Rhodes C. K., 1973. *Appl. Phys. Lett.* **23**, 245.

Hoffman J. M., Hays A. K. and Tisone G. C., 1976. *Appl. Phys. Lett.* **28**, 558.

Hughes W. M., Shannon J. and Hunter R., 1974. *Appl. Phys. Lett.* **24**, 488.

Hung T. Y. and Su C.-S., 1992. *Appl. Opt.* **31**, 4397.

Hutchinson M. H. R., 1980. *Appl. Phys.* **21**, 95.

Ishihara T., Sandstrom R., Reiser C. and Sengupta U., 1992. *Proc. SPIE* **1674**, 473.

Itoh M., Endo A., Kuroda K., Watanabe S. and Ogura I., 1989. *Opt. Commun.* **74**, 253.

Jain K., 1990. *Excimer Laser Lithography*. SPIE Press, Bellingham, Washington.

Johnson T. H., Cartland H. E., Genoni T. C. and Hunter A. M., 1990. *J. Appl. Phys.* **66**, 5707.

Jortner J., Meyer L., Rice S. A. and Wilson E. G., 1965. *J. Chem. Phys.* **42**, 4250.

Kahlert H. J., Sarbach U., Burghardt B. and Klimt B., 1992. *Proc. SPIE* **1835**, 110.

Kannari F., Kimura W. D. and Ewing J. J., 1990. *J. Appl. Phys.* **68**, 2615.

Kannari F., Obara M. and Fujioka T., 1985. *J. Appl. Phys.* **57**, 4309.

Koehler H. A., Ferderber L. J., Redhead D. L. and Ebert P. J., 1972. *Appl. Phys. Lett.* **21**, 198.

Koehler H. A., Ferderber L. J., Redhead D. L. and Ebert P. J., 1975. *Phys. Rev.* **A12**, 968.

Kompa K. L., 1986. *Proc. SPIE* **650**, 75.

Krajnovich D. J., Kulkarni M., Leung W., Tam A. C., Spool A. and York B., 1992. *Appl. Opt.* **31**, 6062.

Leclerc N., Pfleiderer C., Hitzler H., Wolfrum J., Greulich K.-O., Thomas S., Fabian H., Takke R. and English W., 1991. *Opt. Lett.* **16**, 940.

Leung W. P., Kulkarni M., Krajnovich D. and Tam A. C., 1991. *Appl. Phys. Lett.* **58**, 551.

Loree T. R., Sze R. C. and Barker D. L., 1977. *Appl. Phys. Lett.* **30**, 150.

Maiman T. H., 1960. *Nature* **187**, 493.

Mann K., Hopfmüller A., Gorzellik P., Schild R., Stöffler W., Wagner H. and Wolbold G., 1992. *Proc. SPIE* **1834**, 184.

Marling J. B. and Lang D. B., 1977. *Appl. Phys. Lett.* **31**, 181.

McIntyre I. A. and Rhodes C. K., 1991. *J. Appl. Phys.* **69**, R1.

Midorikawa K., Obara M. and Fujioka T., 1984. *IEEE J. Quant. Electron.* **20**, 198.

Newman B. E., 1988. *Proc. SPIE* **738**, 155.

Obara M. and Kannari F., 1991. *Encyclopedia of Lasers and Optical Technology* ed. R. A. Meyers, Academic Press, New York, p. 567.

Pini R., Salimbeni R. and Vannini M., 1987. *Appl. Opt.* **26**, 4185.

Reintjes J., 1980. *Appl. Opt.* **19**, 3889.

Rice, J. K., Hays A. K. and Woodsworth J. R., 1977. *Appl. Phys. Lett.* **31**, 31.

Rice J. K., Tisone G. C. and Patterson E. L., 1980. *IEEE J. Quant. Electron.* **14**, 1315.

Rothschild M. and Sedlacek J. H. C., 1993. *Opt. Eng.* **32**, 2421.

Rumbsy P. T. and Gower M. C., 1991. *Proc. SPIE* **1598**, 36.

Schawlow A. L. and Townes C. H., 1958. *Phys. Rev.* **112**, 1940.

Schermerhorn P., 1992. *Proc. SPIE* **1835**, 70.

Searles S. K. and Hart G. A., 1975. *Appl. Phys. Lett.* **27**, 243.

Sengupta U. K., 1993. *Opt. Eng.* **32**, 2410.

Sercel J. P., 1992. *Proc. SPIE* **1835**, 172.

Serreze H. B. and Chen Y. C., 1993. *Proc. SPIE* **1850**, 397.

Shaw M. J., 1979. *Prog. Quant. Electron.* **6**, 1.

Simon P., Gerhardt H. and Szatmari S., 1989. *Opt. Lett.* **14**, 1207.

Smiley V. N., 1990. *Proc. SPIE* **1225**, 1.

Smith B. W., Gower M. C., Westcott M. and Fuller L. F., 1993. *Proc. SPIE* **1927**, 914.

Szatmari S. and Schäfer F. P., 1987. *J. Opt. Soc. Am.* **B4**, 1943.

Tabat M., O'Keefe T. R. and Ho W., 1992. *Proc. SPIE* **1835**, 144.

Tamura S., Kimura S., Sato Y., Yoshida H. and Yoshida K., 1993. *Thin Solid Films* **228**, 222.

Taylor A. J., Gibson R. B. and Roberts J. P., 1988a. *Appl. Phys. Lett.* **52**, 773.

Taylor A. J., Gibson R. B. and Roberts J. P., 1988b. *Opt. Lett.* **13**, 814.

Toepke I. and Cope D., 1992. *Proc. SPIE* **1835**, 89.

Tsai T. E. and Griscom D. L., 1991. *Phys. Rev. Lett.* **67**, 2517.

Velazco J. E. and Setser D. W., 1975. *J. Chem. Phys.* **62**, 1990.

Wallmeier H. and Zacharias H., 1988. *Appl. Phys.* **B45**, 263.

Wang C. P., Mirels H., Sutton D. G. and Suchard S. N., 1976. *Appl. Phys. Lett.* **28**, 326.

Waynant R. W., 1972. *Phys. Rev. Lett.* **28**, 533.

Waynant R. W. and Ediger M. N., 1993. *Selected Papers on UV, VUV and X-Ray Lasers*, SPIE Press, Bellingham, vol. MS71.

Waynant R. W., Shipman J. D., Elton R. C. and Ali A. W., 1970. *Appl. Phys. Lett.* **17**, 383.

Weber M. J., 1982. *CRC Handbook of Laser Science and Technology*, Vol. IV, Part 2, CRC Press, Boca Raton, Florida.

Wittekoek S., van den Brink M., Poppelaars G. and Reuhman-Huisken M., 1993. *Proc. SPIE* **1927**, 582.

Znotins T. A. and Norris B., 1988. *Proc. SPIE* **894**, 9.

CHAPTER 2

Optical properties of materials at UV wavelengths

2.1 OPTICAL CONSTANTS

The interaction of electromagnetic radiation with condensed matter can be characterized in terms of a complex frequency-dependent dielectric constant

$$\epsilon(w) = \epsilon_1(w) + i\epsilon_2(w) \tag{1}$$

where $\epsilon_1(w)$ and $\epsilon_2(w)$ are related to the complex refractive index, m, as follows:

$$\epsilon_1 = n^2 - k^2 \tag{2}$$

$$\epsilon_2 = 2nk \tag{3}$$

with

$$m = n - ik \tag{4}$$

where n and k are both frequency-dependent.

For an ideal vacuum, $n = 1$ and k is identically zero. The presence of matter causes both n and k to deviate from these values. For example, the refractive index of helium gas is 1.000 086 at 1 atm pressure and at a wavelength $\lambda = 586\,\text{nm}$. This increase in n above unity arises from the presence of electronic transitions in He at much shorter wavelengths.

With condensed matter, the density is many times larger than that of a gas and deviations of n and k from vacuum values are correspondingly larger. Indeed, it is not unusual to have $n, k \gg 1$ over a wide wavelength range in most solids. As an example, the refractive indices for Si at 308 nm are $n = 5.01$ and $k = 3.59$ (Aspnes and Studna 1983).

Physically, the dependence of n on wavelength leads to dispersive effects in optical systems whereas the absorption at a particular wavelength is directly related to k. In fact, it can be shown that the real and imaginary terms either in the dielectric constant or in the refractive index are related through the Kramers–Kronig integrals (Wooten 1972). For the dielectric constant $\epsilon(w)$ these are

$$\epsilon_1(w) = 1 + \frac{2}{\pi} P \int_0^\infty \frac{w^1 \epsilon_2(w^1)}{(w^1)^2 - w^2} \, dw^1 \tag{5}$$

$$\epsilon_2(w) = \frac{-2w}{\pi} P \int_0^\infty \frac{[\epsilon_1(w^1) - 1]}{(w^1)^2 - w^2} \, dw^1 \tag{6}$$

where P is the principal part of the integral. These relations show that knowledge of either ϵ_1 or ϵ_2 over the frequency range $0 < w < \infty$ provides information on the value of the other at a specific frequency w. The Kramers–Kronig relations are often used to verify the consistency of experimental data for ϵ_1 and ϵ_2. Alternatively, the Kramers–Kronig relations can be used to calculate either ϵ_1 or ϵ_2 from laboratory measurements of the other component (Karlsson and Ribbing 1982, Aspnes and Studna 1983, Wooten 1972). The Kramers–Kronig approach is valid for gases as well as for condensed matter.

For wavelengths in the ultraviolet, the attenuation of radiation in materials is due to the excitation of electronic transitions. In metals such excitations involve both intra- and interband transitions whereas in insulators attenuation arises through the excitation of impurity centers and defects as well as band to band transitions. The absorption of light propagating through a medium characterized by refractive indices n and k is given by the Beer–Lambert law (Birks 1970)

$$I(x) = I_0 \, e^{-\alpha x} \tag{7}$$

where I_0 is the intensity at $x = 0$ and $I(x)$ is the intensity after a distance x. The attenuation coefficient

$$\alpha = 4\pi k / \lambda \tag{8}$$

is found to be directly proportional to the imaginary term in the refractive index. At UV wavelengths, a transparent material would have $\alpha \leq 1\,\text{cm}^{-1}$ whereas strong absorbers such as semiconductors or metals would have $\alpha = (2\text{–}3) \times 10^{6}\,\text{cm}^{-1}$. The characteristic penetration depth for radiation under these conditions is then α^{-1}. For the examples given the penetration depth would be about 300–500 nm (metals) and $\geq 1\,\text{cm}$ (transparent media). These values show that surface effects will dominate in the interaction of UV radiation with metals and many semiconductors. Surface roughness and composition are also important in determining the coupling of laser radiation to solids (Barbarino *et al.* 1982, Roos *et al.* 1989, Kinsman and Duley 1993).

Equation (7) is valid only under conditions in which I_0 is much less than the intensity at which non-linear effects may become significant. At excimer laser wavelengths in condensed media this limit implies $I_0 \leq 10^{4}\,\text{W}\,\text{cm}^{-2}$. At incident intensities higher than this value multiphoton processes may become important. When two-photon effects dominate, equation (7) can be used to calculate $I(x)$ with α replaced by $\alpha^{(2)}$ where (Yariv 1989)

$$\alpha^{(2)} = \text{constant} \times I_0 \tag{9}$$

so that the absorption coefficient is intensity-dependent. The value of the constant is determined by such factors as the intrinsic strength of nearby one-photon transitions, the population difference between initial and final two-photon states and the frequency of the transition.

In gases and in molecular solids, it is often useful to replace α in equation (7) by

$$\alpha = \sigma n \tag{10}$$

where σ is the absorption cross-section per absorber and n is the number of absorbers per cm^{3}. A strong absorption band in a solid of density $n = 10^{22}\,\text{cm}^{-3}$ would have $\alpha \simeq 10^{6}\,\text{cm}^{-1}$ and $\sigma = 10^{-16}\,\text{cm}^{2}$. It is important to note that both α and σ are frequency-dependent and that there is a fundamental constraint on the value of the integral

$$\int_{0}^{\infty} \sigma(w)\,\mathrm{d}w = 4.1 \times 10^{-3} fn \tag{11}$$

where f is the oscillator strength.

2.2 METALS

In metals, the complex dielectric constant $\epsilon(w)$ can be written as follows (Abelés 1966)

$$\epsilon(w) = \epsilon^L + \epsilon^D(w) + \epsilon^I(w) \tag{12}$$

where ϵ^L is the contribution due to lattice vibrations, $\epsilon^D(w)$ is the intra-band or Drude term and $\epsilon^I(w)$ is the term due to interband transitions. At UV wavelengths ϵ^L is approximately constant. The Drude term dominates at long (infrared) wavelengths and is given by

$$\epsilon^D(w) = 1 - \frac{w_p^2}{w(w + i/\tau)} \tag{13}$$

where

$$w_p = \left(\frac{n_e e^2}{m^* \epsilon_0}\right)^{1/2} \tag{14}$$

is the plasma frequency, n_e is the electron density and m^* is the electron effective mass. τ is the electron relaxation time and $\epsilon_0 = 8.85 \times 10^{-12}\,\mathrm{N\,m^2\,C^{-2}}$. The Drude term can be rewritten

$$\epsilon^D(w) = \epsilon_1^D + i\epsilon_2^D \tag{15}$$

with

$$\epsilon_1^D(w) = 1 - \frac{w_p^2 \tau^2}{1 + w^2 \tau^2} \tag{16}$$

$$\epsilon_2^D(w) = \frac{w_p^2 \tau}{w(1 + w^2 \tau^2)} \tag{17}$$

Some numerical values for n_e, w_p and m^* for various metals are given in Table 2.1. The plasma frequency $w_p = 2\pi c/\lambda_p$, where c is the speed of light, is in the region of $10^{16}\,\mathrm{rad\,s^{-1}}$. It corresponds to the cutoff frequency for the propagation of electromagnetic radiation through the metal. Light of frequency $w \leq w_p$ will be strongly attenuated in the metal, leading to high reflectivity. Conversely, when $w > w_p$ (or $\lambda < \lambda_p$) the metal can become relatively transparent. This is particularly noticeable in

Table 2.1. *Electron density, n_e, effective mass, m^*, and plasma frequency, w_p, for several elemental metals. λ_p is the plasma cutoff wavelength and m is the mass of a free electron.*

Metal	n_e (10^{28} m^{-3})	m^*/m	w_p (10^{16} rad s^{-1})	λ_p (nm)
Na	2.5	0.98	0.90	209
Ag	5.8	1.07	1.32	142
Au	5.9	1.13	1.29	146
Cu	8.5	2.56	1.03	183
Al	6.0	1.60	1.09	173

thin metal films (Marr 1967) and leads to the use of self-supported metal films as windows for the extreme ultraviolet.

The electron relaxation time in equation (13) arises from a combination of electron–phonon and electron–defect scattering processes. For many metals at 300 K, $\tau \simeq 10^{-14}$–10^{-15} s and is temperature-dependent. In practice, the value of τ in a given metal will also depend on wavelength. The Drude model predicts that

$$\tau = 5.3 \times 10^{-16} \lambda \frac{k^2 - n^2}{2nk} \text{ s} \qquad (18)$$

where λ is in micrometers. The relation between τ and the DC conductivity of the metal, σ_0, is

$$\sigma_0 = \epsilon_0 w_p^2 \tau \qquad (19)$$

Figure 2.1 shows the wavelength and temperature dependences of σ_0 and τ for Fe calculated from the (n, k) data of Shvarev *et al.* (1978).

The rôle of interband transitions in the dielectric response of metals (equation (12)) is most important at UV wavelengths (Ehrenreich and Philipp 1962). Estimation of the contribution of $\epsilon^I(w)$ to $\epsilon(w)$ can only be made from a calculation of band structure and electronic density of states (Wooten 1972, Stohl and Jung 1979, Adachi 1988, Philipp 1966). The effect both of direct and of indirect (i.e. phonon-assisted) electronic transitions must be considered.

At normal incidence, the absorptivity A, of a metal surface is given by (Born and Wolf 1975)

$$A = \frac{4n}{(n+1)^2 + k^2} \qquad (20)$$

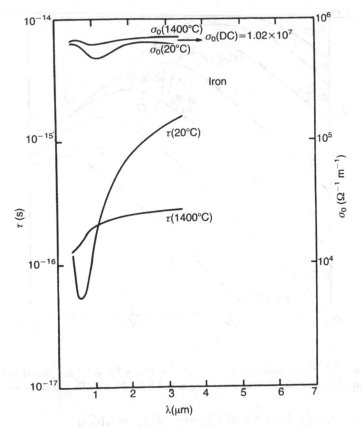

Figure 2.1. Wavelength dependences of σ_0 and τ derived from the Drude model for Fe using n and k from Shvarev *et al.* (1978).

A general solution to this equation is plotted in Figure 2.2, which also shows the wavelength dependence of A for Fe and Ti at room temperature. A plot of $A(\lambda)$ for Fe and Ti (Figure 2.3) shows that the absorptivity rises throughout the visible and near UV spectral regions attaining a value of 0.6–0.75 at excimer laser wavelengths.

This behavior is typical of many metals. These values of A can be contrasted with $A \leq 0.05$–0.1 for metals at $10.6\,\mu m$ (Duley 1976).

Table 2.2 gives a summary of calculated values of A at various excimer wavelengths obtained by using measured refractive indices together with equation (20). These values are likely to be accurate only for low intensity radiation incident on clean, non-oxidized surfaces.

Temperatures can also have a significant effect on absorptivity. This arises owing to volume expansion, changing phonon populations and a shift of the Fermi level (Wooten 1972). Effects are often largest at low

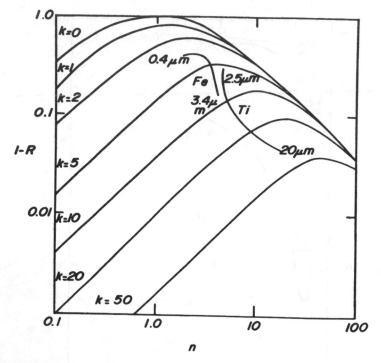

Figure 2.2. The solution to $1 - R = A = 4n/[(n+1)^2 + k^2]$ in generalized form. $A(\lambda)$ calculated from measured n and k for Fe and Ti is also shown.

photon energies (Johnson and Christy 1975), at which

$$A(T) \cong A(20°C)[1 + B(T - 20°C)] \tag{21}$$

where B is a constant and T is temperature. At excimer wavelengths, changes in A with temperature are usually small (cf. Figure 2.4). However, either an increase or a decrease in A with increasing temperature is possible.

There have been few measurements of the reflectivity of liquid metals at UV wavelengths. Gushkin *et al.* (1978) reported n and k values for liquid Ni at 1600°C. They found little change on passing through the melting point and concluded that the absence of long range order in the melt has little effect on the optical properties of the metal.

2.3 SEMICONDUCTORS

There have been extensive studies of the optical properties of semi-conductors (see reviews by Philips 1966, Wooten 1972, Cardona 1969,

Table 2.2. *Absorptivities, A, of some metals at
308, 249 and 193 nm.*

Metal	Wavelength (nm)		
	308	249	193
Ag	0.89	0.74	0.74
Au	0.63	0.66	0.76
Cd	0.18	0.38	0.92
Co	0.51	0.60	0.64
Cr	0.51	0.59	0.61
Cu	0.64	0.62	0.65
Fe	0.59	0.67	0.73
Mn	0.56	0.64	0.71
Mo	0.41	0.31	0.37
Ni	0.58	0.54	0.64
Pd	0.47	0.52	0.61
Sn	0.30	0.32	0.36
Ta	0.60	0.59	0.48
Ti	0.47	0.60	0.64
V	0.45	0.53	0.61
Zn		0.90	0.90

Palik 1985, 1991). The key to obtaining accurate spectral data has been the development of surface etching and cleaning techniques that eliminate many of the surface structural and compositional uncertainties that existed in earlier measurements. This, combined with the development of automatic ellipsometers, has provided the experimenter with real-time diagnostics of surface quality. As a result, optical constants can now be obtained for smooth, clean and undamaged surfaces. Such pristine surfaces are unlikely to be encountered in many practical laser processing applications and so reflectivity and absorptivity are likely to differ from these values measured in the laboratory. With this caveat Tables 2.3 and 2.4 provide some data on A and α for various elemental and compound semiconductors at excimer laser wavelengths. These data were obtained from the measurements of Cardona and Greenaway (1964) and Aspnes and Studna (1983). The wavelength dependences of A for Si and GaAs at 300 K are shown in Figure 2.5. Both elemental and compound semiconductors are characterized by high absorption ($\alpha \simeq 10^6 \, \mathrm{cm}^{-1}$) and large absorptivity at excimer wavelengths. This arises from strong interband absorption at these wavelengths (Adachi 1988).

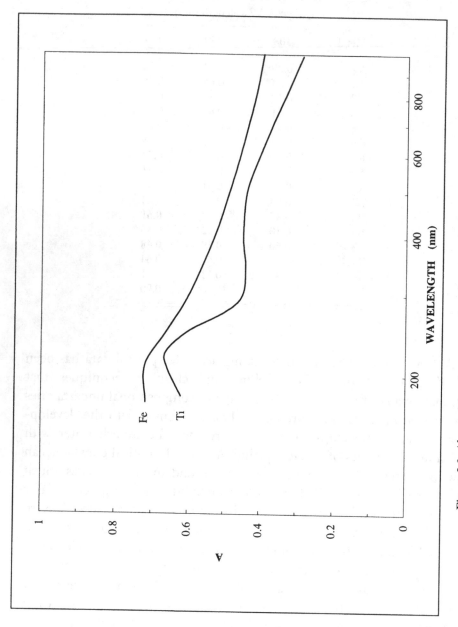

Figure 2.3. Absorptivity A versus wavelength λ (nm) for Fe and Ti.

Table 2.3. *Absorptivities calculated from* (n, k) *for selected semiconductors at 300 K.*

Material	Wavelength (nm)		
	308	248	193
Ge	0.44	0.35	
Si	0.41	0.33	
GaP	0.55	0.42	
GaAs	0.58	0.33	
GaSb	0.42	0.41	
GeTe	0.59	0.75	0.80
InP	0.62	0.39	
InAs	0.61	0.42	
InSb	0.39	0.46	
PbS	0.53	0.62	0.82
PbSe	0.29	0.50	0.77
PbTe	0.59	0.50	0.77
SnTe	0.49	0.66	0.75
CoS_2	0.75	0.85	0.80
NiS_2	0.79	0.83	0.76
Mg_2Si (77 K)	0.35	0.40	0.50

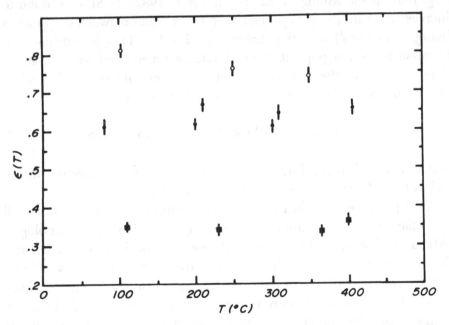

Figure 2.4. Measured values of coupling coefficients $\epsilon(T)$ for 308 nm excimer laser radiation of metals: (O), silver; (●), copper; and (■), aluminum. All data points represent an average of at least four measurements (Kinsman and Duley 1986).

Table 2.4. *Absorption coefficients* α
($10^6\ cm^{-1}$) for various semiconductors at
300 K at two excimer wavelengths.

Material	Wavelength (nm)	
	308	248
Ge	1.35	1.62
Si	1.54	1.81
GaP	0.88	1.84
GaAs	0.78	2.07
GaSb	1.48	1.39
InP	0.70	1.77
InAs	0.73	1.46
InSb	1.50	1.24
PbS	0.92	0.82
PbSe	0.75	0.62
PbTe	0.63	0.52
SnTe	0.80	0.67

The temperature dependence of reflectivity and absorption co-efficients in semiconductors has been studied for a variety of materials (e.g. Jellison and Modine 1982, Bilenko *et al.* 1982). In Si, α is found to increase dramatically with temperature for energies between the indirect bandgap (1.12 eV) and the direct gap (3.4 eV). This is attributed to increasing phonon population at elevated temperatures with an attendant increase in the rate of indirect transitions. Jellison and Modine (1982) find the following relation for $\alpha(T)$ at 308 nm:

$$\alpha(T) = (1.43 \pm 0.01) \times 10^6\ (cm^{-1}) \exp{(T/4680)} \qquad (22)$$

where T is in kelvins. The rôle of increasing α during laser heating of Si has been discussed by Kwong and Kim (1983).

Amorphous semiconductors are of increasing importance in solar cell manufacture, optical memories, xerography and other new technologies (Madan and Shaw 1988). The optical properties of these materials differ from those of their crystalline counterparts because of a lack of long range order and the presence of large defect concentrations. A significant difference is the introduction of a high density of electronic states within the normal bandgap. In fact, the density of these states is sufficiently large that the concept of a forbidden energy gap might not be valid. Under these conditions, the specification of bandgap energy,

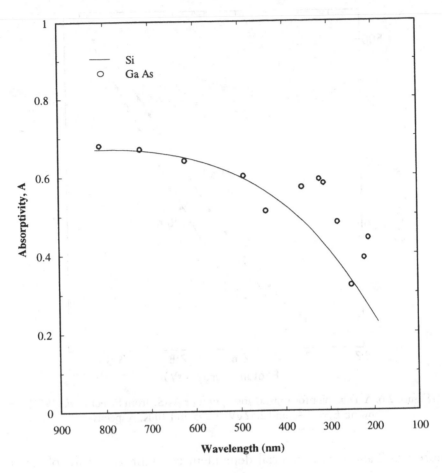

Figure 2.5. Absorptivity A versus λ for Si and GaAs.

E_g, is uncertain. For many amorphous semiconductors the energy dependence of α follows the Urbach (1953) relation

$$\alpha = \alpha_0 \exp\left(\frac{\hbar(w - w_0)}{\beta(T)}\right) \tag{23}$$

when $\alpha \leq 10^5 \text{ cm}^{-1}$. In this expression α_0 is the value of the absorption coefficient at some low frequency w_0 and $\beta(T)$ is the width of the exponential tail. This depends on the degree of structural disorder. At higher photon energy α often exhibits a Tauc (1973) dependence

$$(\alpha\hbar w)^{1/n} = C(\hbar w - E_g) \tag{24}$$

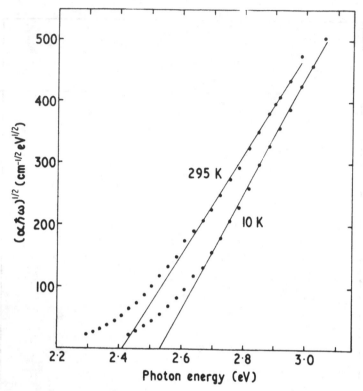

Figure 2.6. A Tauc plot for optical absorption of As_2S_3 from Street *et al.* (1974) showing E_g = 2.41 and 2.52 eV at 295 and 10 K, respectively.

where $n \cong 2$ and C is a material-dependent constant. E_g can be obtained by extrapolation using equation (24). Alternatively, E_g is often taken to be that energy at which $\alpha \simeq 10^4\,\text{cm}^{-1}$. An example of the Tauc dependence of α on photon energy in the vicinity of the bandgap is shown in Figure 2.6 for amorphous As_2S_3.

A comparison of the ultraviolet spectral properties of crystalline and amorphous Si (Ewald *et al.* 1979) prepared by glow discharge deposition shows that the UV reflectivity and optical constants are profoundly affected by substrate temperature. For example, the reflectivity at 250 nm increases from 0.37 to 0.65 as the substrate temperature rises from 27 to 400°C. The reflectivity of samples deposited at 400°C is similar to that of crystalline samples, although fine structure correlated with singularities in the band structure is not reproduced in amorphous samples. A similar effect has been observed in ϵ_1 and ϵ_2 spectra of amorphous GaAs and other amorphous III–V compounds (Stuke and Zimmerer 1972).

There has recently been much interest in the formation and properties of diamond-like amorphous carbon both in elemental and in hydrogenated forms (see reviews by Robertson 1986, Robertson and O'Reilly 1986). These solids are formed by plasma deposition or ion-beam deposition from gaseous hydrocarbons. They can also be formed by direct deposition from atomic and molecular beams. Both a-C and a-C:H differ from glassy carbon by being truly amorphous and semiconducting rather than metallic. The bandgap energy can vary from ≤ 0.5 to $4\,eV$ depending on composition and preparation conditions. The bandgap is characterized by a high density of localized states extending from valence band to conduction band.

Absorption in the region $\hbar w < 5\,eV$ is strongly dependent on the ratio of trigonal (sp^2) to tetrahedral (sp^3) bonding. Solids with a high degree of sp^2 bonding are more graphitic, with smaller E_g and higher absorption over this energy range. Diamond-like solids have a large sp^3 content and low absorption in the visible and near UV. For $\hbar w > E_g$ one has

$$(\alpha \hbar w)^{1/2} \cong 3.7 \times 10^4 (\hbar w - 0.5\,eV) \qquad (25)$$

for a-C and

$$(\alpha \hbar w)^{1/2} \cong 6 \times 10^4 (\hbar w - 1.8\,eV) \qquad (26)$$

for a-C:H (Robertson 1986). A comparison of $(\alpha \hbar w)^{1/2}$ versus E for a-C and a-C:H over a wide energy range is shown in Figure 2.7. The absorptivity of various carbon solids at excimer laser wavelengths is given in Table 2.5.

2.4 INSULATORS

In insulating crystals or wide bandgap semiconductors, absorption at UV wavelengths can derive from interband transitions, excitonic resonances and the presence of impurity and defect centers. The excitation of interband transitions involves the creation of electron–hole pairs. Near the band edge the absorption coefficient for such transitions is about $10^6\,cm^{-1}$.

Excitons are bound electron–hole pairs and contribute discreet absorption resonances at energies near the band edge. The similarity in structure between these bound electron–hole pairs and the H atom

Optical properties of materials at UV wavelengths

Table 2.5. *Absorptivities of carbon solids at excimer wavelengths.*

	Wavelength (nm)			
	308	248	193	Reference
Graphite	0.68	0.48	0.71	Palik (1985)
Glassy carbon	0.88	0.89	0.96	Williams and Arakawa (1972)
a-C	0.82	0.84	0.88	Duley (1984a)
a-C:H	0.86	0.80	0.80	McKenzie *et al.* (1983)
Arc evaporated	0.84	0.87	0.88	Arakawa *et al.* (1985)
Diamond (type IIa)	0.20	0.23	0.28	Walker and Osantowski (1964)

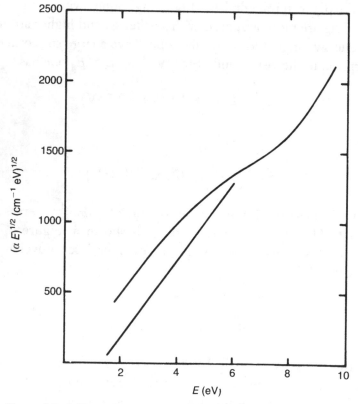

Figure 2.7. A Tauc plot for a-C (Duley 1984a) (upper curve) and for a-C:H (Dischler *et al.* 1983) (lower curve).

results in the production of hydrogenic energy levels in the bandgap just below the bottom of the valence band. A series of discreet, relatively narrow absorption lines can often be observed in spectra of insulating solids near the onset of band–band absorption.

Table 2.6. *Energies and strengths of charge transfer (CT) bands in several oxide and silicate solids.*

Solid	Energy of CT band (eV)	f	Reference
$Fe^{3+}:Mg_2SiO_4$	4.78	0.2	Weeks *et al.* (1974)
	6.0	1.0	
	6.6	0.5	
$Mn^{2+}:Mg_2SiO_4$	6.5	0.1	Weeks *et al.* (1974)
$Fe^{2+}:(MgFe)_2SiO_4$	6.5		Nitsan and Shankland (1976)
$Fe^{3+}:MgO$	4.3	$\simeq 0.1$	Hansler and Segelken (1960)
	5.7	$\simeq 0.1$	
$Cr^{3+}:MgO$	5.9		Hansler and Segelken (1960)
$Fe^{3+}:Al_2O_3$	4.83	0.1	Lehmann (1970)
	6.32	0.9	
$Fe^{3+}:SiO_2$	5.04	0.085	Lehmann (1970)
	6.33	0.13	

The presence of impurities and defects can introduce localized absorption at energies within the bandgap. Absorption features can be very weak, such as for the forbidden d–d transitions of transition metal ions in silicate or oxide lattices. In the ultraviolet, however, the excitation of dipole-allowed charge transfer transitions often dominates. Such transitions can have an oscillator strength $f \simeq 0.1$–1.0. A list of representative charge transfer bands is given in Table 2.6.

Defect centers are present in all crystalline solids. The concentration of such impurities depends on preparation conditions and radiation history. Even nominally clear solids usually have defect concentrations in the parts per million range. Such defects may consist of missing or interstitial atoms, non-equilibrium charge states or isotopic centers. The absorption coefficient due to a concentration n (cm^{-3}) of absorbers is approximately

$$\alpha = \frac{6.9 \times 10^{-17} nf}{\Delta \nu} \tag{27}$$

where f is the oscillator strength of the transition involved and $\Delta \nu$ is the full width at half maximum of the resulting spectral line in electron-volts. For example, for a defect such as a color center in an alkali halide crystal, one would have $\Delta \nu \simeq 1\,eV$, $n \simeq 10^{17}\,cm^{-3}$ and $f \simeq 0.1$. Then $\alpha \simeq 0.7\,cm^{-1}$.

Spectra of the simple cubic oxide MgO (Figure 2.8) illustrate a number of these features. Figure 2.8(a) shows the reflectance R of

(a)

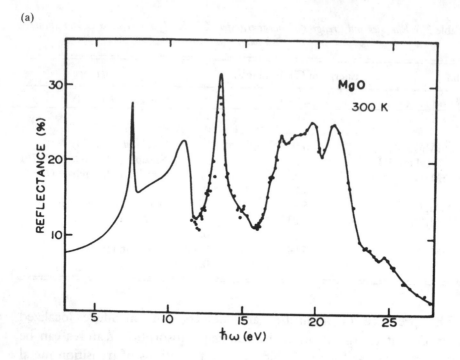

(b)

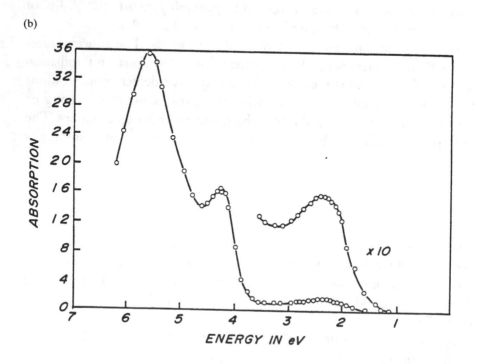

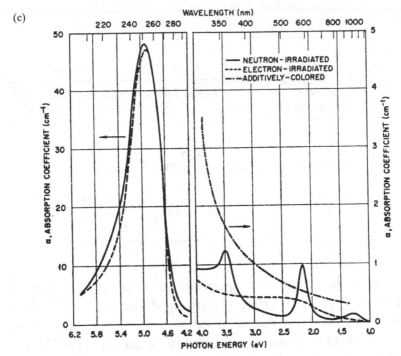

Figure 2.8. (a) A near-normal-incidence (8°) reflectance spectrum typical of a freshly cleaved crystal of MgO at 300 K (Roessler and Walker 1967). (b) Absorption bands due to Fe^{3+} ions in MgO single crystal (Molnar and Hartman 1950). (c) A comparison of absorption spectra of MgO crystals produced by neutron irradiation, electron irradiation, and additive coloration. The 250 nm band for additively colored crystal ($\alpha_{250} = 400 \, cm^{-1}$) is not shown here. (From Chen, Y., Williams, R. T., and Sibley, W. A., *Phys. Rev.*, **182**, 960, 1969.)

nominally pure crystalline MgO over the 4–14 eV region. The most noticeable feature at lower energies is the sharp resonance at 7.60 eV, which is superimposed on a slowly rising reflectivity. This resonance (Cohen *et al.* 1967) is attributed to the creation of excitons. The absorption coefficient is about $0.78 \times 10^6 \, cm^{-1}$ at the peak of this feature (Roessler and Walker 1967). The rising background reflectivity throughout this region is due to an increase in k accompanying the onset of interband absorption.

An absorption spectrum of Fe doped MgO is shown in Figure 2.8(b). Weak d–d transitions with $\alpha \simeq 3 \, cm^{-1}$ are present at low energy. The strong resonances at 4.3 and 5.7 eV are due to charge transfer transitions involving Fe^{3+}. Within these bands $\alpha \simeq (1-2) \times 10^3 \, cm^{-1}$ indicating that these crystals contain about 0.1% of Fe^{3+} (equation (27)).

Table 2.7. *Absorption bands in SiO$_2$ at UV wavelengths.*

Band energy (eV)	Width (eV)	Occurrence	Center	Reference
7.6	0.5	Pure SiO$_2$	Peroxy radical	Antonini *et al.* (1982)
7.1	0.8	Ion irradiated		Antonini *et al.* (1982)
5.8	0.6	Pure SiO$_2$ (irradiated)	Positively charged O vacancy	Rothschild *et al.* (1989)
5.1	0.4	Pure SiO$_2$ (irradiated)	Neutral O vacancy	Levy (1960)
4.7		Pure SiO$_2$		Antonini *et al.* (1982)
4.0	1.2	Doped SiO$_2$	Electron trap	Nassau and Prescott (1975)

Figure 2.8(c) shows the UV spectrum of neutron and electron irradiated MgO (Chen *et al.* 1969). The strong absorption band at about 5 eV in irradiated MgO is due to negative ion vacancy sites. These centers can be annealed out at temperatures exceeding 400°C. It is apparent that such irradiated crystals would be strongly absorbing at the KrF excimer laser wavelength while still being relatively transparent at 308 nm.

In SiO$_2$ the bandgap energy is about 9 eV. As a result, there is little intrinsic absorption in SiO$_2$ at excimer wavelengths above about 160 nm. High purity synthetic fused quartz typically has <10 ppm of OH and ≤1 ppm of metallic impurities. Cl is, however, a common impurity in these materials. It is these impurities, together with intrinsic defects, that effectively limit the transmission of quartz at wavelengths above the bandgap. Table 2.7 provides a summary of impurity and defect absorption bands in SiO$_2$ in the region $\hbar w > 3$ eV. Further data on the occurrence and properties of these spectral features can be found in the literature (Levy 1960, Nelson and Weeks 1961, Nassau and Prescott 1975, Friebele *et al.* 1979, Antonini *et al.* 1982, Jones and Embree 1976, Stathis and Kastner 1987, Rothschild *et al.* 1989, Williams and Friebele 1986).

Because of the use of silica for UV window materials and for UV transmitting optical fibers, there have been a number of studies of the effect of excimer laser radiation on this material (Stathis and Kastner 1984, Mizunami and Takagi 1988, Taylor *et al.* 1988a,b, Tsai *et al.* 1988, Arai *et al.* 1988, Devine 1989, Rothschild *et al.* 1989). This work

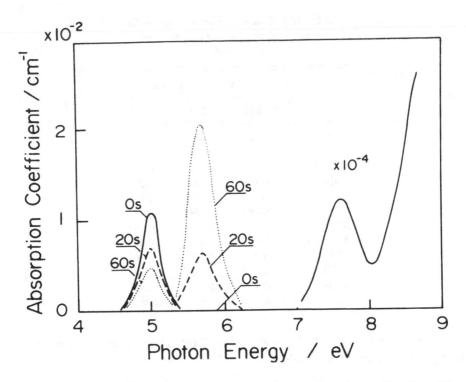

Figure 2.9. Optical absorption characteristics of dehydrated synthetic silica glass. The 7.6 and 5.0 eV bands are due to the intrinsic oxygen-deficient defects. The growth of the 5.8 eV band due to the E′ center, and the decrease of the 5.0 eV band, were observed after irradiation by the ArF laser of $45\,mJ\,cm^{-2}$ per pulse at 30 Hz with irradiation times shown (broken line). Thicknesses of samples used were 0.6 mm for the 7.6 eV band and 30 mm for the 5.0 and 5.8 eV bands (Arai *et al.* 1988).

has shown that a variety of defects, similar to those produced by ionizing radiation, can be produced by excimer laser light even when the photon energies involved are much less than that required for interband transitions. An example of defect absorption features induced by ArF laser radiation in hydroxyl-free silica is shown in Figure 2.9 (Arai *et al.* 1988). The decrease in the amplitude of the 5.1 eV feature due to neutral O vacancies is accompanied by an increase in the amplitude of the 5.8 eV band attributable to positively charged O vacancies. The density of 5.8 eV absorbers was found to depend on the square of the incident laser fluence ($J\,cm^{-2}$) suggesting that two-photon excitation was responsible for the creation of these defects (Arai *et al.* 1988). It appears that these centers are created when excitons generated via two-photon excitation decay non-radiatively at precursor sites (Tsai *et al.* 1988, Devine 1989).

Table 2.8. *Reflectance at normal incidence for various insulating solids at excimer laser wavelengths.*

	Wavelength (nm)		
	308	248	193
Al_2O_3	0.088	0.092	0.10
BN	0.08	0.18	
$BaTiO_3$	0.26	0.26	0.12
GaN	0.18	0.20	0.27
LiF	0.028	0.029	0.031
MgO	0.08	0.094	0.115
MgF_2	0.026	0.028	0.031
Mg_2SiO_4	0.063	0.07	0.084
SiO_2	0.051	0.057	0.061
$SrTiO_3$	0.25	0.26	0.12
TiO_2	0.41	0.38	0.30

A summary of data on the normal incidence reflectance of some insulating solids is given in Table 2.8.

2.5 ORGANIC MEDIA

The tight bonding of electrons to molecular groups in most organic systems results in electronic spectra that can be classified according to the molecular bonds present in the system. Strong absorption in organic systems is due to electronic transitions between molecular states. The energy required for such transitions corresponds approximately to bond energies (several electronvolts) with the result that most electronic transitions of organic molecules occur at UV wavelengths.

In general the electronic levels involved may be classified as bonding, non-bonding or antibonding (Murrell 1963). Wave functions are classified in terms of their angular momentum so that states with $L = 0, 1$ and 2 correspond to σ, π and δ orbitals, respectively (Flygare 1978).

Figure 2.10 shows a schematic representation of such energy levels in a hypothetical molecule. In terms of energy, bonding σ and π orbitals lie lowest. Antibonding orbitals occur at high energy since they represent an activated state of the molecule. Non-bonding orbitals are observed at intermediate energies. Depending on the symmetry of the molecule, electronic transitions may be strongly allowed between these

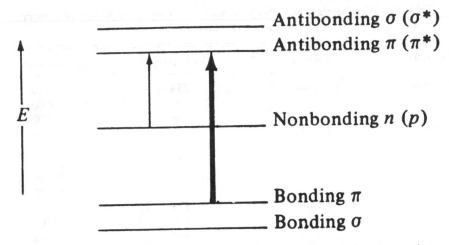

Figure 2.10. A summary of electronic energy levels. Both $n \rightarrow \pi^*$ and $\pi \rightarrow \pi^*$ (heavy arrow) transitions are represented.

electronic states. The intensity of a molecular absorption band is usually expressed in terms of the molar absorptivity ϵ, where

$$\epsilon = \frac{0.434\alpha}{c} \tag{28}$$

and c is the solute concentration with α the absorption coefficient (cm^{-1}). Thus a strong absorption band in an organic system would have $\alpha \simeq 10^6 \, cm^{-1}$ and $\epsilon \simeq 10^5 \, cm^2 \, mol^{-1}$.

Extensive listing of band type, energy and absorptivity can be found in the following works: Birks (1970), Clar (1964) and Murrell (1963). Table 2.9 summarizes some of these data for representative aromatic molecules as given by Birks (1970). These spectra are characterized by the presence of three bands increasing in intensity as wavelength decreases. These bands move to longer wavelength with increasing substitution or increasing number of rings. The result is that most aromatic molecules will be moderate to strong absorbers at excimer laser wavelengths. The strongest of these absorption bands will involve $\pi \rightarrow \pi^*$ transitions of ring electrons.

Spectral structure associated with individual molecular electronic groups in solution is also apparent in the spectra of extended systems. Figure 2.11 shows the energy dependence of the absorption coefficient of polystyrene. The strong absorption peak at 6.4 eV can be correlated with a $\pi \rightarrow \pi^*$ ring excitation in the monomer. A similar resonance is seen in the spectrum of the polyimide Kapton[TM] near 5.5 eV, where

Table 2.9. *Wavelengths of peak absorption and molar absorptivities for selected aromatic molecules (from Birks 1970).*

Molecule	Solvent	Peak (nm)	ϵ_{max} (cm^2 mol^{-1})
Benzene	Hexane	254	250
		204	8800
		184	68 000
Naphthalene	Hexane	301	270
		275	5000
		221	117 000
		190	10 000
		167	30 000
Anthracene	Hexane	375	8500
		252	220 000
		221	11 400
		186	32 000
Toluene	Hexane	262	260
		208	7900
		189	55 000
o-Xylene	Heptane	263	260
		210	8900
		191	56 000
Phenol	Hexane	270	1450
		210	6200
Styrene	Alcohol	282	450
		244	12 000
Biphenyl	Alcohol	246	20 000
Furan	Cyclohexane	200	10 000

$\alpha = 4.2 \times 10^5$ cm^{-1}. The monomer in this compound has the structure

with two types of aromatic ring. It is likely that the occurrence of two chemically distinct six-membered rings in Kapton is responsible in part for the structure in α between 4 and 7 eV and that all three peaks are due to $\pi \rightarrow \pi^*$ transitions (Brannon *et al.* 1985).

Figure 2.11 shows that several organic polymers are only weakly absorbing at XeCl and KrF photon energies. Polyethylene, in particular, is relatively transparent at 4.02 eV although impurities can increase absorption in this region by a significant amount.

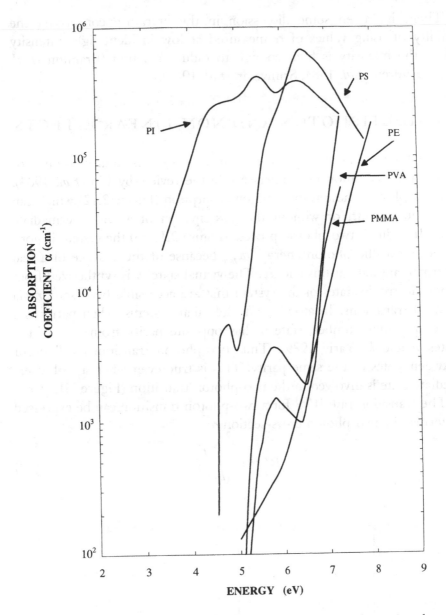

Figure 2.11. Absorption coefficient versus photon energy for several organic polymers: PE, polyethylene (Painter *et al.* 1980); PVA, polyvinyl acetate; PMMA, polymethylmethacrylate; PI, polyimide (Philipp *et al.* 1986); and PS, polystyrene (Inagaki *et al.* 1977).

There has been some discussion in the literature concerning the validity of using values of α measured at low incident light intensity when the intensity is high enough to induce ablation (Brannon *et al.* 1985, Andrew *et al.* 1983, Srinivasan *et al.* 1987).

2.6 MULTIPHOTON AND NON-LINEAR EFFECTS

At the high intensities available from laser sources, multiphoton processes are a common phenomenon (see the review by Lin *et al.* 1984). The simplest of these is two-photon absorption (Figure 2.12) which can occur either with or without the participation of a real intermediate state. In a direct two-photon process (Figure 2.12(b)) the system is non-absorbing at the photon energy $\hbar w_{mn}$ because of the absence of a real intermediate state at this energy. The virtual state m is synthesized from other electronic states of the system that are accessible from state n via allowed transitions. In atomic and molecular systems, when parity is a good quantum number, state m has opposite parity from that of the states n and k (Yariv 1989). Thus two-photon transitions will occur between states of the same parity. This is true even when a real intermediate state is involved in the two-photon transition (Figure 2.12(c)).

The transition rate $W^{(2)}$ for a two-photon transition can be expressed in terms of a two-photon cross-section $\sigma^{(2)}$

$$W^{(2)} = \frac{\sigma^{(2)} I^2}{(\hbar w)^2} \qquad (29)$$

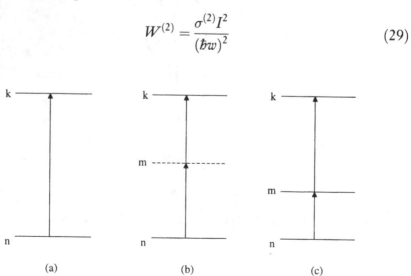

Figure 2.12. (a) One-photon transition at photon energy $\hbar w_{kn}$. (b) Two-photon transition via virtual intermediate state m. Each photon has energy $\hbar w_{kn}/2$. (c) Two-photon transition via real intermediate state m. The transition energy is $\hbar w_{kn} = \hbar w_{mn} + \hbar w_{km}$.

where $\sigma^{(2)}$ will depend on the location of all other electronic states of the system. In addition, the two-photon absorption coefficient

$$\alpha^{(2)} = \frac{\sigma^{(2)} n I}{\hbar w} \tag{30}$$

where n is the population of the ground electronic state. For a general N-photon process these expressions become

$$W^{(N)} = \frac{\sigma^{(N)} I^N}{(\hbar w)^N} \tag{31}$$

$$\alpha^{(N)} = \frac{\sigma^{(N)} n I^{N-1}}{(\hbar w)^{N-1}} \tag{32}$$

with units $\sigma^{(N)} \equiv cm^{2N} s^{N-1}$, $I \equiv W$ and $\hbar w \equiv J$. An important result is that

$$\log W^{(N)} = N \log I + K \tag{33}$$

where K is a constant. Thus N-photon processes will exhibit a slope N on a log–log plot. Examples of this dependence can be seen in the laser-induced multiphoton ionization of benzene (Boesl et al. 1981), in the creation of defect centers in fused quartz with KrF or ArF laser radiation (Arai et al. 1988, Tsai et al. 1988) and in the excitation of luminescence from MgO and MgO:Al crystals with 308 nm XeCl laser radiation (Duley 1984b, Dunphy and Duley 1988). It has also been suggested that multiphoton processes may be important in excimer ablation of organic polymers (Srinivasan et al. 1987, Duley 1987).

When real intermediate states are involved (Figure 2.12(c)), multi-photon transitions can be excited at quite low intensity. For example, with a one-photon cross-section $\sigma_{mn} \simeq 10^{-16} cm^2$ (as appropriate for a moderately strong molecular transition)

$$W_{mn}^{(1)} = \frac{10^{-16} I}{\hbar w} \tag{34}$$

If $W_{mn}^{(1)}$ is now taken to be equal to a typical radiative relaxation rate (i.e. $10^8 s^{-1}$), then $I \simeq 10^6 W cm^{-2}$ for $\hbar w = 10^{-18} J$. Under these conditions, the population of the intermediate state, m, will approximate that of the ground state (Svelto 1982) so that the probability of resonant absorption of a second photon becomes large. In the example shown in Figure 2.12(c) this second photon would have an energy $\hbar w_{km}$.

The cross-section $\sigma^{(2)}$ for a direct two-photon transition involving two photons of the same energy is a strong function of photon energy and available molecular states. To illustrate the magnitude of the effects involved, we will take $\sigma^{(2)} = 10^{-46}\,\mathrm{cm^4\,s^{-1}}$. Then

$$W^{(2)} = \frac{10^{-46}I^2}{(\hbar w)^2} \tag{35}$$

$$= 10^{-10}I^2 \tag{36}$$

when $\hbar w = 10^{-18}\,\mathrm{J}$. An excitation rate $W^{(2)} = 10^6\,\mathrm{s^{-1}}$ could then be attained when $I = 10^8\,\mathrm{W\,cm^{-2}}$. This intensity would be produced with a fluence of $1\,\mathrm{J\,cm^{-2}}$ in a $10\,\mathrm{ns}$ pulse. With $\sigma^{(1)} = 10^{-16}\,\mathrm{cm^2}$, $\sigma^{(2)} = 10^{-46}\,\mathrm{cm^4\,s^{-1}}$ and $\hbar w = 10^{-18}\,\mathrm{J}$, $I \simeq 10^{12}\,\mathrm{W\,cm^{-2}}$ is seen to make $W^{(1)} = W^{(2)}$.

Two-photon absorption has been shown to limit the intensity of UV laser radiation that can be transmitted through UV window materials (Taylor *et al.* 1988b) and optical fibers (Mizunami and Takagi 1988, Taylor *et al.* 1988a, Brimacombe *et al.* 1989). These effects are observed to occur in materials such as fused silica, MgF_2 and LiF at intensities in the 10^{10}–$10^{11}\,\mathrm{W\,cm^{-2}}$ range (Taylor *et al.* 1988b). Table 2.10 lists measured two-photon absorption coefficients β at UV wavelengths from the work of Lin *et al.* (1979) and Mizunami and Takagi (1988). β is related to $\alpha^{(2)}$ (equation (32)) as follows:

$$\beta = \alpha^{(2)}I^{-1} = \sigma^{(2)}n(\hbar w)^{-1} \tag{37}$$

where β has the units $\mathrm{cm\,W^{-1}}$.

The differential attenuation of light, dI, over a pathlength dx when both one- and two-photon absorption terms are present is then

$$\frac{dI}{dn} = -\alpha I - \beta I^2 \tag{38}$$

The transmittance $I(x)/I_0$ becomes

$$\frac{I(x)}{I_0} = \left[\left(1 + \frac{\beta I_0}{\alpha}\right)\exp\left(\alpha x\right) - \frac{\beta I_0}{\alpha}\right]^{-1} \tag{39}$$

Table 2.10. *Two-photon absorption coefficients, β (cm W^{-1}) at UV wavelengths.*

Material	Band gap (eV)	Wavelength (nm)				
		355[a]	282[b]	266[a]	248[c]	193[d]
Fused silica	7.8	$<1.3 \times 10^{-12}$	5×10^{-12}	1.7×10^{-11}	4.5×10^{-11}	2.2×10^{-9}
BaF$_2$	9.1			$<4.0 \times 10^{-12}$	1.1×10^{-10}	
SrF$_2$	9.6			$<5.4 \times 10^{-12}$	1.1×10^{-11}	
CaF$_2$	10.0			$<2.0 \times 10^{-11}$	8.3×10^{-12}	
LiF	11.6			$<2.0 \times 10^{-11}$	$<1.3 \times 10^{-12}$	
MgF$_2$	11.8			$<2.8 \times 10^{-12}$	$<1.3 \times 10^{-12}$	

[a] Lin *et al.* (1978, 1979).
[b] Mizunami and Takagi (1988).
[c] Taylor *et al.* (1988a).
[d] Brimacombe *et al.* (1989).

Applying this to two fused silica fibers, Mizunami and Takagi (1988) found that $\beta = 5 \times 10^{-12} \pm 20\%$ cm W^{-1}. The corresponding low intensity attenuation coefficient α was found to be $(3.7-7.8) \times 10^{-4}$ cm^{-1} or $(160-340)$ dB km^{-1}, respectively.

A comprehensive study of the non-linear transmission of UV optical fibers induced by ArF, KrF, XeCl and XeF laser radiation has been reported by Brimacombe *et al.* (1989). They found that irradiation at 193 or 248 nm leads to the production of a 215 nm absorption band similar to that observed in samples subjected to radiation from conventional sources. The formation of the color centers involved was not found to depend significantly on the OH$^-$ content of the fibers or the nature of the cladding. Onset of strong absorption at 215 nm was observed after a dosage of about 0.1 J (200 pulses at 0.5 mJ per pulse) had been passed down the fiber. With a fiber diameter of 400 µm, the fluence under these conditions was about 0.4 J cm^{-2}.

Figure 2.13 shows the non-linear transmission of two quartz fibers at 193 nm. It can be seen that the threshold for appreciable non-linear effects occurs at an intensity of 10^6 W cm^{-2}. Subsequent increases in intensity result in a rapid decrease in fiber transmission. A similar effect was observed at 248 nm for intensities exceeding 10^7 W cm^{-2}. Data obtained at 308 nm showed that fiber transmission remained high even at intensities near 10^9 W cm^{-2}, although surface damage could be produced at laser intensities in excess of this value. A study of the threshold for front surface damage in fused silica fibers in

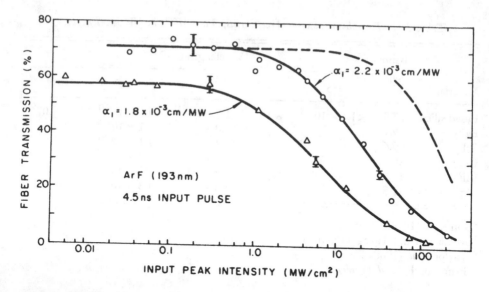

Figure 2.13. Non-linear transmission at 193 nm of $I = 101\,\mathrm{W\,cm^{-2}}$ Diaguide ($\triangle$) and $I = 39\,\mathrm{W\,cm^{-2}}$ Fiberguide GUV (O) fiber. The dashed curve represents the small signal transmission for the GUV fiber obtained after each high intensity shot. The solid curves are theoretical predictions using the values of α_1 indicated in the figure (Brimacombe *et al.* 1989).

relation to laser pulse duration at 308 nm has been reported by Taylor *et al.* (1987).

2.7 SMALL PARTICLES AND THIN FILMS

The optical properties of particles and thin films with characteristic dimensions $\leq \lambda$, where λ is the wavelength, have been widely discussed and reviewed. For general reviews of this subject the reader is referred to the articles by van de Hulst (1957), Kerker (1969), Ruppin and Englman (1970), Bohren and Huffman (1983), Perenboom *et al.* (1981), Ruppin (1982) and Kreibig and Genzel (1985). In general, the optical properties of small particles and thin films can differ from those of bulk solids because of quantum size effects and the excitation of collective resonances. Quantum size effects lead to terms in the dielectric constants ϵ_1 and ϵ_2 that are size-dependent and are not usually obvious from measurements on bulk samples. When ϵ_1 and ϵ_2 (or n and k) are known, however, the optical properties of small particles and thin films can be predicted from classical electromagnetic theory. For small particles, the solution of Maxwell's equations was first derived by Mie (1908) and Debye (1909). The resulting Mie theory, as it is generally known, yields a description of the optical response of a small particle irradiated with a monochromatic plane wave. In practice such solutions are possible only for simple particle shapes. An alternate approach based on the discrete dipole array has been used to study the scattering properties of a wider range of particle geometries (de Voe 1965, Purcell and Pennypacher 1973, Lakhtakia 1992).

The attenuation of radiation by small particles is produced by a combination of absorption and scattering. Absorption of incident radiation leads to internal heating while scattering redirects radiation away from the original direction of propagation. The attenuation coefficient for a monodisperse aerosol of small particles is

$$\alpha = Q\pi a^2 n \tag{40}$$

where each particle is assumed to be spherical and of radius a. n is the number of particles per cm^3 in the aerosol and Q is the efficiency factor for extinction. Mie theory shows that $Q \ll 1$ for particles of radius $a \ll \lambda$. For particles with $a \gg \lambda$, $Q = 2$. Thus particles can look much smaller than their geometrical size at certain wavelengths while appearing several times larger than their geometrical cross-section at other wavelengths.

The result of a Mie scattering calculation for 0.15 and 0.015 μm radius glassy carbon spheres is shown in Figure 2.14. Refractive indices were obtained from the work of Williams and Arakawa (1972). It is apparent that for 0.15 μm radius particles Q shows little spectral structure and is close to $Q = 2$. When $a = 0.015$ μm, $Q < 1$ over the entire energy range. Notice that Q has a similar energy dependence to that of k for $\hbar w > 4$ eV. It is also significant that the extinction cross-section per unit mass, $Q\pi a^2/V$, increases with decreasing particle volume.

Since Q is due to both scattering and absorption one can write

$$Q = Q_{sca} + Q_{ab} \tag{41}$$

where Q_{sca} is the efficiency factor for scattering and Q_{ab} is that for absorption. In the Rayleigh limit where $a \ll \lambda$ and $|m|2\pi a/\lambda \ll 1$ (van

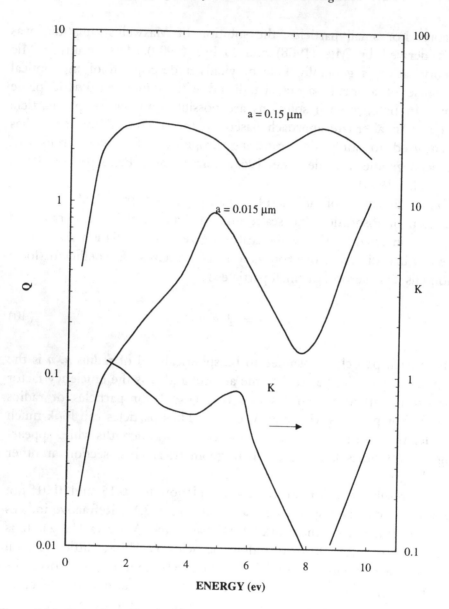

Figure 2.14. Extinction efficiency Q versus photon energy for 0.015 and 0.15 μm radius particles of glassy carbon. The energy dependence of k is also shown (Williams and Arakawa 1972).

de Hulst 1957)

$$Q_{abs} = \frac{-8\pi a}{\lambda} Im\left(\frac{m^2 - 1}{m^2 + 2}\right) \tag{42}$$

$$Q_{sca} = \frac{8}{3}\left(\frac{2\pi a}{\lambda}\right)^4 \left|\frac{m^2 - 1}{m^2 + 2}\right|^2 \tag{43}$$

In the Rayleigh limit similar simple analytical expressions are available for ellipsoids. For example,

$$C^j = \frac{2\pi}{\lambda} V \frac{\epsilon_m \epsilon_2}{[L_j\epsilon_1 + \epsilon_m(1 - L_j)]^2 + (L_j\epsilon_2)^2} \tag{44}$$

where C^j is the extinction cross-section (cm^2), V is the ellipsoid volume, ϵ_m is the dielectric constant of the surrounding medium, and $4\pi L_j$ is the depolarization factor for the jth ($j = 1, 2, 3$) axis of the ellipsoid. ϵ_1 and ϵ_2 are the components of the complex dielectric constant of the particle. L_j can take values between 0 and 1.

It is apparent that there will be a resonance in the extinction produced by these particles whenever

$$\epsilon_1 = -\epsilon_m\left(\frac{1}{L_j} - 1\right) \tag{45}$$

For spheres, where $L_j = \frac{1}{3}$, this condition reduces to

$$\epsilon_1 = -2\epsilon_m \tag{46}$$

or for a particle in vacuum, $\epsilon_1 = -2$.

This resonance is unique to small spheres and is due to the collective motion of ions (IR) or electrons (UV). In metals, ϵ_1 is generally negative for $\hbar w < \hbar w_p$, where w_p is the plasma frequency. The resulting surface plasmon resonance arises from the collective motion of electrons within the particle.

The effect of shape on the value of ϵ_1 at which the surface resonance occurs is shown in Figure 2.15 (Huffman 1977). One can see that, although $\epsilon_1 = -2\epsilon_m$ for a sphere, $\epsilon_1 = -17\epsilon_m$ for long cylinders (electric

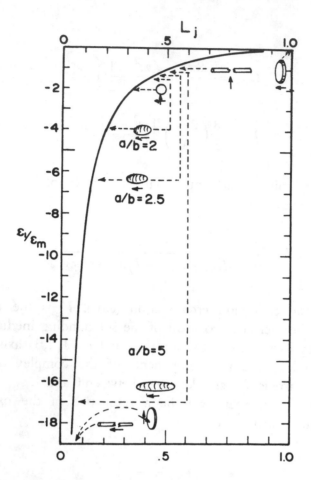

Figure 2.15. The effect of particle shape on the lowest order surface mode resonance as calculated in the Rayleigh approximation for ellipsoids. The heavy line is a plot of the function $\epsilon_1/\epsilon_m = -(1/L_j - 1)$, where $4\pi L_j$ is known as a depolarization factor. Solid arrows next to the various shapes designate the orientation of the electric field vector with respect to the particle axis (Huffman 1977).

field parallel to axis) and $\epsilon_1 = -\epsilon_m$ for long cylinders with the electric field perpendicular to the long axis. Values of $\hbar w$ at which $\epsilon_1 = -2$ are given in Table 2.11 for several metals.

Figure 2.16 shows the evolution of the extinction produced by Au particles as one progresses from isolated particles to a thin Au film (Kreibig 1986). The plasmon resonance at $19\,000\,\text{cm}^{-1}$ (about 2.4 eV), which is clearly present in the spectrum of isolated particles, broadens and then disappears as increasing clustering occurs. A similar effect has been observed for Ag particles in aqueous and solid suspension in a

Table 2.11. *The photon energy $\hbar w$ at which $\epsilon_1 = -2$ for several metals. This would be the surface plasmon resonance in small spherical particles.*

Metal	Energy $\hbar w$ for $\epsilon_1 = -2$ (eV)
Ag	3.4
Al	8.8
Au	2.5
Co	4.7
Mg	6.2
Mo	6.5
Ni	5.9
Sn	7.3
Ta	7.2
V	4.1

gelatin matrix (Kreibig *et al.* 1989). The optical properties of very thin granular films have been studied both theoretically (Hunderi 1989, Berthier and Driss-Khodja 1989, Bedeaux and Vlieger 1983) and experimentally (Bieganski *et al.* 1989, Jebari *et al.* 1989, Abelés *et al.* 1975). In metal films that can be considered to be porous, extra absorption at UV wavelengths appears due to the excitation of resonances within cavities that exist in the film (Jebari *et al.* 1989).

An example of the way in which structural disorder may affect electronic structure in thin insulating films can be seen in Figure 2.17. This plot shows the wavelength dependence of α in thin amorphous MgO films deposited by CO_2 laser sublimation of bulk MgO (MacLean and Duley 1984a). The absence of the 160 nm excitonic resonance is apparent in these thin film spectra. One sees instead an enhancement of absorption in the 180–240 nm region. An increase in α over this wavelength range can be associated with the presence of O^{2-} ions in states of low coordination (Zecchina *et al.* 1975). Such low coordination sites are also observed in finely divided powders of MgO and other oxides (Tench and Pott 1974). When these defects are present, the effect is to introduce states within the bandgap. Electronic transitions involving these defects can be seen in photo-excitation spectra of MgO particles as well as from bulk MgO that has been mechanically abraded (Figure 2.18) (MacLean and Duley 1984b).

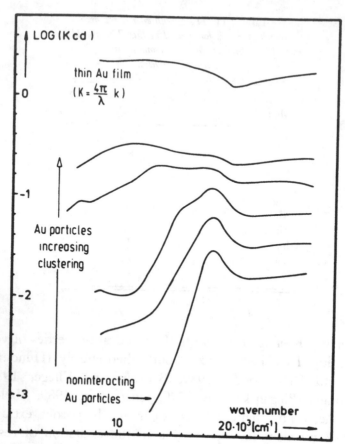

Figure 2.16. Measured extinction spectra of Au particle systems with varying amount of aggregation. Particle size: $2R = 10\,\text{nm}$. The computed absorption spectrum of a thin Au film is plotted for comparison (upper curve). Kcd is the extinction constant. The aggregation (clustering) increases from bottom to top (Kreibig 1986).

2.8 COMPOSITES

The optical response of a heterogeneous system containing metal (i.e. conducting) particles in a matrix of insulating material can be approximated using the Maxwell-Garnett (1904, 1906) theory. In this approximation, the dielectric constant of the medium as a whole can be expressed

$$\epsilon^{MG} = \frac{\epsilon_m[1 + 2f(\epsilon - \epsilon_m)/(\epsilon + 2\epsilon_m)]}{[1 - f(\epsilon - \epsilon_m)/(\epsilon + 2\epsilon_m)]} \tag{47}$$

where ϵ is the complex dielectric constant of the metal and ϵ_m is that of the insulating matrix. f is the volume fraction of metal and must be $\ll 1$

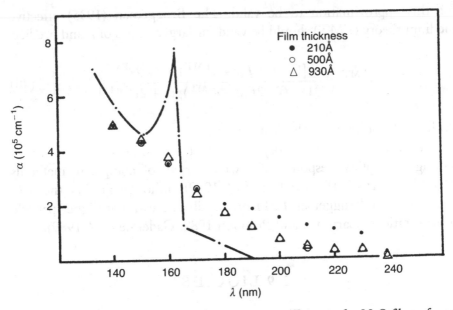

Figure 2.17. Wavelength dependence of absorption coefficient α for MgO films of various thickness. The dash–dot line is α calculated from bulk optical constants (MacLean and Duley 1984a).

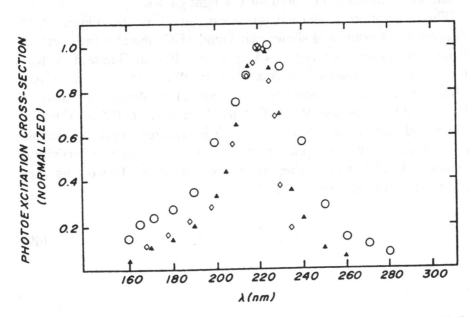

Figure 2.18. Normalized photoexcitation curves obtained with $\lambda_L = 380\,nm$: ($\bigcirc$), MgO from $MgCO_3$; other symbols, absorption seen in various MgO crystals (MacLean and Duley 1984b).

for this approximation to be valid. The Bruggeman (1935) effective medium theory (EMT) should be valid for larger values of f and predicts

$$\epsilon^{EMT} = \frac{\epsilon_m[1 - f + f(\epsilon - \epsilon^{EMT})/(\epsilon + 2\epsilon^{EMT})]}{[1 - f - 2f(\epsilon - \epsilon^{EMT})/(\epsilon + 2\epsilon^{EMT})]} \tag{48}$$

Both results pertain to spherical particles.

The Maxwell-Garnett theory has been shown to be successful in predicting the optical response of a wide variety of composite materials (Perenboom *et al.* 1981, Ashrit *et al.* 1989), although neither the MG theory nor the Bruggeman EMT are totally successful at higher particle concentrations (Carmona and Prudhon 1989, Gadenne *et al.* 1989).

2.9 LIQUIDS

There has been little work done on the interaction of intense UV laser radiation with liquids. A number of organic and inorganic liquids are quite transparent to 308 and 248 nm excimer laser radiation, making them effective media for the propagation of intense UV laser light. As a result, liquid media can be used for UV light guides.

The absorption coefficient for clear liquid water (Figure 2.19) illustrates this point and shows that liquid H_2O provides little attenuation at all excimer wavelengths longer than 193 nm. These data show that the e^{-1} propagation length is 50 cm at 308 nm and 5 cm at 193 nm. Over this wavelength range, the near normal incidence reflectivity of liquid H_2O is generally $R = 0.03$–0.04 (Painter *et al.* 1968). Although the introduction of dissolved compounds is expected to greatly increase k and hence α, the absorption coefficient, little change in reflectivity is produced by all but the highest solute concentrations. This occurs since the normal incidence reflectivity

$$R = \frac{(n - 1)^2 + k^2}{(n + 1)^2 + k^2} \tag{49}$$

reduces to

$$R \simeq \frac{(n - 1)^2}{(n + 1)^2} \tag{50}$$

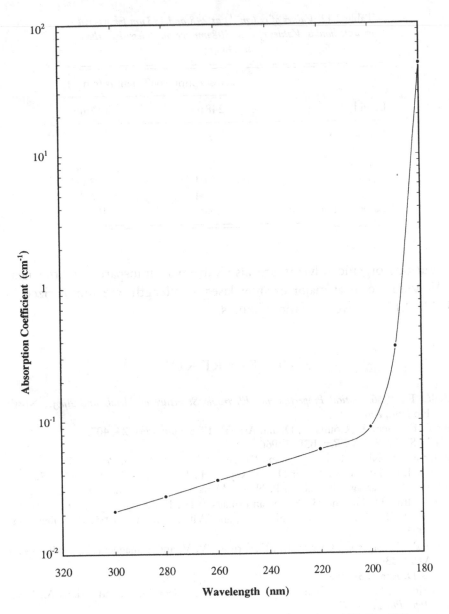

Figure 2.19. The absorption coefficient α (cm^{-1}) for liquid H_2O at UV wavelengths (data from Painter *et al.* 1968).

when $k \ll 1$ whereas changes in n occur only at high solute concentrations. This effect can be seen in the study of the UV optical constants of aqueous solutions of various organic molecules reported by Anderson *et al.* (1978).

Table 2.12. *Values of* α *(cm^{-1}) at 248 and 193 nm for several organic liquids. Values of* α *at 308 nm are much less than those at 248 nm.*

	Absorption coefficient α (cm^{-1})	
Liquid	248 nm	193 nm
Cyclohexane	0.01	3.0
Hexane	<0.01	2.2
Heptane	<0.01	2.08
Methanol	0.04	23.6
Ethanol	0.04	16.4
Carbon tetrachloride	<0.01	>100

Various organic solvents are also extremely transparent in the near UV. Values of α at major excimer laser wavelengths are summarized in Table 2.12 for several of these liquids.

2.10 REFERENCES

Abelés F., 1966. *Optical Properties and Electronic Structure of Metals and Alloys*, North Holland, Amsterdam.

Abelés F., Sheng P., Coutts M. D. and Arie Y., 1975. *Adv. Phys.* **24**, 407.

Adachi S., 1988. *Phys. Rev.* **B38**, 12966.

Andersen M., Nir S., Heller J. M. and Painter L. R., 1978. *Radiat. Res.* **76**, 493.

Andrew J. E., Dyer P. E., Foster D. and Key P. H., 1983. *Appl. Phys. Lett.* **43**, 717.

Antonini M., Camagni P., Gibson P. N. and Manara A., 1982. *Radiat. Effects* **65**, 41.

Arai K., Imai H., Hosono H., Abe Y. and Imagawa H., 1988. *Appl. Phys. Lett.* **53**, 1891.

Arakawa E. T., Dolfini S. M., Ashley J. C. and Williams M. W., 1985. *Phys. Rev.* **B31**, 8097.

Ashrit P. V., Bader G., Gironard F. E., Truong V. V. and Yamaguchi T., 1989. *Physica* **A157**, 333.

Aspnes D. E. and Studna A. A., 1983. *Phys. Rev.* **B27**, 985.

Barbarino S., Grasso F., Guerriera G., Musumeci F., Scordino A. and Triglia A., 1982. *Appl. Phys.* **A29**, 77.

Bedeaux D. and Vlieger J., 1983. *J. Physique Coll.* **C10**, 379.

Berthier S. and Driss-Khodja K., 1989. *Physica* **A157**, 356.

Bieganski P., Dobierzsewska-Mozrzymas E. and Piesert J., 1989. *Physica* **A57**, 371.

Bilenko D. I., Belobrovaya O. Y., Dvokin B. A. and Tsiporukha V. D., 1982. *Opt. Spectrosk.* **53**, 276.

Birks J. B., 1970. *Photophysics of Aromatic Molecules*, Wiley-Interscience, London.

Boesl U., Neusser H. J. and Schlag E. W., 1981. *Chem. Phys.* **55**, 193.

Bohren C. F. and Huffman D. R., 1983. *Absorption and Scattering of Light by Small Particles*, John Wiley, New York.

Born M. and Wolf E., 1975. *Principles of Optics*, Pergamon, New York.

Brannon, J. H., Lankard J. R., Baise A. I., Burns F. and Kaufman J., 1985. *J. Appl. Phys.* **58**, 2036.

Brimacombe R. K., Taylor R. S., Bechtel K. E., Bloembergen N. and Adhav R. S., 1989. *Phys. Rev.* **B17**, 4620.

Bruggeman D. A., 1935. *Ann. Phys. (Leipzig)* **24**, 636.

Cardona M. 1969. *Solid State. Phys. Suppl.* 11, ed. F. Seitz, D. Turnbull and H. Ehrenreich, Academic Press, New York.

Cardona M. and Greenaway D. L., 1964. *Phys. Rev.* **A133**, 1685.

Carmona F. and Prudhon P., 1989. *Physica* **A157**, 328.

Chen Y. R., Williams T. and Sibley W. A., 1969. *Phys. Rev.* **A182**, 960.

Clar E. 1964. *Polycyclic Hydrocarbons: Their Synthesis and Reactions*, Vol. 1, Academic Press, New York.

Cohen M. L., Lin P. J., Roessler D. M. and Walker W. C., 1967. *Phys. Rev.* **155**, 992.

Debye P., 1909. *Ann. Phys. (Leipzig)* **30**, 57.

Devine R. A. B., 1989. *Phys. Rev. Lett.* **62**, 340.

de Voe H., 1965. *J. Chem. Phys.* **53**, 3199.

Dischler B., Bubenzer A. and Koidl P., 1983. *Appl. Phys. Lett.* **42**, 636.

Duley W. W., 1976. *CO₂ Laser: Effects and Applications*, Academic Press, New York.

Duley W. W., 1984a. *Ap. J.* **287**, 694.

Duley W. W., 1984b. *Opt. Commun.* **51**, 160.

Duley W. W., 1987. *Proc. LAMP'87*, High Temperature Society of Japan, Osaka, p. 585.

Dunphy K. and Duley W. W., 1988. *Phys. Stat. Sol.* **B148**, 729.

Ehrenreich H. and Phillip H. R., 1962. *Phys. Rev.* **128**, 1622.

Ewald D., Milleville M. and Weiser G., 1979. *Phil. Mag.* **B40**, 291.

Flygare W. H., 1978. *Theory of the Electronic Spectra of Organic Molecules*, Methuen, London.

Friebele E. J., Griscom D. L., Stapelbrock M. and Weeks R. A., 1979. *Phys. Rev. Lett.* **42**, 1346.

Gadenne M., Lafait J. and Gadenne P., 1989. *Opt. Commun.* **71**, 273.

Gushkin V. S., Shvarev K. M., Baum B. A. and Geld P. V., 1978. *Sov. Phys. Sol. State* **20**, 948.

Hansler R. L. and Segelken W. G., 1960. *J. Phys. Chem. Solids* **13**, 124.

Huffman, D. R., 1977. *Adv. Phys.* **26**, 129.

Hunderi O., 1989 *Physica* **A157**, 309.

Inagaki T., Arakawa E. T., Hamm R. N. and Williams M. W., 1977. *Phys. Rev.* **B15**, 3243.

Jebari M., Borensztein Y. and Vuye G., 1989. *Physica* **A157**, 371.

Jellison G. E. and Modine F. A., 1982. *Appl. Phys. Lett.* **41**, 180.

Johnson P. B. and Christy R. W., 1975. *Phys. Rev.* **B11**, 1315.

Jones C. E. and Embree D., 1976. *J. Appl. Phys.* **47**, 5366.

Karlsson B. and Ribbing C. G., 1982. *J. Appl. Phys.* **53**, 6340.

Kerker M., 1969. *The Scattering of Light*, Academic Press, New York.

Kinsman G. and Duley W. W., 1986. *Proc. SPIE* **668**, 19.

Kinsman G. and Duley W. W., 1993. *Appl. Opt.* **32**, 7462.

Kreibig U., 1986. *Z. Phys.* **D3**, 239.

Kreibig U. and Genzel L., 1985. *Surf. Sci.* **156**, 678.

Kreibig U., Quinten M. and Schoenauer D., 1989. *Physica* **A157**, 244.

Kwong D. L. and Kim D. M., 1983. *J. Appl. Phys.* **54**, 366.

Lakhtakia A., 1992. *Astrophys. J.* **394**, 494.

Lehmann G., 1970. *Phys. Chem. Neue Folge* **72**, 279.

Levy P. W., 1960. *J. Phys. Chem. Solids* **13**, 287.

Lin S. H., Fujimura Y., Neusser H. J. and Schlag E. W., 1984. *Multiphoton Spectroscopy of Molecules*, Academic Press, New York.

Lin S. H., Smith W. L., Lotem H., Bechtel J. H., Bloembergen N. and Adhav R. S., 1978. *Phys. Rev.* B17, 4620.

Lin S. H., Yen R. and Bloembergen N., 1979. *Appl. Opt.* 18, 1015.

MacLean S. and Duley W. W., 1984a. *J. Phys. Chem. Solids* 45, 223.

MacLean S. and Duley W. W., 1984b. *J. Phys. Chem. Solids* 45, 227.

Madan A. and Shaw M. P., 1988. *The Physics and Applications of Amorphous Semiconductors*, Academic Press, New York.

Marr G. V., 1967. *Photoionization Processes in Gases*, Academic Press, New York.

Maxwell-Garnett J. C., 1904. *Phil. Trans. R. Soc. Lond.* 203, 835.

Maxwell-Garnett J. C., 1906. *Phil. Trans. R. Soc. Lond.* 205, 237.

McKenzie D. R., McPhedran R. C., Savuides N., and Botten L. C., 1983. *Phil. Mag.* B48, 341.

Mie G., 1908. *Ann Phys.* 25, 377.

Mizunami T. and Takagi F., 1988. *Opt. Commun.* 68, 223.

Molnar J. P. and Hartman C. D., 1950. *Phys. Rev.* 79, 1015.

Murrell J. N., 1963. *Theory of the Electronic Spectra of Organic Molecules*, Methuen, London.

Nassau K. and Prescott B. E., 1975. *Phys. Stat. Sol.* A30, 659.

Nelson C. M. and Weeks R. A., 1961. *J. Appl. Phys.* 32, 883.

Nitsan U. and Shankland T. J., 1976. *Geophys. J. R. Astr. Soc.* 45, 59.

Painter L. R., Arakawa E. T., Williams M. W. and Ashley J. C., 1980. *Radiat. Res.* 83, 1.

Painter L. R., Hamm R., Arakawa E. T. and Birkhoff R. D., 1968. *Phys. Rev. Lett.* 21, 282.

Palik E. D., 1985. *Handbook of Optical Constants of Solids*, Vol. I, Academic Press, New York.

Palik E. D., 1991. *Handbook of Optical Constants of Solids*, Vol. II, Academic Press, New York.

Perenboom J., Wyder P. and Maier F., 1981. *Phys. Rep.* 78, 173.

Philips J. C., 1966. *Solid State Phys.* 18, 55.

Phillipp H. R., Cole H. S., Lin Y. S. and Sitnik T. A., 1986. *Appl. Phys. Lett.* 48, 192.

Purcell E. M. and Pennypacker C. R., 1973. *Astrophys. J.* 186, 705.

Robertson J., 1986. *Adv. Phys.* 35, 317.

Robertson J. and O'Reilly E. P., 1986. *Phys. Rev.* B35, 2946.

Roessler D. M. and Walker W. C., 1967. *Phys. Rev.* 159, 733.

Roos A., Bergkvist M. and Ribbing C. G., 1989. *Appl. Opt.* 28, 1360.

Rothschild M., Ehrlick D. J. and Shaver D. C., 1989. *Appl. Phys. Lett.* 55, 1276.

Ruppin R., 1982. *Electromagnetic Surface Modes*, Wiley, New York.

Ruppin R. and Englman R., 1970. *Rep. Prog. Phys.* 33, 149.

Shvarev K. M., Gushkin V. S. and Baum B. A., 1978. *High Temp. Res.* 16, 1441.

Srinivasan R., Braren B. and Dreyfus R. W., 1987. *J. Appl. Phys.* 61, 372.

Stathis J. H. and Kastner M. A., 1984. *Phys. Rev.* B29, 7079.

Stathis J. H. and Kastner M. A., 1987. *Phys. Rev.* B35, 2972.

Stohl M. P. I. and Jung C., 1979. *J. Phys. F: Met. Phys.* 9, 2491.

Street R. A., Searle T. M., Austin I. G. and Sussmann R. S., 1974. *J. Phys. C. Solid State Phys.* 7, 1582.

Stuke J. and Zimmerer G., 1972. *Phys. Stat. Sol.* B49, 513.

Svelto O., 1982. *Principles of Lasers*, 2nd Edition, Plenum Press, New York.

Tauc J., 1973. *Amorphous Semiconductors*, Plenum Press, New York.

Taylor R. S., Gibbson R. B. and Roberts J. P., 1988a. *Opt. Lett.* 13, 814.

Taylor R. S., Leopold K. E., Brimacombe R. K. and Mihailov S., 1988b. *Appl. Opt.* **27**, 3124.

Taylor R. S., Leopold K. E., Mihailov S. and Brimacombe R. K., 1987. *Opt. Commun.* **63**, 26.

Tench A. J. and Pott G. T., 1974. *Chem. Phys. Lett.* **26**, 590.

Tsai T. E., Griscom D. L. and Friebele E. J., 1988. *Phys. Rev. Lett.* **61**, 444.

Urbach F., 1953. *Phys. Rev.* **92**, 1134.

van de Hulst J. C., 1957. *Light Scattering by Small Particles*, Wiley, New York.

Walker W. C. and Osantowski J., 1964. *Phys. Rev.* **A134**, 153.

Weeks R. A., Pigg J. C. and Finch C. B., 1974. *Am. Mineral.* **59**, 1259.

Williams R. W. and Arakawa E. T., 1972. *J. Appl. Phys.* **43**, 3460.

Williams R. and Friebele E., 1986. *CRC Handbook of Laser Sciences, III. Optical Materials* ed. M. Weber, CRC Press, Boca Raton.

Wooten F., 1972. *Optical Properties of Solids*, Academic Press, New York.

Yariv A., 1989. *Quantum Electronics*, 3rd Edition, J. Wiley, New York.

Zecchina A., Lofthouse M. G. and Stone F. S., 1975. *J. Chem. Soc. Faraday Trans. I.* **71**, 1476.

CHAPTER 3

Photochemical and photothermal effects

3.1 INTRODUCTION

Ultraviolet laser sources can initiate both photochemical and photothermal effects in condensed media. The relative importance of these two effects depends on a variety of factors including laser wavelength, pulse duration, intensity and the photochemical/photothermal response of the irradiated material. In addition, exposure to UV laser radiation can result in radiation conditioning or hardening, such that the response of the medium to subsequent irradiation may be quite different from its initial response.

This chapter explores some of the fundamental limitations of materials processing with lasers as they relate to the physical and chemical response of the irradiated medium. Some general constraints on the relative rate of ablation in photochemical and photothermal regimes are also discussed. The question of radiation resistance is shown to exhibit both geometrical and physico-chemical characteristics.

3.2 FUNDAMENTAL LIMITATIONS IN LASER MATERIALS PROCESSING

At the intensities customarily used in laser processing of materials, the irradiated sample is exposed to an intense radiative environment that is generally far from the equilibrium state of the ambient medium. The thermal or physical change in the irradiated medium is then driven by an attempt to approach a new equilibrium in the applied radiation field.

In general, even at intensities that may be as large as $10^8\,\mathrm{W\,cm^{-2}}$, the response function of an irradiated medium is usually described using classical heat transfer theory. There are, however, implicit limitations to the validity of this theory as well as assumptions implied by the adoption of this description of the thermal response that may be relevant at high incident laser intensities or short pulse durations (Harrington 1967, Duley 1976). There are also some fundamental limitations that are quantum mechanical in nature.

Consider a medium in which the cohesive energy is U J per atom. Then the energy density stored in bonding within this medium is

$$U_v = \frac{U}{V}\,\mathrm{J\,m^{-3}} \tag{1}$$

where V is the volume of the unit cell and

$$V = \frac{m}{\rho} \tag{2}$$

where m (kg) is the atomic weight and ρ $(\mathrm{kg\,m^{-3}})$ is the density. The applied radiation field with intensity I $(\mathrm{W\,m^{-2}})$ has an energy density

$$u = \frac{I}{c}\,\mathrm{J\,m^{-3}} \tag{3}$$

where c is the speed of light in the medium, $c = 3 \times 10^8/n$ $(\mathrm{m\,s^{-1}})$, where n is the real refractive index. The equality $u = U_v$ then implies a fundamental upper limit to I of

$$I_\mathrm{m} = \frac{3 \times 10^8}{mn}\rho U\ \mathrm{W\,m^{-2}} \tag{4}$$

I_m is typically 10^{18}–$10^{19}\,\mathrm{W\,m^{-2}}$ for most materials and represents an unattainable upper limit at which the electromagnetic energy density becomes equal to the cohesive energy. The electromagnetic electric field strength for $I = 10^{18}\,\mathrm{W\,m^{-2}}$ is $E = 2 \times 10^{10}\,\mathrm{V\,m^{-1}}$. This is comparable to the electric field which exists within an atom.

Another fundamental limitation to the maximum intensity is provided by the Heisenberg uncertainty principle, which can be written as follows:

$$\Delta t = \frac{h}{\Delta(h\nu)} \tag{5}$$

where $h\nu$ is the photon energy. For incident photon flux $F = I/(h\nu)$ and an absorption cross-section σ, the absorption rate is $\sigma F\,\mathrm{s}^{-1}$. If this is then equated to $(\Delta t)^{-1}$, one obtains

$$\frac{1}{\sigma F} = \frac{h\nu}{\sigma I} = \frac{h}{\Delta(h\nu)} \tag{6}$$

or

$$I = \frac{\nu\Delta(h\nu)}{\sigma} \tag{7}$$

This tells us that excitation at a high rate, i.e. at high laser intensity, yields an uncertainty $\Delta(h\nu)$ in the energy of the transition. Under normal conditions $\Delta(h\nu)$ is usually quite small and its effect on σ can safely be ignored. For example, with $I = 10^{12}\,\mathrm{W\,m}^{-2}$, $h\nu = 10^{15}\,\mathrm{Hz}$ and $\sigma = 10^{-20}\,\mathrm{m}^2$, $\Delta(h\nu) = 10^{-23}\,\mathrm{J}$ or $6.25 \times 10^{-5}\,\mathrm{eV}$.

At higher intensities, however, $\Delta(h\nu)$ can become significant, so that the absorption cross-section σ can no longer be specified with accuracy. One can define a limit to I based on the requirement that $\Delta\nu \lesssim \nu$. Then

$$I_{\mathrm{H}} \leq \frac{h\nu^2}{\sigma} \tag{8}$$

For our example, $I_{\mathrm{H}} \leq 6.6 \times 10^{16}\,\mathrm{W\,m}^{-2}$ at $\nu = 10^{15}\,\mathrm{Hz}$. This intensity reduces to $I_{\mathrm{H}} \leq 6.6 \times 10^{12}\,\mathrm{W\,m}^{-2}$ at $\nu = 10^{13}\,\mathrm{Hz}$.

In practice, both I_{m} and I_{H} are several orders of magnitude larger than the threshold for plasma formation and therefore may not be attainable under normal processing conditions. The plasma threshold intensity for opaque materials such as metals ranges from about $5 \times 10^{10}\,\mathrm{W\,m}^{-2}$ at $\lambda = 10.6\,\mu\mathrm{m}$ to about $5 \times 10^{12}\,\mathrm{W\,m}^{-2}$ at excimer wavelengths.

When only *thermal* effects are considered, then laser radiation is effective for materials processing only if the absorbed laser intensity is greater than, or comparable to, the radiative cooling rate of the material under ambient conditions. This requirement can be expressed as follows:

$$\epsilon_\lambda^T I_\lambda \gtrsim \epsilon^T \sigma_{\mathrm{b}} T^4 \tag{9}$$

where ϵ_λ^T is the absorptivity of the surface at temperature T (K), ϵ^T is the total hemispherical emissivity of the surface at T and

Table 3.1. *Values of I_λ at different temperatures for a blackbody.*

T (K)	I_λ (W m^{-2})
10^4	5.7×10^8
3×10^3	4.6×10^6
10^3	5.7×10^4
3×10^2	4.6×10^2
3×10^1	4.6×10^{-2}

$\sigma_b = 5.67 \times 10^{-8}\,\mathrm{m}^{-2}\,\mathrm{s}^{-1}\,\mathrm{K}^{-4}$ is the Stefan–Boltzmann constant. For a blackbody one has the values in Table 3.1.

Thus, at $T = 3000\,\mathrm{K}$, $4.6 \times 10^6\,\mathrm{W\,m}^{-2}$ is required to match the equilibrium thermal radiative emission rate. It is interesting that, at cryogenic temperatures (e.g. $T = 30\,\mathrm{K}$), significant thermal effects may be possible with very low incident laser intensities.

Thermal effects are customarily described in terms of a temperature, T, which is related to the entropy S of the system (Reif 1965)

$$\frac{1}{T} = \frac{\partial S}{\partial U} \tag{10}$$

where U is the internal energy. The concept of temperature is valid as long as the system is in thermodynamic equilibrium and of macroscopic size. Specifically, the requirement is that fluctuations in the internal energy $\Delta U \ll U$. For a sample with N atoms, $\Delta U/U$ becomes

$$\frac{\Delta U}{U} \simeq \frac{1}{(3N)^{1/2}} \tag{11}$$

which is ordinarily a very small number for macroscopic samples. If we consider the subvolume of a material defined by the laser focal area ($\simeq \lambda^2$) and the penetration depth for radiation $\delta \simeq \alpha^{-1}$ where α is the absorption coefficient,

$$N = \frac{\rho \lambda^2}{m \alpha} \tag{12}$$

$$\frac{\Delta U}{U} = \left(\frac{m \alpha}{3 \rho \lambda^2}\right)^{1/2} \tag{13}$$

For a material such as amorphous carbon, $m = 12 \times 1.67 \times 10^{-27}$ kg, $\alpha = 10^8$ m^{-1}, and $\rho = 1.6 \times 10^3$ kg m^{-3}. Taking $\lambda = 248$ nm, we have

$$\frac{\Delta U}{U} = 8.2 \times 10^{-5} \tag{14}$$

This is still sufficiently small that the system can be considered to be macroscopic so that thermal quantities such as U, S, etc. can be defined accurately during such an interaction.

The concept of temperature can be used to describe the system within the laser focus, but T itself is to be associated with particular statistical distributions of particles among available energy levels. For example, assignment of a temperature to the electron gas implies a Fermi–Dirac distribution of populations. Vibrational temperature can only be assigned when a local thermodynamic equilibrium exists between vibrational energy level populations as given by the Boltzmann function. Strong radiative interactions over selected wavelength ranges can influence these population distributions, leading to deviations from thermal equilibrium. Under such conditions, the concept of temperature loses its validity, although an effective or excitation temperature is often assigned to systems under these conditions. This temperature, T_x, is often defined by the ratio of the population of two energy levels according to the Boltzmann relation

$$\frac{N_2}{N_1} = \exp\left[-\frac{E_2 - E_1}{kT_x}\right] \tag{15}$$

where N_1 and N_2 are the populations of states with energies E_1 and E_2, respectively, and $E_2 > E_1$. It is important to realize that T_x defined in this way may be quite different from any thermal equilibrium temperature, if indeed a thermal equilibrium temperature can be defined. Super-heating can occur under conditions of high radiative excitation. In metals this has the effect of decoupling the temperature of the electron gas from that of the lattice (see Chapter 4). The specific heat of the electron gas in a metal (Omar 1975) is

$$C_v^e = \gamma T \tag{16}$$

where γ is typically 10^2 J m^{-3}. Thus, with E (J m^{-2}) absorbed over a depth α^{-1} (m), the temperature excursion of the electron gas is

$$\Delta T = \frac{2E\alpha}{\gamma} \tag{17}$$

in the absence of any transfer to the lattice. Taking $E = 10^2\,\mathrm{J\,m^{-2}}$, $\alpha = 10^8\,\mathrm{m^{-1}}$ and $\gamma = 10^2\,\mathrm{J\,m^{-3}}$, $\Delta T = 2 \times 10^8\,\mathrm{K}$. In fact, ΔT is very much less than this because of the rapid ($10^{-14}\,\mathrm{s}$) transfer of energy from the electron gas to the lattice via electron–phonon scattering. Propagation of heat away from the point of excitation is due to the diffusion of phonons.

The thermal diffusivity, κ, is related to the phonon speed v and mean free path, l, as follows:

$$\kappa = vl \tag{18}$$

For metals, $l \simeq 10\,\mathrm{nm}$ at $300\,\mathrm{K}$. This reduces to $1\text{--}2\,\mathrm{nm}$ at elevated temperatures. In wide band gap insulators, where $\kappa \simeq 10^{-6}\,\mathrm{m^2\,s^{-1}}$, $l \simeq 1\text{--}2\,\mathrm{nm}$ at $300\,\mathrm{K}$. Harrington (1967) has pointed out that l effectively limits the range of validity of the classical heat transfer equations since they cannot be expected to apply over distances that are less than or comparable to l. For two isothermal planes, one at temperature T_1 and the other at temperature T_2, conductive heat transfer is given by

$$\kappa \left[\frac{\Delta T}{\Delta z}\bigg|_1 - \frac{\Delta T}{\Delta z}\bigg|_2 \right] = \Delta z \frac{\Delta T}{\Delta t} \tag{19}$$

Mathematically, $\Delta z \to 0$ to form the derivatives. However, this would imply that $l \to 0$, which is unreasonable from a physical point of view. As a consequence, one expects that there will be a range of Δz and Δt over which equation (19) will not be strictly applicable.

When $l \simeq \alpha^{-1}$, as in metals, laser heating occurs over a distance of about one electron mean free path. Under these conditions, the temperature should be closely related to the variation of incident intensity with distance away from the surface (Harrington 1968):

$$T(z,t) = T_0(t)\exp(-\alpha z) \tag{20}$$

where $T_0(t) = \epsilon_\lambda \alpha I t/(\rho C_p)$. Because $\alpha^{-1} = l$, this becomes

$$T(z,t) = T_0(t)\exp(-z/l) \tag{21}$$

Numerically, with $\epsilon_\lambda = 1$, $\alpha = 10^8\,\mathrm{m^{-1}}$, $I = 10^{10}\,\mathrm{W\,m^{-2}}$, $t = 10^{-9}\,\mathrm{s}$ and $\rho C_p = 10^4\,\mathrm{J\,kg^{-1}\,{}^\circ C^{-1}}$, one obtains $T_0 = 10^5\,{}^\circ\mathrm{C}$. Classical heat transfer theory for a surface source with the same incident intensity predicts

$$T_0(t) = \frac{2\epsilon_\lambda I}{\rho C_p}\left(\frac{t}{\kappa\pi}\right)^{1/2} \tag{22}$$

$$= \frac{35.7}{\sqrt{\kappa}} \tag{23}$$

The large difference between $T_0(t)$ calculated from equations (23) and (21) can be reconciled if part of κ changes dramatically within the layer on the metal surface that is directly heated by laser radiation. Harrington (1968) discussed this point and concluded that K (and hence κ) decreases by a factor of 10^{-5}–10^{-6} in this layer.

An enhanced T within the surface layer would result in rapid vaporization as well as in the emission of charged particles. The kinetic temperature of vaporized atoms and charged species would therefore be considerably higher than that predicted from a calculation of surface temperature. Particles with kinetic energies many times larger than kT are observed in the ejecta from surfaces heated with intense excimer laser pulses (Dyer *et al.* 1988), although it is not clear whether such suprathermal particles are ejected from the surface with this energy, or whether they acquire this energy through interaction with the laser plasma or the incident radiative flux in the gas in front of the surface.

A useful additional insight into the limitations of laser material processing can be obtained by assuming the standard solution for the temperature at the center of a Gaussian focal spot on the surface of an infinite half-space. This is

$$T(r=0, t) = \frac{\epsilon_\lambda I_0 d}{K\sqrt{\pi}}\tan^{-1}\left(\frac{4\kappa t}{d^2}\right)^{1/2} \tag{24}$$

for a source with

$$I(r, t) = I_0(t)\exp\left(-\frac{r^2}{d^2}\right) \tag{25}$$

Ignoring any effect of melting, the time required for this temperature to rise to the vaporization temperature T_v is obtained from the equation

$$\tan\left(\frac{KT_v\sqrt{\pi}}{\epsilon_\lambda I_0\lambda}\right) = \left(\frac{4\kappa t_v}{\lambda^2}\right)^{1/2} \tag{26}$$

where the Gaussian radius has been taken to be $d = \lambda$. A lower limit to t_v would have $t_v = \nu^{-1}$, where ν is the laser frequency. This would require the surface temperature to rise to T_v in only one cycle of the incident radiative flux. A pulse width of this duration would be highly non-monochromatic since $t_v \nu = 1$ and the assumption of focusing to a spot with radius λ is not strictly valid. However, the solution to equation (26) provides another limit to maximum practical values of I_0^m. Then

$$\tan\left(\frac{KT_v\sqrt{\pi}}{\epsilon_\lambda I_0^m \lambda}\right) = \left(\frac{4\kappa}{c\lambda}\right)^{1/2} \tag{27}$$

where c is the speed of light. For metals, at $\lambda = 0.248\,\mu m$ and with $T_v = 2300°C$, $\epsilon_\lambda I_0^m \simeq 5 \times 10^{14}\,W\,m^{-2}$.

A lower limit to ϵI for effective utilization of laser radiation for thermal processing can be obtained from the solution to equation (26) as $t \to \infty$. With $d = \lambda$, the equilibrium temperature excursion ΔT at the center of a Gaussian focus at the sample surface is

$$\Delta T = \frac{\epsilon_\lambda I_0 \lambda \sqrt{\pi}}{2K} \tag{28}$$

where the assumption is made that the material is opaque at λ. Taking $\Delta T = 2300°C$, $\lambda = 0.248\,\mu m$ and $K = 50\,W\,m^{-1}\,°C^{-1}$, $\epsilon_\lambda I_0 = 5.2 \times 10^{11}\,W\,m^{-2}$. At $\lambda = 10.6\,\mu m$, $\epsilon_\lambda I_0 = 1.2 \times 10^{10}\,W\,m^{-2}$.

Another limit is provided by the cooling rate R_v due to evaporation at the normal vaporization temperature. This rate is

$$R_v = pU\left(\frac{1}{2\bar{m}kT_v}\right)^{1/2} \tag{29}$$

whereas the rate for a surface vaporizing at the speed of sound is

$$R_s = v_s UN \tag{30}$$

where p is the pressure, $\bar{m}$ is the average mass of vaporizing atoms, U is the vaporization energy per atom, v_s is the speed of sound, and N is the atomic density. Using values appropriate to Fe, these rates become

$$R_v \simeq 4.1 \times 10^8\,W\,m^{-2}$$

$$R_s \simeq 2.8 \times 10^{14}\,W\,m^{-2}$$

Vaporization inevitably leads to plasma formation when the incident laser intensity exceeds some threshold. This threshold intensity depends on wavelength, the ionization potential of vaporizing atoms and geometric factors such as beam area. When the laser intensity exceeds this threshold value, rapid avalanche ionization occurs and a plasma is generated, which is then heated further by inverse *Bremsstrahlung*. Thomas (1975) has defined the threshold in terms of two collision times, τ_1 and τ_{EA}, where

$$\tau_1^{-1} = \frac{2\alpha_{EA}I}{3(kT_E n_E)} \tag{31}$$

$$\tau_{EA}^{-1} = 2\left(\frac{M_E}{M_a}\right)n_a\Omega \tag{32}$$

τ_1 is then the electron heating time where α_{EA} is the inverse *Bremsstrahlung* absorption coefficient, T_E is the electron temperature and n_E and n_a are the electron and atomic densities, respectively.

τ_{EA} is the relaxation time for energy transfer in electron–atom collisions, where M_E and M_a are electron and atom masses, respectively. Ω is the average $\langle \sigma_{TR} V_E \rangle$ over the Maxwellian electron velocity distribution, σ_{TR} is the collision cross-section and V_E is the electron velocity.

Non-equilibrium heating of the plasma occurs when $\tau_1 \ll \tau_{EA}$. This effect, which neglects multiphoton ionization, yields a scaling law for a characteristic intensity I_c of

$$I_c = 1.1 \times 10^9 \frac{T_v \ (\text{K})}{\bar{m}(\lambda \ (\mu\text{m}))^2} \ \text{W m}^{-2} \tag{33}$$

where $\bar{m}$ is the average atomic mass. With $T_v = 2500\,\text{K}$, $\bar{m} = 50$ and $\lambda = 0.248$, $I_c = 8.9 \times 10^{11}\,\text{W m}^{-2}$.

The plasma angular frequency, w_p, defines a limit to the lowest frequency of laser radiation that can be transmitted through a plasma

$$w_p = \left(\frac{e^2 n_E}{\epsilon_0 m_E}\right)^{1/2} \tag{34}$$

where ϵ_0 is the permittivity of free space. Numerically,

$$\nu_p = \frac{w_p}{2\pi} = 8.98(n_E)^{1/2} \tag{35}$$

Then only those laser wavelengths with

$$\lambda < \frac{c}{8.98(n_E)^{1/2}} \tag{36}$$

will be transmitted by a plasma with electron density n_E.

The electron density in front of a workpiece during laser processing is typically $10^{18}-10^{24}\,\mathrm{m}^{-3}$ (Poprawe et al. 1984). At $\lambda = 0.248\,\mu\mathrm{m}$, reflection occurs only for plasmas with $n_E > 1.8 \times 10^{28}\,\mathrm{m}^{-3}$. At $\lambda = 10.6\,\mu\mathrm{m}$ this limit is $n_E > 9.9 \times 10^{24}\,\mathrm{m}^{-3}$. Thus plasmas over materials during laser processing are generally non-reflecting and incident radiation may penetrate into the plasma. However, they still act to attenuate incident radiation through a variety of absorption, refractive, and scattering processes. As noted earlier, the energy absorbed from the laser acts to heat the plasma and may result in rapid expansion.

When $w < w_\mathrm{p}$, the real refractive index, n, of the plasma can be written (Corson and Lorrain 1962)

$$n = \left[1 - \left(\frac{w_\mathrm{p}}{w}\right)^2\right]^{1/2} \tag{37}$$

$$= \left(1 - \frac{3180 n_E}{w^2}\right)^{1/2} \tag{38}$$

For $n_E = 10^{24}\,\mathrm{m}^{-3}$ and $\lambda = 0.248\,\mu\mathrm{m}$, $n = [1 - (5.5 \times 10^{-5})]^{1/2}$ and refraction can be safely ignored. However, at $\lambda = 10.6\,\mu\mathrm{m}$, $n = (1 - 0.1)^{1/2}$ under the same conditions.

The absorption in laser-produced plasmas is dominated by inverse *Bremsstrahlung* at high electron densities (Herziger and Kreutz 1986). Following Raizer (1965), the absorption coefficient due to this process can be written

$$\alpha = \frac{3.1 \times 10^{-31} Z^3 n^2 g}{[T\,(\mathrm{K})]^{3/2} [h\nu\,(\mathrm{eV})]^2} \quad \mathrm{cm}^{-1} \tag{39}$$

with

$$g = 0.55 \ln\left(\frac{2.4 \times 10^3 T\,(\mathrm{K})}{Z^{4/3} n^{1/3}}\right) \tag{40}$$

where n is the gas density (cm^{-3}), and Z is the effective ionic charge. With $T = 2 \times 10^4\,\mathrm{K}$, $Z \simeq 1$, $n = 3 \times 10^{19}\,\mathrm{cm}^{-3}$ and $h\nu = 5\,\mathrm{eV}$, one

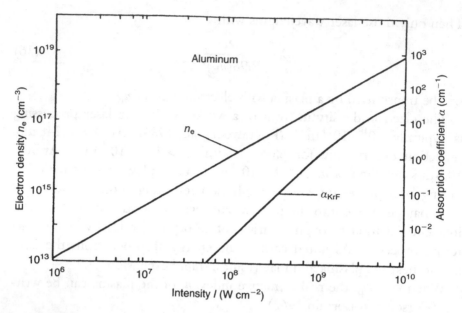

Figure 3.1. Electron density and absorption coefficient for an aluminum
plasma at 0.248 nm versus laser intensity (Poprawe *et al.* 1984).

obtains $g = 1.51$ and $\alpha = 5.95\,\text{cm}^{-1}$. Thus the attenuation length in a
plasma with these parameters would be $\alpha^{-1} = 0.17\,\text{cm}$. The free–free
absorption coefficient with constant plasma parameters scales as ν^{-2}.
Thus energy transfer from a propagating UV laser beam is much less
efficient than from an IR beam of the same total intensity. However,
heating due to multiphoton ionization is greatly enhanced at short laser
wavelengths (Tozer 1965, Poprawe and Herziger 1986). Figure 3.1 shows
electron density and α for 0.248 μm KrF laser radiation in an aluminum
plasma at various incident laser intensities (Poprawe *et al.* 1984).

Heating of the plasma formed in front of a vaporizing surface leads to
a variety of shock phenomena. Herziger and Kreutz (1986) identified
several interaction regimes classified by incident intensity for vaporization. If I_c is the threshold for rapid plasma heating (equation (33)) and
I_D is the threshold for a laser detonation wave, then, for $I_v < I < I_c$, the
surface vaporizes significantly but the plasma produced is of low density
and high transparency. When $I_c < I < I_D$ strong heating of the laser
plasma results in a laser-supported combustion (LSC) wave with the
geometry shown in Figure 3.2(a). This wave is preceded by a shock
wave that propagates away from the surface. However, laser radiation
can still penetrate the plasma to reach the surface. A reverse gas flow
impinges from the shock wave on to the surface.

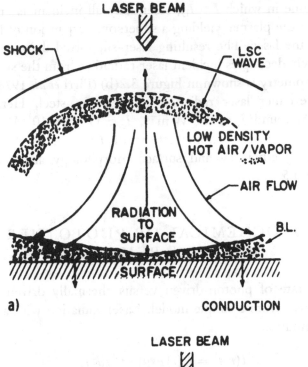

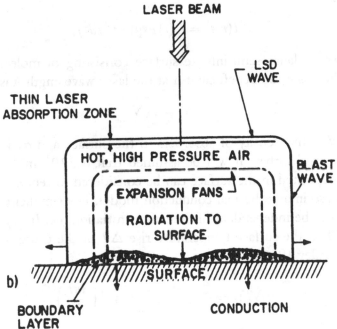

Figure 3.2. (a) LSC wave plasma dynamics. (b) LSD wave plasma dynamics (Pirri *et al.* 1978).

In the regime in which $I > I_D$, essentially all incident laser radiation is absorbed in the plasma, yielding a supersonic expansion of heated gas back toward the laser. The resulting laser-supported detonation (LSD) wave effectively decouples incident laser radiation from the surface. The LSD wave geometry is shown in Figure 3.2(b) (Pirri *et al.* 1978).

For KrF excimer laser radiation incident on steel, Herziger and Kreutz (1986) found $I_v \simeq 10^{10}\,\mathrm{W\,m^{-2}}$, $I_c \simeq (1\text{--}2) \times 10^{11}\,\mathrm{W\,m^{-2}}$, and $I_D \simeq (2\text{--}4) \times 10^{12}\,\mathrm{W\,m^{-2}}$. Some effects of LSC and LSD waves on laser machining efficiency and surface morphology are discussed in Chapters 4 and 5.

3.3 PHOTOCHEMICAL OR PHOTOTHERMAL MATERIAL REMOVAL

The relative rate of photon-driven versus thermally driven processes can be estimated from a simple model. Laser radiation with a Gaussian intensity distribution

$$I(r,t) = I_0(t)\exp(-r^2/d^2) \tag{41}$$

irradiates a planar semi-infinite surface consisting of molecular absorbers. The absorption coefficient α at the laser wavelength λ is

$$\alpha = \sigma_\lambda N \tag{42}$$

where N is the density of absorbers in the surface and σ_λ is the cross-section per absorber. It will be assumed that $\alpha \gtrsim 10^4\,\mathrm{cm^{-1}}$ so that the energy deposited in the surface can be considered to represent a surface heat source in any thermal conduction model. For simplicity $I(r,t)$ will be taken to be independent of time after initiation, i.e. $I(r,t) = I(r)$, for $t > 0$. Then the surface temperature rise ΔT at the center of the focal spot is (Duley 1976)

$$\Delta T(r=0,t) = \frac{\epsilon I_0 d}{K\sqrt{\pi}}\tan^{-1}\left(\frac{4\kappa t}{d^2}\right) \tag{43}$$

where ϵ is the absorptivity, K is the thermal conductivity and κ is the thermal diffusivity. For a laser pulse of infinite duration

$$\Delta T(r=0,\infty) = \frac{\epsilon I_0 d\sqrt{\pi}}{2K} \tag{44}$$

The rate of a thermally induced reaction at the focus will be driven by $\Delta T(t)$. Assuming a first-order rate constant of the Arrhenius form (Busch *et al.* 1978)

$$\delta = A \exp[-\Delta E/(kT)] \tag{45}$$

where A (s^{-1}) is the preexponential frequency factor, ΔE is the activation energy, k is Boltzmann's constant and T is in kelvins,

$$\delta(t) = A \exp\{-\Delta E/[k(T_0 + \Delta T(t))]\} \tag{46}$$

where T_0 (K) is the surface temperature at $t = 0$.

The rate of a photon-driven reaction (e.g. photodissociation, photo-rearrangement) is

$$\rho_\lambda^{(n)}(t) = \sigma_\lambda^{(n)} F_\lambda^{(n)}(t) \tag{47}$$

where $F_\lambda(t)$ is the photon flux, and $\sigma_\lambda^{(n)}$ is the nth-order absorption cross-section. At the center of the focal spot

$$F_\lambda(t) = \frac{\lambda \epsilon I_0(t)}{hc} \tag{48}$$

where h is the Planck constant and c is the speed of light. With $I_0(t)$ constant $\rho_\lambda^{(n)}$ is independent of time. Then, the ratio

$$R_\lambda^{(n)}(t) = \frac{\rho_\lambda^{(n)}}{\delta(t)} \tag{49}$$

yields a quantitative estimate of the relative importance of an nth-order photo-process and a thermally driven reaction with an activation energy ΔE. Taking $d = \lambda$, $R_\lambda^{(n)}(t)$ becomes

$$R_\lambda^{(n)}(t) = \frac{\sigma^{(n)}}{A} F_\lambda^n \exp\left(\frac{T^*}{[T_0 + \Delta T(t)]}\right) \tag{50}$$

where $T^* = \Delta E/k$ and

$$\Delta T(t) = \frac{\epsilon I_0 \lambda}{K\sqrt{\pi}} \tan^{-1}\left(\frac{4\kappa t}{\lambda^2}\right)^{1/2} \tag{51}$$

or

$$\beta = \tan^{-1}(\eta)^{1/2} \tag{52}$$

where $\beta = K\sqrt{\pi}\,\Delta T/(\epsilon I_0 \lambda)$ and $\eta = 4\kappa t/\lambda^2$. In normalized form equation (50) becomes

$$R_\lambda^{(n)}(t) = \gamma^{(n)} \exp\left(\frac{T^*}{(T_0 + \beta I_\lambda')}\right) \tag{53}$$

where

$$\gamma^{(n)} = \frac{\sigma^{(n)}}{A} F_\lambda^m \tag{54}$$

$$I_\lambda' = \frac{\epsilon I_0 \lambda}{K\sqrt{\pi}} \tag{55}$$

As $t \to \infty$; $\beta = \pi/2$ and

$$R_\lambda^{(n)}(\infty) = \gamma^{(n)} \exp\left(\frac{T^*}{T_0 + \epsilon I_0 \lambda\sqrt{\pi}/(2K)}\right) \tag{56}$$

Thus, assuming constant photon flux, thermal effects become more dominant as time increases. Conversely, photon-dominated reaction rates are most significant for $t < \lambda^2/(4\kappa)$.

Figure 3.3 shows a plot of $R_\lambda^{(n)}(t)/\gamma^{(n)}$ versus η for various values of I_λ' and with $T^* = 5000\,\mathrm{K}$. Since the activation energy, ΔE, is typically 20% of the bond energy (Busch *et al.* 1978), $T^* = 5000\,\mathrm{K}$ would correspond to a bond energy of 2.16 eV and the thermal reaction rate or dissociation rate would involve breaking a bond with this energy. The strong dependence of $R_\lambda^{(n)}/\gamma^{(n)}$ on η at small η clearly shows the dominance of photon effects at short times.

If we consider that the solid is a collection of molecular units that absorb photons, redistribute this energy among internal vibrational modes, and then transfer this energy over a longer timescale to the surrounding medium, then the rate of thermal dissociation becomes

$$\delta_\mathrm{m} = \nu_\mathrm{v} \exp[-D/(kT)] \tag{57}$$

where D is the dissociation energy for a bond with frequency ν_v. In this equation, T is to be interpreted as the intramolecular vibrational

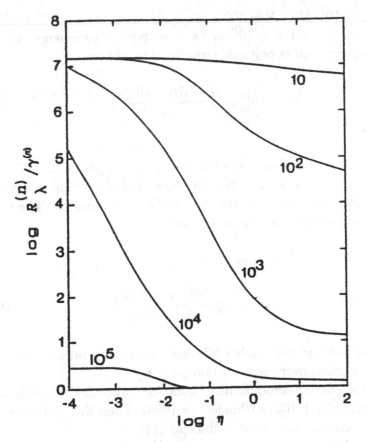

Figure 3.3. A plot of $R_\lambda^{(n)}/\gamma^{(n)}$ versus η for various values of I_λ' and assuming that $T^* = 5000\,\mathrm{K}$ and $T_0 = 293\,\mathrm{K}$.

temperature following absorption of one or more photons. For a molecule with N atoms,

$$T = \frac{q h \nu}{k(3N - 6)} \tag{58}$$

where q is the number of photons absorbed by the molecule over a timescale corresponding to the intermolecular energy delocalization time, τ_R. Assuming constant incident photon flux,

$$q = n\sigma^{(n)} F^n \tau_R \tag{59}$$

$$\delta_m = \nu_v \exp\left(\frac{-(3N - 6)D}{h\nu n\sigma^{(n)} F^n \tau_R}\right) \tag{60}$$

Taking $\nu_v = 10^{13}$ Hz, $N = 100$, $D = 3$ eV, $h\nu = 5$ eV, $\sigma^{(n)} = \sigma^{(1)} = 10^{-21}$ m^2, $n = 1$, and $F = 10^{29}$ m^{-2} s^{-1} as typical parameters for an organic polymer such as polyimide irradiated at 248 nm,

$$\delta_m = 10^{13} \exp\left(\frac{-1.76 \times 10^{-6}}{\tau_R}\right) \tag{61}$$

The value to be assigned to τ_R will depend on the surrounding medium and its state of aggregation. For a free molecule in a gas under collision-less conditions, τ_R will be either an IR radiative lifetime (about 1–10 ms) or the laser pulse length τ_p, whichever is shorter. Then the probability of thermal dissociation per pulse is

$$p^{(n)} = \delta_m \tau_R \tag{62}$$

$$= \nu_v \tau_R \exp\left(\frac{-(3N - 6)D}{nh\nu\sigma^{(n)}F^n\tau_p}\right) \tag{63}$$

since photons are absorbed only while laser radiation is present, whereas dissociation persists over the relaxation time, τ_R.

Vibrational energy delocalization times for molecules in condensed media are typically 1–100 ps (Bondeybey 1984). Then the probability of thermal dissociation per n-photon absorption is

$$p_1^{(n)} = \nu_v \tau_R \exp\left(\frac{-(3N - 6)D}{nh\nu}\right) \tag{64}$$

where $\tau_R = 1$–100 ps. The overall dissociation probability for a laser pulse with $\tau_p > \tau_R$ is approximately

$$p^{(n)} = \sigma^{(n)}F^n\tau_p p_1^{(n)} \tag{65}$$

For $\nu_v = 10^{13}$ Hz, $\tau_R = 100$ ps, $N = 100$, $D = 3$ eV, $h\nu = 5$ eV and $n = 1$, $p_1^{(1)} = 10^3 \exp(-176.4)$, which is vanishingly small. However, under the same conditions, but with $N = 10$, $p_1^{(1)} = 5.6 \times 10^{-4}$ per event. Taking $\sigma = 10^{-21}$ m^2, $F = 10^{29}$ m^{-2} s^{-1} and $\tau_p = 100$ ns, $p^{(1)} = 5.6 \times 10^{-3}$ per pulse. This simple estimate of dissociation probability ignores any statistical effects in the photon flux or redistribution of vibrational energy within the excited molecule. Further discussion of photodissociation is given in Chapter 5.

3.4 RADIATION RESISTANCE

In many applications of lasers for material processing, the material to be worked is subjected to a number of overlapping laser pulses or to an extended dwelling time within the laser beam. It is a general principle in such an instance that the material will attempt to adapt to its radiative environment. This adaptation can take the following forms.

(i)   Development of a surface that minimizes exposure to incident laser radiation, for example by assuming an orientation at oblique incidence with respect to the beam.

(ii)   Development of a shielding plasma or gas flow that acts to intercept the incident beam.

(iii)   Development of a mass flow that provides strong convective cooling of the laser-heated surface.

(iv)   Structural and/or chemical modification of the absorbers exposed to laser radiation.

Mechanism (i) is geometrical in nature and is readily observed in the formation of cone-like structures in solids heavily irradiated with excimer laser radiation (Krajnovich and Vazquez 1993). It also occurs in the formation of a keyhole during penetration welding of metals with high power CO_2 laser radiation (Duley 1983).

The second mechanism is observed when processing materials such as metals at laser intensities exceeding about $10^{10}\,\mathrm{W\,m^{-2}}$ at 10.6 µm and about $10^{11}\,\mathrm{W\,m^{-2}}$ at excimer laser wavelengths. It arises when rapidly vaporized material is ionized by multiphoton absorption or by inverse *Bremsstrahlung* followed by avalanche ionization.

Rapid heat loss due to convection occurs in the flow of liquid metal around the laser keyhole and in conduction welding (Mazumder and Kar 1993).

The last adaptation mechanism to an intense radiative environment involves physical and chemical changes in irradiated material that allow it to withstand the applied radiative flux without further substantial alteration. A simple example is the graphitization of amorphous carbon films upon exposure to laser radiation (Armeyev *et al.* 1989). In this case, adaptation involves a change in bonding from the sp^3 hybridized orbitals characteristic of diamond-like carbon to the sp^2 bonding associated with graphite. The effect of radiation is then to increase the proportion of sp^2 bonded material, thus decreasing the bandgap energy and increasing the conductivity (Robertson and O'Reilly 1987). This permits heat to be more readily conducted away from the laser focus.

Mechanism (iv) can involve both photochemical and photothermal effects. On a molecular basis, weak bonds are the first to be broken, leading to the evolution of molecular fragments. The parent molecule then undergoes a rearrangement, or a reaction with its environment, to yield a product with fewer bonds that are susceptible to dissociation under the ambient radiative conditions. Eventually, the dissociation probability per photochemical or photothermal event is reduced to the point at which dissociation does not occur in this ambient radiative field over the timescale of the interaction.

In the special case of organic materials, this process initially involves the loss of volatile species such as H, OH, CO, etc. Generally this can lead to a significant dehydrogenation of the parent material. Since bond energies for ring carbons are typically 11 eV, whereas peripheral functional groups are bonded to these rings by $\lesssim 5$ eV (Busch *et al.* 1978), dehydrogenation and the loss of other peripheral functional groups will facilitate the formation of extended aromatic ring structures. Such compact aromatic ring structures are known to be radiation-resistant (Gonzales-Hernandez *et al.* 1988).

Compositional and morphological modifications leading to enhanced radiation resistance have also been observed on the surface of $YBa_2Cu_3O_{7-x}$ films exposed to many overlapping 308 nm excimer laser pulses (Foltyn *et al.* 1991). After high levels of irradiation, the exposed surface was found to have evolved a columnar cone-like structure oriented in the direction of the incident laser beam. The chemical composition of these cones was observed to differ significantly from that of the un-irradiated Y–B–C–O surface. The overall effect of these compositional and structural changes was to reduce the evaporation rate/pulse to about 10% of its original value.

3.5 SOME RULES FOR LASER PROCESSING

The adaptation effects outlined in the previous section suggest some general rules for laser materials processing. These rules apply for processing at all laser wavelengths and can be summarized as follows.

1. All physical systems, when exposed to intense laser radiation, will transform so as to minimize their interaction with the incident radiation field.

2. The response of a physical system in adapting to an applied laser field occurs sequentially through a variety of discrete stages.

3. A system will transform so as to attempt to establish an equilibrium between excitation and deexcitation at the point of highest intensity.

3.6 REFERENCES

Armeyev V. Y., Chapliev N. I., Konov V. I., Ralchenko V. G., Strelinitsky V. E. and Volkov Y. Y., 1989. *Proc. SPIE* **1352**, 200.

Bondeybey V., 1984. *Ann. Rev. Phys. Chem.* **35**, 591.

Busch D. H., Shull H. and Conley R. T., 1978. *Chemistry*, Allyn and Bacon Inc., Boston.

Corson D. and Lorraine P., 1962. *Introduction to Electromagnetic Fields and Waves*, W. H. Freeman, San Francisco.

Duley W. W., 1976. *CO_2 Lasers: Effects and Applications*, Academic Press, New York.

Duley W. W., 1983. *Laser Processing and Analysis of Materials*, Plenum Press, New York.

Dyer P. E., Greenough R. D., Issa A. and Key P. H., 1988. *Appl. Phys. Lett.* **53**, 534.

Foltyn S. R., Dye R. C., Ott K. C., Peterson E., Hubbard K. M., Hutchinson W., Muenchausen R. E., Esther R. C. and Wu C. D., 1991. *Appl. Phys. Lett.* **59**, 594.

Gonzalez-Hernandez J., Asomoza R., Mena A. R., Rickards C. J., Chao S. S. and Pawlik D., 1988. *J. Vac. Sci. Technol.* **A6**, 1798.

Harrington R. E., 1967. *J. Appl. Phys.* **38**, 3266.

Harrington R. E., 1968. *J. Appl. Phys.* **39**, 3699.

Herziger G. and Kreutz E. W., 1986. *Phys. Scripta* **T13**, 139.

Krajnovich D. J. and Vazquez J. E., 1993. *J. Appl. Phys.* **73**, 3001.

Mazumder J. and Kar A., 1993. *Proc. NATO ASI Laser Applications for Mechanical Industry* ed. S. Martellucci, A. N. Chester and A. M. Scheggi, Kluwer Academic, Dordrecht, p. 47.

Omar M. A., 1975. *Elementary Solid State Physics*, Addison-Wesley, Reading, Massachusetts.

Pirri A. N., Root R. G. and Wu P. K. S., 1978. *AIAA J.* **16**, 1296.

Poprawe R., Beyer E. and Herziger G., 1984. *Inst. Phys. Conf. Ser.* **72**, 67.

Poprawe R. and Herziger G., 1986. *IEEE J. Quant. Electron.* **22**, 590.

Raizer Y. P., 1965. *Sov. Phys. JETP* **21**, 1009.

Reif F., 1965. *Fundamentals of Statistical and Thermal Physics*, McGraw-Hill, New York.

Robertson J. and O'Reilly E. P., 1987. *Phys. Rev.* **B35**, 2946.

Thomas, P. D., 1975. *AIAA J.* **13**, 1279.

Tozer B. A., 1965. *Phys. Rev.* **A137**, 1665.

CHAPTER 4

Interaction of UV laser radiation with metals

4.1 ABSORPTION AND LATTICE HEATING

The initial stage in the conversion of laser radiation to heat during irradiation involves the excitation of electrons to states of higher energy. For this process to occur, vacant states have to be available to accept excited electrons. When the photon energy $h\nu$ is small, as for example when 10.6 μm laser radiation is absorbed, only electrons with energies within a narrow range $h\nu$ near the Fermi energy, ϵ_F, can participate in absorption. At 0 K, the highest energy reached upon absorption is $\epsilon_F + h\nu$.

At higher temperatures, electrons occupy a range of states given by the Fermi–Dirac distribution (Omar 1975). This reduces to a Boltzmann function for electron energies ϵ such that $\epsilon - \epsilon_F \gg kT$, where T is the metal temperature. Absorption of photons then populates those states with energy $\epsilon + h\nu$. Since ϵ_F is usually several electronvolts, whereas $h\nu = 0.117$ eV for CO_2 laser photons, absorption of IR laser radiation then acts to redistribute electrons among states close to those on the Fermi surface.

This situation is different at excimer laser wavelengths, since $h\nu$ is then comparable to or larger than the work function, φ, of many metals. When $h\nu > \varphi$, electrons may be directly excited from states near the Fermi surface to continuum states associated with the ejection of an electron from the metal. These electrons will originate from levels within the skin depth, δ. Those electrons that are not ejected will dissipate their excess energy as heat within the skin depth. Photoelectrons, as they leave the surface with kinetic energy about $h\nu + kT - \varphi$, will *cool* the surface.

Figure 4.1 shows a plot of photoelectron current density versus laser intensity for 248 nm KrF laser radiation incident on several metals

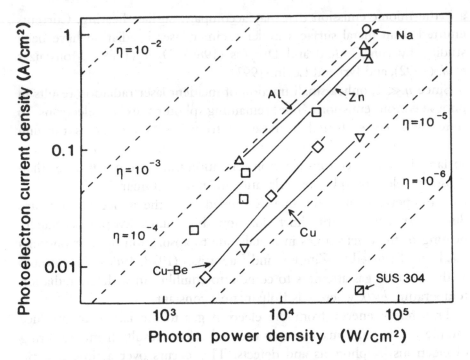

Figure 4.1. Quantum efficiencies for photoelectron emission from various metals subjected to 248 nm KrF laser radiation (Kawamura *et al.* 1984).

(Kawamura *et al.* 1984). The current density $\mathcal{J}$ (A cm^{-2}) can be found from the incident laser intensity I (W cm^{-2}) through the relation

$$\mathcal{J} = \eta \frac{Ie}{h\nu} \tag{1}$$

where e is the electron charge and η is the emission probability. The experimental data suggests $\eta = 10^{-4}$–10^{-5} for those metals investigated. There is no indication of any dependence of η on laser intensity at the relatively low intensities used in this study. For a fluence E (J cm^{-2}) and a pulse duration (τ_p) equation (1) can be rewritten as

$$\mathcal{J} = \eta \frac{Ee}{\tau_p h\nu} \tag{2}$$

Then for $E = 2\,\mathrm{J\,cm^{-2}}$ and $h\nu = 5\,\mathrm{eV}$ (KrF photons) one obtains $\mathcal{J} = 1.6 \times 10^7 \eta\,\mathrm{A\,cm^{-2}}$ when $\tau_p = 25$ ns. A current pulse $\mathcal{J} \simeq 10^2$–$10^3\,\mathrm{A\,cm^{-2}}$ due to photoelectrons would then flow under these conditions. This current is coincident with the incident laser pulse and is distinct from any thermionic emission that may accompany surface heating. Currents emitted from metal surfaces under excimer laser irradiation have been studied by von Gutfeld and Dreyfus (1989), Dyer (1989), Moustaizas *et al.* (1992), and Ivri and Levin (1993).

Since $\eta \ll 1$, only a small fraction of incident laser radiation results in photoelectron emission. The remaining photons excite electrons to states with $\epsilon = \epsilon_F + h\nu$ but these electrons lose their excess through electron–electron and electron–phonon scattering prior to leaving the surface. Because electron–electron relaxation times are much faster than those for electron–phonon relaxation in metals (Omar 1975) the electron temperature may become decoupled from the lattice temperature during irradiation with short duration, high intensity pulses. Superheating of the electron gas in metals due to absorption of laser photons has been discussed by Zinovev and Lugovskoi (1978, 1980). This occurs as the electron gas attempts to come into equilibrium with the radiation temperature $h\nu/k$ where k is Boltzmann's constant.

Transfer of energy from the electron gas to the lattice to produce sample heating and macroscopic thermal effects results from scattering of electrons by phonons and defects. This occurs over a timescale τ_R, which is both temperature- and wavelength-dependent. For DC fields τ_R is typically 10^{-14} s at 300 K. It is apparent then that absorption of photons by electrons in metals results in rapid conversion of excess electronic energy to lattice heat. The distribution of this heat in response to a radiative source with defined spatial and temporal properties can then be calculated using the heat equation (Carslaw and Jaeger 1976)

$$\nabla^2 T(\bar{r},t) - \frac{1}{\kappa}\frac{\partial T(\bar{r},t)}{\partial t} = -\frac{A(\bar{r},t)}{K} \tag{3}$$

where κ is the thermal diffusivity ($\mathrm{cm^{-2}\,s^{-1}}$) and K is the thermal conductivity ($\mathrm{W\,cm^{-1}\,{}^\circ C^{-1}}$). $A(\bar{r},t)$ is the position-dependent rate of heat production per unit time per unit volume ($\mathrm{W\,cm^{-3}}$). Equation (3) assumes that K and κ are independent of temperature and do not vary across the sample.

4.2 SOLUTION OF THE HEAT EQUATION

As has been discussed, the deposition of heat under excimer irradiation of opaque samples occurs over a depth defined by the single-photon absorption coefficient α (cm^{-1}). In metals, at optical frequencies this dimension is typically about 10^{-6} cm. Since α^{-1} is usually much smaller than the lateral spatial extent of a focused excimer beam the heat conduction equation (equation (3)) can be linearized. Then

$$\frac{\partial^2 T}{\partial x^2}(x,t) - \frac{1}{\kappa}\frac{\partial T}{\partial t}(x,t) = -\frac{A(x,t)}{K} \tag{4}$$

where x is a coordinate extending from the sample surface into the material. Because $A(x,t)$ is a volume heat source, it must be evaluated over some incremental length Δx located at x. The usual approximation is to assume that

$$A(x,t) = (1-R)I_0(t)\alpha\exp(-\alpha x) \tag{5}$$

where R is the surface reflectivity and $I_0(t)$ is the time-dependent laser intensity incident at the surface. Then

$$\int_0^\infty A(x,t)\,\mathrm{d}x = (1-R)I_0(t) \tag{6}$$

with

$$\frac{\partial^2 T(x,t)}{\partial x^2} - \frac{1}{\kappa}\frac{\partial T(x,t)}{\partial t} = -\frac{1-R}{K}I_0(t)\alpha\exp(-\alpha x) \tag{7}$$

Equation (7) then describes the solution for the temperature profile produced inside a semi-infinite half space exposed to a uniform surface heat source of intensity $(I-R)I_0(t)$ distributed as $\exp(-\alpha x)$ with depth. Standard analytical solutions to this equation are available for several types of time-dependent source, $I_0(t)$ (Carslaw and Jaeger 1976, Ready 1971, Duley 1976). Some approximate solutions are as follows.

4.2.1 $(\kappa t)^{1/2} \gg \alpha^{-1}$; $I_0(t) = I_0$

In this situation, which may be encountered with strongly absorbing media such as metals, the temperature profile is given by

$$T(x,t) = T_0 + \frac{(1-R)I_0(\kappa t)^{1/2}}{K}\,\mathrm{ierfc}\left(\frac{x}{2(\kappa t)^{1/2}}\right) \tag{8}$$

where T_0 is the initial temperature, and ierfc is the integral of the error function. Then

$$T(0,t) = T_0 + 2\frac{1-R}{K}I_0\left(\frac{\kappa t}{\pi}\right)^{1/2} \tag{9}$$

For example, taking $R = 0.5$, $K = 1.0\,\mathrm{W\,cm^{-1}\,K^{-1}}$, $\kappa = 0.3\,\mathrm{cm^2\,s^{-1}}$ and $t = 20 \times 10^{-9}\,\mathrm{s}$ the surface temperature calculated from equation (9), is $T(0, 2 \times 10^{-8}) \cong 4.35 \times 10^{-5}I_0\,\mathrm{K}$. Thus surface temperatures approaching several thousand kelvins can be produced by excimer laser pulses with $I_0 \simeq 10^8\,\mathrm{W\,cm^{-2}}$. This temperature will increase with pulse duration. For a pulse with fluence $E = I_0 t$, equation (9) becomes

$$T(0,t) = T_0 + 2\frac{1-R}{K}E\left(\frac{\kappa}{\pi t}\right)^{1/2} \tag{10}$$

4.2.2 $(\kappa t)^{1/2} \gg \alpha^{-1}$; $I_0(t) = I_0$, $0 < t < T$; $I_0(t) = 0$, $t > T$

Here equations (8) and (9) describe the temperature profile for $x > 0$ during the laser pulse, i.e. $0 < t < T$. Subsequent to the cessation of the pulse the temperature varies as follows:

$$T(x,t) = T_0 + 2\frac{1-R}{K}\kappa^{1/2}$$

$$\times \left(t^{1/2}\,\mathrm{ierfc}\frac{x}{2(\kappa t)^{1/2}} - (t-T)^{1/2}\,\mathrm{ierfc}\frac{x}{2[\kappa(t-T)]^{1/2}}\right) \tag{11}$$

where ierfc is the integral of the error function erfc:

$$\mathrm{ierfc}\,y = \int_y^\infty \mathrm{erfc}\,s\,ds \tag{12}$$

$$\mathrm{erfc}\,y = \left(\frac{2}{\pi}\right)^{1/2}\int_y^\infty e^{-s^2}\,ds \tag{13}$$

4.2.3 $I_0(t) = I_0$

Here the approximation that $(\kappa t)^{1/2} \gg \alpha^{-1}$ is not necessary and the temperature profile becomes (Carslaw and Jaeger 1976)

$$T(x,t) = T_0 + \frac{1-R}{K\alpha}I_0 f(x,t) \tag{14}$$

where $f(x,t)$ is

$$f(x,t) = 2\alpha(\kappa t)^{1/2} \operatorname{ierfc}\left(\frac{x}{2(\kappa t)^{1/2}}\right) - \exp(-\alpha x)$$

$$+ \tfrac{1}{2}\exp(\kappa\alpha^2 t + \alpha x)\operatorname{erfc}\left(\alpha(\kappa t)^{1/2} + \frac{x}{2(\kappa t)^{1/2}}\right)$$

$$+ \tfrac{1}{2}\exp(\kappa\alpha^2 t - \alpha x)\operatorname{erfc}\left(\alpha(\kappa t)^{1/2} - \frac{x}{2(\kappa t)^{1/2}}\right) \qquad (15)$$

4.2.4 A general pulse shape

To accommodate a general pulse shape we first rewrite the source function $A(x,t)$ (equation (5))

$$A(x,t) = (1 - R)I_0(t)\alpha\exp(-\alpha x)\,\beta(t) \qquad (16)$$

where $\beta(t)$ is a Heaviside unit function

$$\beta(t) = \begin{cases} 1 & t > 0 \\ 0 & t < 0 \end{cases}$$

Then the general pulse shape can be synthesized from a linear combination of step-on, step-off Heaviside functions such that

$$T(x,t) = T^0 + \frac{(1-R)I_0}{K\alpha}[f(x,t) + f(x,t - \tau_{1\,\text{on}}) + f(x,t - \tau_{2\,\text{on}}) + \cdots$$

$$- f(x,t - \tau_{1\,\text{off}}) - f(x,t - \tau_{2\,\text{off}}) - \cdots] \qquad (17)$$

Figure 4.2 shows how the source function for an excimer laser pulse can be approximated in this way (Mihailov 1992). Note that only functions with $t > \tau_i$ would be included in equation (17) for the source function shown in Figure 4.2. τ_I is therefore the specific Heaviside function to be incorporated into equation (16).

4.3 MELTING AND VAPORIZATION

As the surface temperature rises during the initial stages of heating with an excimer laser pulse, the temperature is given by equation (17), which

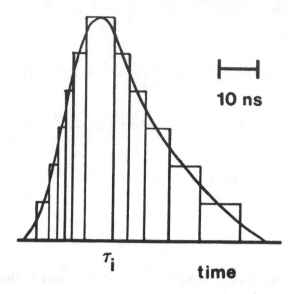

10 ns

τ_i

time

Figure 4.2. A representation of an excimer laser pulse as a linear combination
of step-on, step-off Heaviside functions (Mihailov 1992).

can be approximated by equation (14) if a constant intensity is assumed.
When the melting temperature is reached, further heating supplies the
latent heat of fusion, L_f (J cm^{-3}). The effect of this is to reduce dT/dx at
points within the sample. In the simple one-dimensional model this delay
can be approximated by assigning a fictitious equivalent temperature
increment ΔT to the latent heat, where $\Delta T = L_f/C$ and C is the heat
capacity (J cm^{-3} °C^{-1}) (Rosen *et al.* 1982). Then using equation (14)

$$T(x,t)\,|\,t > t_m = T_0 - \Delta T + \frac{(1-R)I_0 f(x,t)}{K\alpha} \qquad (19)$$

where t_m is the time to reach the melting temperature. In Al,
$\Delta T = 44$°C whereas $\Delta T = 598$°C for Fe.

With the approximation that I_0 is constant and that $(\kappa t)^{1/2} \gg \alpha^{-1}$,

$$T(x,t)\,|\,t > t_m = T_0 - \Delta T + \frac{(1-R)I_0}{K}(\kappa t)^{1/2}\,\mathrm{ierfc}\left(\frac{x}{2(\kappa t)^{1/2}}\right) \qquad (20)$$

If strong vaporization is assumed to occur at some temperature T_v then

$$\frac{t_v}{t_m} = \left(\frac{T_v - T_0 + \Delta T}{T_m - T_0}\right)^2 \qquad (21)$$

Table 4.1. *Mass evaporation rate, β, and linear evaporation rate, ν_B, for several metals at their normal boiling temperature.*

Metal	$\beta\,(T_B)$ $(\mathrm{kg\,m^{-2}\,s^{-1}})$	ν_B $(\mathrm{m\,s^{-1}})$
Al	43.3	1.6×10^{-2}
C	21.2	1.1×10^{-2}
Cu	65.3	0.73×10^{-2}
Fe	57.4	0.73×10^{-2}
Mo	56.2	0.55×10^{-2}
W	75.6	0.39×10^{-2}

is the ratio of the time interval to achieve vaporization to that to reach the melting temperature, T_m. In Al, $t_v/t_m = 2.3$. This ratio does not depend on laser intensity. However, the assumption has been made that K, R and κ are temperature- and phase-independent.

Vaporization occurs at all temperatures but the vaporization pressure and rate becomes larger only near the boiling temperature, $T_v = T_B$. Since the vapor pressure, p, is known at T_B and is equal to 1 atm (or 1.01×10^5 Pa), the Clausius–Clapeyron equation may be used to find p at other temperatures. Then

$$p(T) = p(T_B) \exp\left[\frac{L_v}{k}\left(\frac{1}{T_B} - \frac{1}{T}\right)\right] \tag{22}$$

This result assumes that liquid and vapor are in thermodynamic equilibrium (Sussman 1972).

The mass evaporation rate can be calculated from $p(T)$ assuming equilibrium vaporization conditions

$$\beta(T) = p(T)\left(\frac{\bar{m}}{2\pi kT}\right)^{1/2} \tag{23}$$

where $\bar{m}$ is an average mass of evaporating species. $\beta(T)$ has the units $\mathrm{kg\,m^{-2}\,s^{-1}}$. For a solid of density ρ $(\mathrm{kg\,m^{-3}})$ the linear evaporation rate ν $(\mathrm{m\,s^{-1}})$

$$\nu = \beta(T)/\rho \tag{24}$$

$\beta(T_B)$ and ν_B are listed for several metals in Table 4.1. It is apparent that these quantities are not strongly material-dependent. Typical linear evaporation rates are about $10^{-2}\,\mathrm{m\,s^{-1}}$, implying a vaporization depth of about 0.1 nm for a 10 ns laser pulse. Thus, the depth of vaporization is comparable to one lattice spacing, assuming no shielding of incident laser radiation by ablating material.

In practice the ablation depth, X, should be obtained from the integral

$$X = \int_0^\infty \nu(t)\, dt \tag{25}$$

where $\nu(t) = \beta(T(t))p^{-1}$. Kelly and Rothenberg (1985) have calculated X for a square pulse of duration τ using the approximation (equation (9)) that $(T(t) - T_0) \propto t^{1/2}$. Their result is

$$X = \frac{\bar{p}}{\rho}\left(\frac{\bar{m}}{2\pi k \bar{T}}\right)^{1/2} \tau_{\text{eff}} \tag{26}$$

where

$$\tau_{\text{eff}} = \frac{4k\bar{T}\tau}{L_v} \tag{27}$$

and $\bar{T}$ is the maximum surface temperature at which pressure $\bar{p}$ is obtained. If $\bar{T} \simeq 3 \times 10^3\,\text{K}$, then $\tau_{\text{eff}}/\tau \simeq 0.1$–$0.4$ for many metals, implying that significant vaporization occurs only near the end of the laser pulse.

The total number of atoms removed under these conditions is

$$n_\tau = \frac{\rho X}{\bar{m}} \tag{28}$$

Using values for Al, $\bar{T} = 3000\,\text{K}$ and $\bar{p} = 10^5\,\text{Pa}$, one obtains $X = 5.2 \times 10^{-3}\tau$ from equation (26). Thus with $\tau = 20\,\text{ns}$, $X = 0.1\,\text{nm}$ per pulse. Under these conditions, $n_\tau = 6.1 \times 10^{18}$ atoms m^{-2} per pulse. This can be compared with $n_\tau = 1.95 \times 10^{19}\,\text{m}^{-2}$ per pulse using X determined from equations (23) and (24). This result suggests that equilibrium thermal heating followed by vaporization in the time available during the laser pulse will evaporate less than a monolayer from the metal surface.

This situation may be complicated by the possibility of subsurface heating effects. Rykalin and Uglov (1971) have discussed the rôle played by inclusions in the removal of material under these conditions. These inclusions can act as boiling nuclei in the liquid surface layer, yielding an enhancement in the rate of material removal. Singh *et al.* (1990) recently showed how surface cooling due to vaporization together with the absorption of incident radiation over a length $\delta = \alpha^{-1}$, where α is the absorption coefficient at the laser wavelength, can lead to subsurface heating of metals irradiated with excimer laser pulses.

Equations (23)–(28) were derived under the assumption that vaporization is kinetically limited. It is of interest to examine the solution for the ablation depth per pulse obtained using a conservation of energy equation. In this case, the linear evaporation rate is obtained from

$$\nu = \frac{(1 - R)I}{L_f + L_v + C(\bar{T} - T_0)} \tag{29}$$

where L_f is the latent heat of fusion ($J\,m^{-3}$), $\bar{T}$ is the effective vaporization temperature, L_v is the latent heat of vaporization ($J\,m^{-3}$), and C is the heat capacity ($J\,m^{-3}\,K^{-1}$). Note that ν scales directly with incident intensity I rather than through the dependence of $\bar{T}$ on I (equations (8)–(14)).

For Al, taking $L_f + L_v = 3.36 \times 10^{10}\,J\,m^{-3}$ and $C = 3.6 \times 10^6\,J\,m^{-3}\,K^{-1}$ with $\bar{T} - T_0 = 2494\,K$, one obtains

$$\nu = 2.35 \times 10^{-11}(1 - R)I \tag{30}$$

With $I = 10^{12}\,W\,m^{-2}$, $\nu = 23.5(1 - R)\,m\,s^{-1}$, which should be compared with the value $\nu_v = 1.6 \times 10^{-2}\,m\,s^{-1}$ given in Table 4.1. Since $\nu \gg \nu_v$, kinetic considerations rather than energy conservation must provide the effective limit to ablation depths in Al under these conditions. A similar conclusion can be reached when equations (24) and (29) are compared for other metals.

When steady state vaporization exists (equation (29)) the temperature distribution within the sample has the following simple form (Harrach 1977, Rosen *et al.* 1982):

$$T(x') = \left(T_0 - \frac{L_f}{C}\right) + \left(\bar{T} - T_0 + \frac{L_f}{C}\right)\exp\left(-\frac{x'}{x_0}\right) \tag{31}$$

where $x_0 = K(C\nu)^{-1}$ and x' is measured from the instantaneous position of the ablating surface. $\bar{T}$ can be obtained by equating the evaporation rate ν calculated from equation (24) to that of equation (29):

$$\frac{\beta(\bar{T})}{\rho} = \frac{(1 - R)I}{L_f + L_v + C(\bar{T} - T_0)} \tag{32}$$

The solution to this equation will be the equilibrium surface temperature $\bar{T}$, which scales with absorbed laser intensity $(1 - R)I$. In the absence of plasma shielding effects, $\bar{T}$ can increase to temperatures well above the normal boiling temperature of the material. However, the ablation speed can never exceed that given by the limit provided by kinetic factors (equation (24)).

With longer laser pulses and lower incident intensity, the evaporation rate may be calculated from equation (29) (Duley 1976) and $\bar{T}$ will be close to the normal vaporization temperature T_v. This situation occurs under CO_2 laser irradiation of metals with pulse durations $\tau \geq 1$ ms and $(1 - R)I \leq 10^9 \, \mathrm{W \, m^{-2}}$.

To produce vaporization with a short excimer laser pulse an absorbed intensity $(1 - R)I \geq 10^{12} \, \mathrm{W \, m^{-2}}$ is required (equation (9)). With an intensity this high the solution to equation (29) then immediately implies evaporation rates that are in excess of those that are possible kinetically. As a result, irradiation of metals with short duration (< 100 ns) excimer laser pulses leads to relatively little direct vaporization arising from equilibrium thermal heating. This conclusion was drawn by Kelly and Rothenberg (1985) from their experimental study of the etching of metals with 12 ns, 248 nm excimer laser pulses at fluences near the threshold for vaporization in Al, Au and Pt (about 2.3 J cm^{-2}). They concluded that hydrodynamic effects dominate in the removal of material under these conditions.

4.4 ABLATION RATES IN METALS

In general, the ablation rates observed under excimer laser irradiation greatly exceed those predicted from the theoretical estimates of the previous section. A selection of data obtained for 308 nm ablation of Al and Cu in air (Kinsman 1991) is shown in Figure 4.3. In Cu, X varies from about 0.2 µm per pulse at low fluence to about 0.8 µm per pulse at a fluence of 24 J cm^{-2}. Data for Al show a clear threshold for ablation of 3 J cm^{-2} but much larger values of X at high fluence. Although these data exhibit some scatter, a linear relation exists between X and E with a slope of about $2.5 \times 10^{-2} \, \mu\mathrm{m \, J^{-1} \, cm^2}$ in Cu and $0.28 \, \mu\mathrm{m \, J^{-1} \, cm^2}$ for Al. More scatter is observed when the focal area is reduced (Figure 4.3(a)).

Measurements of ablation rates for other metals in vacuum have been reported by Stafast and von Przychowski (1989). In general, a

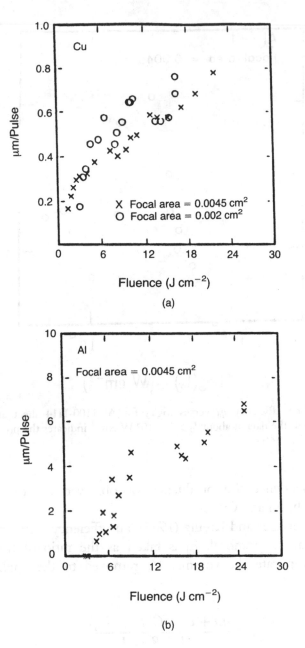

Figure 4.3. Depth per pulse versus fluence at 308 nm for (a) electrolytic Cu and (b) Al 1100-H14 (Kinsman 1991). The laser pulse duration was 31 ns.

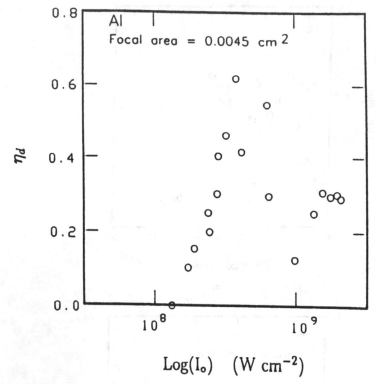

Figure 4.4. Drilling efficiency η_d versus $\log I_0$ for Al 1100-H14 sheet at 308 nm. Increasing scatter in the data at about $I_0 > 5 \times 10^8 \, \mathrm{W \, cm^{-2}}$ indicates the presence of an LSD wave (Kinsman 1991).

non-linear dependence of X on fluence was observed over the range 0–13 J cm^{-2} for Pb, Ni and Cr.

Following Herziger and Kreutz (1984), an efficiency factor η_d can be defined for material removal. η_d is taken as the ratio of the energy required to evaporate the volume, V, removed to the applied laser energy. Then

$$\eta_d = \frac{[L_f + L_v + C(T_v - T_0]V}{(1 - R)I\tau A} \tag{33}$$

$$= \frac{[L_f + L_v + C(T_v - T_0)]X}{(1 - R)I\tau} \tag{34}$$

where A is the focal area. Figure 4.4 shows a plot of η_d versus I for 308 nm laser ablation of Al (Kinsman 1991). This plot shows that η_d first

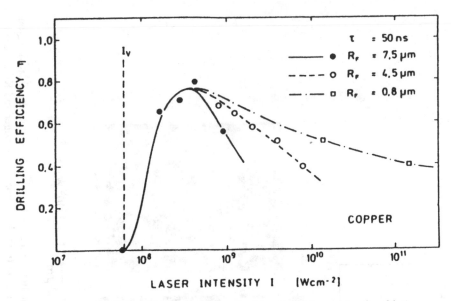

Figure 4.5. Drilling efficiency η_d versus incident laser intensities for ablation of copper at 248 nm (Poprawe *et al.* 1984).

increases with I above threshold to a maximum value of about 0.6 at $4 \times 10^8 \, \text{W cm}^{-2}$. At higher intensities η_d decreases dramatically because much of the incident laser intensity is dissipated in heating the gas in front of the focus. This heating leads to the formation of a laser-supported detonation (LSD) wave that separates from the surface and results in poor coupling of laser radiation into the target area (Herziger and Kreutz 1984, Rykalin *et al.* 1988, Poprawe *et al.* 1984, 1985). Similar behavior has been reported for 248 nm laser ablation of copper (Poprawe *et al.* 1984). A plot of η_d versus laser intensity for constant pulse duration (50 ns) and different focal radii is shown in Figure 4.5. η_d increases from zero at the threshold for vaporization, I_v, to about 0.8 at $I \simeq 6 \times 10^8 \, \text{W cm}^{-2}$. η_d is largest at $I > 6 \times 10^8 \, \text{W cm}^{-2}$ for the tightest focus.

The changing hole morphology associated with an increase in incident intensity through the range within which η_d varies through its maximum value is shown in Figure 4.6 (Herziger and Kreutz 1984). Near threshold, when the efficiency is low but increasing, the ablated area is equal to the area of the incident beam. When η_d is a maximum, the area of material removal has increased somewhat with evidence for redeposited material adjacent to the laser focus. For intensities much in excess of $10^8 \, \text{W cm}^{-2}$, the ablated area has increased to several times that of the laser focus. This increase in spot diameter arises from the reaction of the LSD wave on molten material present on the surface.

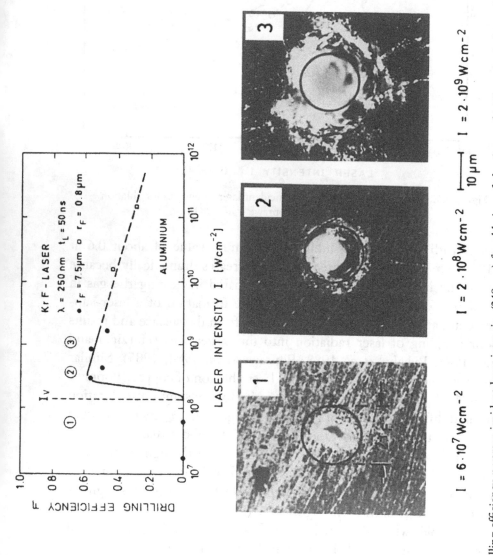

Figure 4.6. (a) Drilling efficiency η versus incident laser intensity (248 nm) for ablation of aluminum. (b) The morphology of the ablated region for intensities extending from threshold (left-hand image) to well above the detonation threshold (right-hand image) (Herziger and Kreutz 1984).

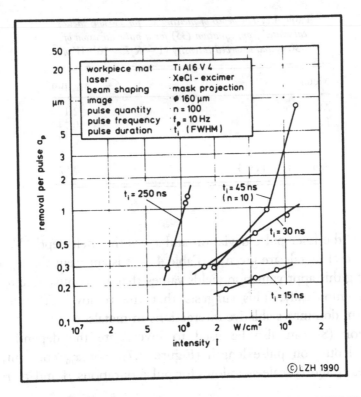

Figure 4.7. The influence of pulse duration on the removal of TiAl6V4
with excimer laser irradiation (Bütje 1990).

The effect of excimer pulse length on the ablation depth per pulse
has been investigated by Bütje (1990). Some data are shown in Figure
4.7 for the alloy TiAl6V4 irradiated at 308 nm. These show that an
ablation depth of 1 μm per pulse can be obtained at an intensity of
$10^8 \, \mathrm{W \, cm^{-2}}$ with a 250 ns pulse. The same ablation depth requires an
intensity of $10^9 \, \mathrm{W \, cm^{-2}}$ with a 30 ns pulse.

Because the ablation depths observed (e.g. Figure 4.3) greatly exceed
those which would be possible from thermal vaporization, material
removal must proceed in another way. This would appear to be via
mechanical removal of liquid in response to the back reaction of the
laser-induced plume on the heated surface. In this case, the ablation
depth per pulse should be related to the depth of melting. This can be
estimated from equations (8) and (19)

$$T_{\mathrm{m}}(x_{\mathrm{m}}, \tau) = T_0 - \Delta T + \frac{(1 - R)I(\kappa\tau)^{1/2}}{K} \mathrm{ierfc}\left(\frac{x_{\mathrm{m}}}{2(\kappa\tau)^{1/2}}\right) \qquad (35)$$

Table 4.2. *The depth of melting* x_m *for various metals calculated from equation (35) for a pulse duration of* $\tau = 30$ *ns and absorbed intensity* $(1 - R)I = 10^{12}\ Wm^{-2}$.

Metal	x_m (μm)
Al	2.1
Fe	0.74
Ti	0.75
W	0.07
Stainless steel (304)	0.57

where τ is the laser pulse deviation and x_m is the melt depth. Values of x_m for several metals are given in Table 4.2. It is apparent that x_m calculated using this approximation yields values that are comparable to those obtained empirically. This suggests that the removal of material in liquid form dominates ablative interactions in metals.

Equation (8) can also be used to investigate the dependence of ablation depths on pulse length (Figure 4.7). For x_m constant, laser intensities I_1 and I_2 scale as follows for pulse durations τ_1 and τ_2, respectively:

$$\frac{I_1}{I_2} = \left(\frac{\tau_2}{\tau_1}\right)^{1/2} \frac{\mathrm{ierfc}\left(\dfrac{x_m}{2(\kappa\tau_1)^{1/2}}\right)}{\mathrm{ierfc}\left(\dfrac{x_m}{2(\kappa\tau_2)^{1/2}}\right)} \tag{36}$$

assuming that R is independent of I.

4.5 SURFACE MORPHOLOGY

The surfaces of metals ablated with excimer laser radiation are characterized by structures that are typical of surface melting accompanied by hydrodynamic interactions. An example of this morphology on the surface of copper can be seen in Figure 4.8 (Kinsman 1991). There is evidence for sputtering from defects, inclusions or subsurface sites as well as extensive recasting of molten material adjacent to sputtering sites. Similar sputtering has been reported in other metals irradiated at 248 nm (Kelly and Rothenberg 1985).

Figure 4.9 shows the surface of copper after exposure to 50 and 500 overlapping 308 nm pulses at an intensity of $1.3 \times 10^9 \, \mathrm{W \, cm^{-2}}$. Here hydrodynamic effects are more apparent, with liquid flowing to form recast material around the periphery of the laser focal area.

At still lower intensity (Figure 4.10) a smoother surface is obtained but radial striations trace the flow of liquid in response to the blast wave created at the center of the focus, where the laser intensity is highest. This is the area over which evaporation will first occur. The laser-supported detonation wave will spread out from this region to interact with melted material farther away from the center of the hole. At higher intensities this detonation wave covers more of the laser focus, accounting for the circular structures observed at the periphery of the hole (Figure 4.9).

For surfaces heavily irradiated in air, oxidation accompanies roughening. This oxide can take the form of small particles that appear to be material redeposited from the laser plume. The primary mechanism would appear to involve the formation of surface asperities which become detached from the metal surface after exposure to several overlapping pulses (Kelly and Rothenberg 1985). Particle sizes of 1–10 μm are predicted for Cu and Al irradiated with excimer laser pulses. SEM photographs of heavily irradiated Cu (Figure 4.11(a)) show that particles in this size range are abundant. Etching of this surface (Figure 4.11(b)) to eliminate oxide reveals a heavily disturbed metal surface with evidence for liquid flow. Figure 4.12 shows the structure of a heavily irradiated Cu surface at high magnification. This morphology is characteristic of turbulent liquid flow.

Kelly and Rothenberg (1985) consider that a turbulent resolidified surface is formed when surface asperities (Figure 4.13) are accelerated away from the liquid surface during each laser pulse due to a volume change on melting followed by thermal expansion of the liquid. Assuming an ideal spherical shape, the restoring force on a sphere of radius r and surface tension σ is

$$f = -\frac{\partial}{\partial r}(4\pi r^2 \sigma) \tag{37}$$

$$= -8\pi r \sigma \tag{38}$$

The change in height of an asperity on going from a solid to liquid is

$$\Delta L = 2r\gamma \, \Delta T + 2r \frac{\rho_s - \rho_l}{3\rho_s} \tag{39}$$

(a)

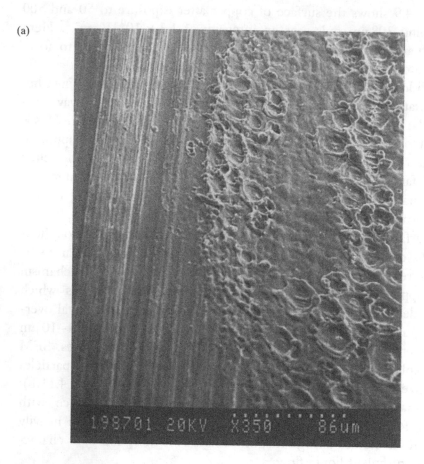

Figure 4.8. SEM micrographs of a Cu surface showing the effect of one pulse at $0.308\,\mu m$ with $I_0 = 1.3 \times 10^9\,\mathrm{W\,cm^{-2}}$. (a) The edge of the irradiation zone. (b) Greater magnification showing evidence of bulk vaporization (Kinsman 1991).

where the first term is the thermal expansion of the liquid on heating to $\Delta T = T - T_m$, and the second term is the change due to conversion from solid to liquid. γ is the linear expansion coefficient of the liquid, and ρ_s and ρ_l are the densities of solid and liquid, respectively. Droplet separation can occur when

$$\frac{4}{3}\pi\rho_l r^3 \frac{\Delta L}{\Delta t} > 8\pi r\sigma\,\Delta t \tag{40}$$

where Δt is the time within the laser pulse after substrate melting has occurred. Solution of these equations for metals such as Al and Cu

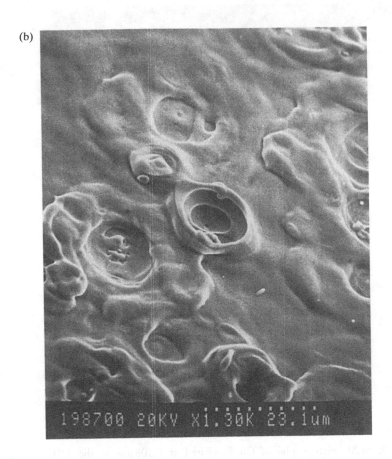

(b)

198700 20KV X1.30K 23.1µm

predicts (Kelly and Rothenberg 1985) a minimum radius for ejected droplets between 0.25 and 1.0 µm.

Organized surface structures consisting of ripples, waves and cones can be produced on the surface of metals at excimer laser intensities that are just above the threshold for surface melting (Ursu *et al.* 1985). For copper, this threshold was found to be $5 \times 10^7 \, \text{W cm}^{-2}$. The nature of the structures produced depended on intensity, the number of overlapping laser pulses, surface preparation and the nature of the ambient gas. Unorganized, monodimensional, and bidimensional structures were observed, with each of these types of structure occurring over a specific intensity versus pulse-number regime (Figure 4.14). The threshold intensity for plasma formation $I^*(N)$ was found to scale linearly with $\log N$ (Figure 4.14).

(a)

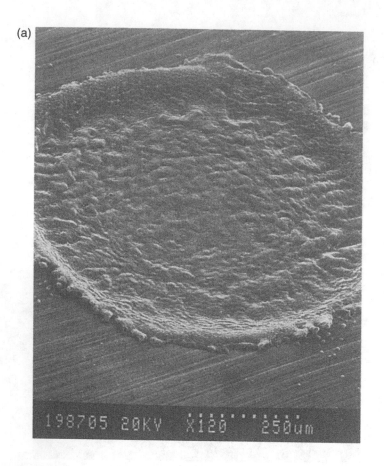

198705 20KV X120 250µm

Figure 4.9. SEM micrographs of Cu irradiated at 0.308 µm in the LSD regime at $I_0 \approx 1.3 \times 10^9$ W cm^{-2} with the target held stationary: (a) 50 pulses and (b) 500 pulses (Kinsman 1991).

In region I, only unorganized structure was observed, with the presence of a random array of 2–5 µm droplets. The substrate exhibited signs of rapid thermal cycling (i.e. cracking). This behavior would appear to follow the model of Kelly and Rothenberg (1985) involving the cycling and ejection of microscopic asperities growing out of a surface melt.

Region II shows a combination of unorganized and monodimensional structures, with the latter following the crystallographic orientation of individual surface grains.

Region III extends over a broad range on the I versus N plot (Figure 4.14) and generally occurs above the threshold for plasma formation.

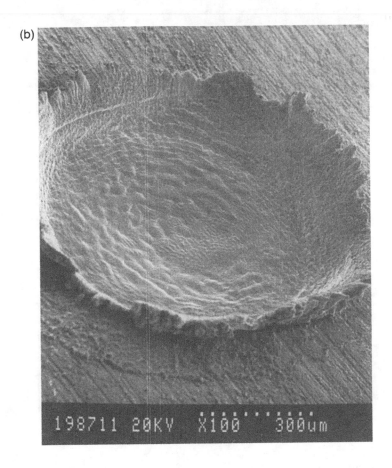

(b)

198711 20KV X100 300µm

The structure consists of regularly spaced conical structures, typically with a circular cross-section of 2–5 µm diameter, accompanied by free particles of about the same diameter.

The crystalline structure of metal surfaces irradiated with XeCl laser pulses has been studied by Pedraza (1987), Juckenath *et al.* (1988) and Bergmann *et al.* (1988). They found that grain structure in the resolidification layer is either the same as that in the substrate or exhibits no detectable changes. In f.c.c. metals such as copper the recrystallized surface layer showed deformation twins whereas those metals with b.c.c. structure such as iron showed no twinning. The grain size was similar to that of the substrate with evidence for epitaxial growth. Materials that ordinarily undergo a martensitic transformation on cooling also show this effect even though cooling rates under these conditions are estimated to be 10^9–10^{11} K s^{-1}. Juckenath *et al.* (1988) estimated that

(a)

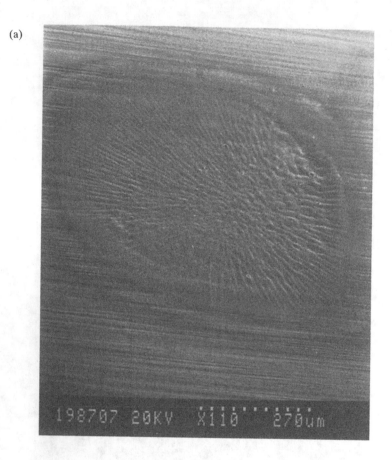

Figure 4.10. SEM micrographs of Cu irradiated at $0.308\,\mu m$ in the LSC regime at $I_0 \approx 1.3 \times 10^8\,W\,cm^{-2}$ with the target held stationary: (a) 50 pulses and (b) 500 pulses (Kinsman 1991).

temperature gradients of 10^5–$10^6\,K\,mm^{-1}$ are present under these conditions and that resolidification speeds exceed $1\,m\,s^{-1}$.

The resulting roughness of the treated surface was found to depend on fluence as well as on the initial roughness of the material. A threshold for enhanced roughness was observed at the laser intensity at which a surface plasma could be initiated (10–$30\,mJ\,mm^{-2}$). Typically, the roughness was $< 1\,\mu m$ below this threshold and 1–$2\,\mu m$ above threshold. The overall roughness was linearly related to initial surface roughness.

Tosto *et al.* (1988) have studied the microhardness of Fe, Ti and TiAl6V4 alloy exposed to XeCl laser pulses in air. They found an

(b)

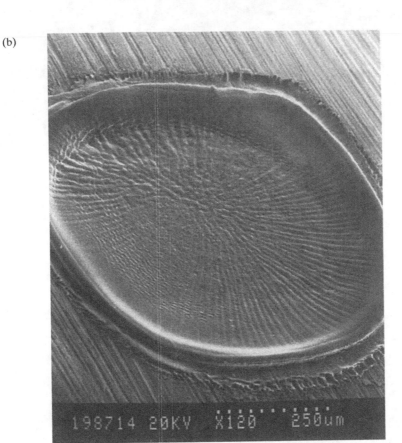

198714 20KV X120 250um

increase in microhardness in all three materials, with the formation of finely dispersed surface precipitates. Precipitate formation was especially noticeable in TiAl6V4 alloy, arising probably via reaction with ambient atmospheric species during irradiation.

Jervis *et al.* (1991) found that martensite in excimer-laser-treated TiAl6V4 is softer than in the mechanically polished alloy. They concluded that surface hardness increases after multiple pulses and is related to interstitial solute hardening.

Surface stress measurements performed on AISI 316, Al and Cu (Tosto *et al.* 1988) indicated that resolidified layers are under tensile stress at close to the yield strength. These materials also showed no increase in hardness in this surface layer. Microcracks were evident in Al.

(a)

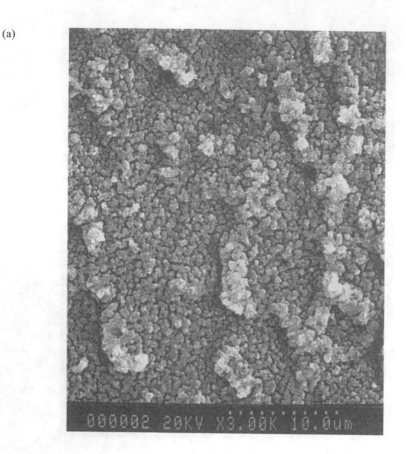

Figure 4.11. SEM micrographs of Cu after irradiation at 0.308 μm with a fluence of $50\,J\,cm^{-2}$. Irradiation proceeded using continuous scanning at $0.3\,cm\,s^{-1}$ so that each focal area was subjected to about 25 overlapping pulses. (a) Shows the resultant oxidized surface. (b) As in (a) but after dissolving surface oxide.

Excimer laser radiation has also been used to mix thin layers of Ti into 304 stainless steel (Jervis *et al.* 1991) to form an amorphous layer with Ti substituting for Fe on a one-to-one basis.

4.6 VAPORIZATION PRODUCTS

Species ejected from metal surfaces irradiated with excimer laser pulses have been studied using mass spectrometry (Viswanathan and Hussla 1986, Kools and Dieleman 1993), ion probe measurements (von Gutfeld

(b)

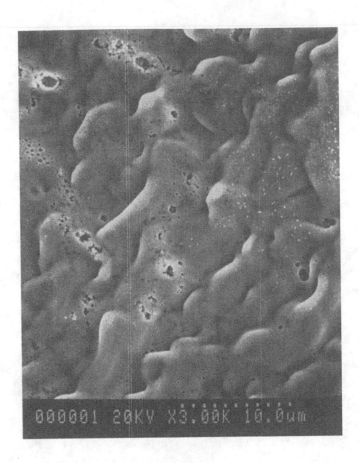

000001 20KV X3.00K 10.0μm

and Dreyfus 1989, Dyer 1989), time-resolved beam deflection (Guo and Lou 1990, Guo *et al.* 1993), emission spectroscopy (Chen and Mazumder 1990), resonant absorption spectroscopy (Gilgenbach and Ventzek 1991), laser spectroscopy (Katsuragawa *et al.* 1989, Brosda *et al.* 1990, Wang *et al.* 1991, Dreyfus 1991), shadowgraphy and *schlieren* photography (Callies *et al.* 1994) and planar laser-induced fluorescence (PLIF) imaging (Sappey and Gamble 1992).

The plume, as imaged in the Al $3\,^2P_{1/2}-4\,^2S_{1/2}$ transition at 394 nm over a solid Al target irradiated with 248 nm laser radiation, shows an expansion at velocities between 0.5×10^6 and $3.4 \times 10^6\,\mathrm{cm\,s^{-1}}$ (Gilgenbach and Ventzek 1991) in vacuum. The highest expansion speed was seen at fluences $E \cong 7.2\,\mathrm{J\,cm^{-2}}$. In the presence of argon or air, the expansion of the Al plume was found to be severely curtailed. However, a shock wave was observed to propagate into the surrounding medium.

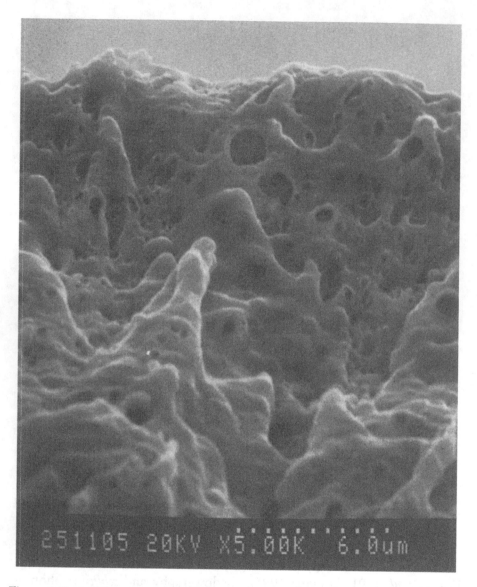

Figure 4.12. A SEM micrograph of Cu irradiated at 0.308 µm with 750 pulses each at a fluence of 75 J cm^{-2} with target held stationary. The micrograph (×5000) shows evidence of turbulent flow at the end of the crater rim.

Gilgenbach and Ventzek (1991) found evidence for kinetic energies of 1–2 eV for neutral Al atoms in the plume initially created over Al in vacuum. Similar kinetic energies have been observed in 193 nm ablation of Cu (Kools and Dieleman 1993). Chen and Mazumder (1990) have recorded emission spectra of some simple diatomic molecules (C$_2$ and

(a)   **(b)**

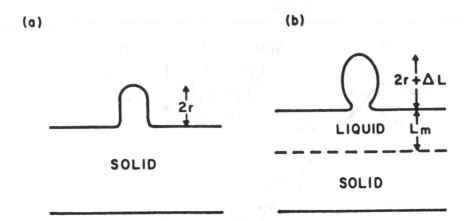

Figure 4.13. A sketch relevant to hydrodynamical sputtering. (a) A solidified asperity with a height of $2r$ on a laser-bombarded metal surface. (b) The same asperity while melted during a subsequent laser pulse. The melt depth of the liquid (L_m) is assumed to be more or less uniform since τ_{thermal} is much less than the laser pulse length. This in turn means that the asperity will show additional total expansion (ΔL) relative to the liquid substrate (Kelly and Rothenberg 1985).

CN) in the plume created by irradiating graphite with 248 nm excimer laser pulses. The intensity at the focus was estimated to be $7 \times 10^8 \, \text{W cm}^{-2}$. They find evidence for vibrational temperatures as high as $(12–15) \times 10^3 \, \text{K}$ from an analysis of relative emission intensity in vibronic lines of C_2.

This analysis assumes an equilibrium Boltzmann distribution, which may not be present. Molecular concentrations over the irradiated surface were estimated to be about $5 \times 10^{14} \, \text{cm}^{-3}$ for C_2 and $2 \times 10^{14} \, \text{cm}^{-3}$ for CN. The mass removal rate under these conditions was approximately $1.5 \times 10^{16} \, \text{cm}^{-3}$ at the laser focus. These values are all much greater than those reported by Katsuragawa *et al.* (1989) over Al and Cu surfaces irradiated at 308 nm. Resonant ionization spectroscopy was used to infer plasma temperatures as high as $10^8 \, \text{K}$. Such temperatures, if confirmed, could only arise through multiphoton excitation of vaporized species by excimer radiation.

Viswanathan and Hussla (1986) have performed a careful study of surface preparation and incident intensity on vaporized species using a variety of diagnostic techniques. They find that, for a clean Cu substrate in a UHV environment, the nature of the species leaving the surface is strongly intensity-dependent. For example, they find that, although neutral Cu monomers are dominant at $3 \times 10^8 \, \text{W cm}^{-2}$, Cu_2 dimers

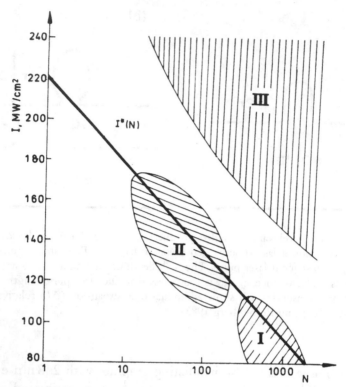

Figure 4.14. An intensity versus pulse-number diagram for electrolytic copper in ambient air (Ursu *et al.* 1985). See text for explanation of regions I–III.

become apparent at $3.6 \times 10^8 \, \text{W cm}^{-2}$. Cu_3 appears at about $5 \times 10^8 \, \text{W cm}^{-2}$, well before the onset of strong plasma formation at about $5 \times 10^9 \, \text{W cm}^{-2}$. Cu^+ ions were first detected at $5 \times 10^8 \, \text{W cm}^{-2}$ with translational energies as high as $7 \, \text{eV}$. Cu atoms and Cu_2 have been detected using laser-induced fluorescence in the ablation of copper using 308 nm laser pulses at a fluence of about $10 \, \text{J cm}^{-2}$ but Cu_2 was not detected under these conditions (Sappey and Gamble 1992).

The presence of adsorbed CO on the Cu surface was found to raise the intensity threshold for emission of neutral Cu and for the appearance of a strong ionized plume (Viswanathan and Hussla 1986). It was suggested that collisional quenching of Cu excitation by CO was responsible for this increasing threshold.

Probe measurements of charged species emitted from Cu irradiated at 248 nm (Dyer 1989, von Gutfeld and Dreyfus 1989) also show that electron temperatures can exceed several electronvolts and that ion energies may be as large as 200 eV. Dyer (1989) found that such ions were emitted with a $\cos^2 \theta$ angular distribution, where θ is the angle between the emitted ion velocity vector and the normal to the surface. The ion velocity was found to decrease substantially as θ increased. The angular dependence of emission has been reviewed by Saenger (1994).

Dyer (1989) proposed that the detection of Cu ions with about 200 eV energies can be understood if these ions are the products of the recombination of highly charged ions created near the metal surface. The limiting ion expansion energy E_i is (Puell 1970)

$$E_i = 5(Z+1)kT_e \qquad (41)$$

where Z is the ionic charge and kT_e is the electron energy. If, as indicated, Z is as large as 4 in this surface region, then $kT_e = 7$ eV yields $E_i = 175$ eV. This energy would be retained when Cu^{4+} ions recombine with electrons to yield Cu^+ and eventually Cu^0.

Wang et al. (1991) used laser-induced fluorescence and Doppler-free spectroscopy to study the state of Al atoms evaporating from Al surfaces excited at 193, 248 and 351 nm. Al atoms with kinetic energies of 6.2 eV were detected, in agreement with other measurements on Al (Gilgenbach and Ventzek 1991) and Cu (Dyer 1989). The velocity distribution was found to be highly non-Maxwellian. If thermal emission from a heated surface is the dominant mechanism, then the velocity distribution should be Maxwellian with only positive velocities. Kelly and Dreyfus (1988) have pointed out that, at high vaporization rates, collisions in the evolving vapor may act to restore this to a full Maxwellian distribution. The surface temperature T_S is then related to atom kinetic energy E as follows:

$$T_S = \frac{E}{\eta k} \qquad (42)$$

where $\eta = 2.52$ for a monatomic gas and 3.28 for a molecule with many internal degrees of freedom. Since $E/k \simeq 72\,000$ K for Al atoms vaporizing under excimer irradiation, $\eta = 2.52$ would imply $T_S \simeq 29\,000$ K, which is clearly incompatible with the critical temperature of Al (3940 K). Thus a model in which highly charged ions created in the immediate vicinity of the metal surface are accelerated by space charge effects and subsequently recombine to yield neutrals may be indicated. Spectroscopic studies of emission from this highly ionized surface layer should be possible and would provide information on the nature and ionization state of the species that exist in this region (Mehlman et al. 1993).

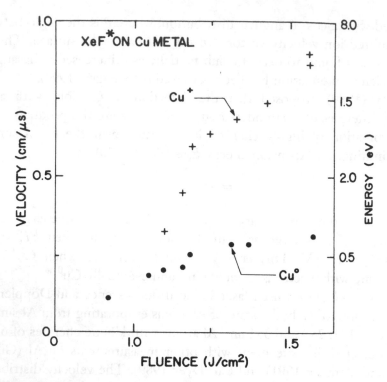

Figure 4.15. The velocity of the leading edge (as defined by the LIF signal reduced to half its maximum) versus fluence. This leading edge velocity is twice the Boltzmann velocity and hence represents an energy four times the Boltzmann energy (Dreyfus 1991).

Dreyfus (1991) used laser-induced fluorescence to obtain the relative concentrations of Cu, Cu^+ and Cu_2 vaporized from solid Cu with 193 and 351 nm pulses at fluences 0.6–12.5 J cm^{-2} and intensities up to 1.2×10^9 W cm^{-2}. Figure 4.15 shows the fluence dependence of particle energy (and velocity) for 351 nm radiation. The velocity of neutral Cu atoms is not strongly dependent on fluence, rising to about 3×10^5 cm s^{-1} at $E = 1.6$ J cm^{-2}. On the other hand, there is a sharp threshold at $E = 1.1$ J cm^{-2} at which the velocity of Cu^+ increases rapidly to 6×10^5 cm s^{-1} and then to 8×10^5 cm s^{-1} at $E = 1.6$ J cm^{-2}. This implies Cu^+ kinetic energies in the 15–22 eV range, some 8–10 times smaller than those observed by Dyer (1989). A comparison of the slopes of plots of the laser-induced fluorescence signals from Cu^+ and Cu^0 versus fluence shows that Cu^+ may arise in a three-photon ionization (at 351 nm) of Cu. Dreyfus (1991) concluded that Cu is produced by thermal evaporation of the parent surface at temperatures near

4000 K. The instantaneous vapor pressure is estimated to be about 30 atm. Multiphoton ionization then leads to a rapid enhancement in the Cu^+ concentration as the fluence is increased. This is followed by avalanche ionization of neutral Cu by electrons heated by inverse *Bremsstrahlung*. Cu_2 is initially formed in the high density environment over the front surface. However, the relative concentration of Cu_2 diminishes as multiphoton ionization becomes more important at high fluences. With plasma heating at high fluence, ion energies may increase into the range measured by Dyer (1989).

The morphology and temporal evolution of the blast wave emanating from metal surfaces irradiated with excimer laser pulses (Sappey and Gamble 1992, Gilgenbach and Ventzek 1991, Callies *et al.* 1994) shows a complex development (Figure 4.16). The structure of an evolving blast wave is summarized in Figure 4.17 (Callies *et al.* 1994) and shows five distinct regions. The shock front US1 precedes an ionization front US2 and expands away from the surface as described by the Sedov (1959) theory for the blast wave accompanying an explosion. The radius, r_2, of the shock front, which is assumed to be spherical, is

$$r_2 = \lambda_0 \left(\frac{E_0}{\rho_1} \right)^{1/5} t^{2/5} \qquad (43)$$

where E_0 is the energy deposited in the explosion ($J\,cm^{-2}$), ρ_1 is the density of the ambient medium and λ_0 is a dimensionless constant, which is 1.033 for a spherical explosion but which takes the value 1.1863 under the conditions of Figure 4.16. Equation (43) has been found to accurately represent the expansion of the shock front accompanying excimer laser ablation of Cu and Al if E_0 is interpreted as the energy deposition up to time t. Using E_0 as an empirical parameter, Callies *et al.* (1994) found that the fraction of the total pulse energy that is deposited in the shock wave by the termination of the laser pulse is a strong function of pulse energy and can increase to about 80% of the incident pulse energies at high fluences.

The discontinuity US2 (Figure 4.17) is a region of luminosity that accompanies the absorption of laser radiation in an ionization front. This front appears about halfway through the laser pulse and subsequently expands at the same speed as US1.

US3 is a contact front between vaporized material and the shocked ambient gas. It starts off as a planar expansion and then evolves after some time t_0 into a spherical expansion. After this time, the radius r_{cf} of

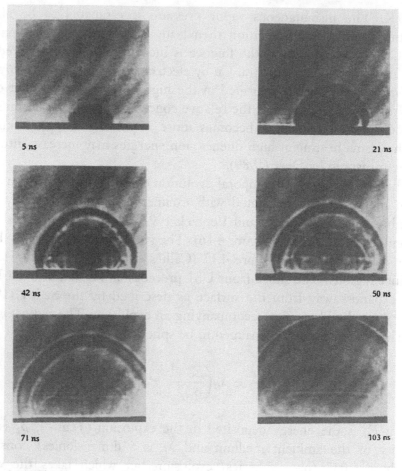

5 ns

21 ns

42 ns

50 ns

71 ns

103 ns

Figure 4.16. The laser-induced density variation detected with the shadowgraph method over a Cu target irradiated with a 248 nm KrF pulse at a fluence $E = 30\,\mathrm{J\,cm^{-2}}$ (Callies *et al.* 1994).

the contact front and r_2 are related as follows:

$$r_{\mathrm{cf}} = \frac{2}{(\gamma + 1)} r_2 \left(\frac{t}{t_0}\right)^{2(1-\gamma)/[5(1+\gamma)]} \tag{44}$$

where γ is the adiabatic coefficient. Equation (44) has been shown to accurately model r_{cf} during the laser pulse (Callies *et al.* 1994).

The lateral expansion giving rise to the discontinuity US4 (Figure 4.17) can be attributed to the stagnation produced by plasma ignition (US5) near the surface. This yields a rapid radial surface expansion of vaporized material localized along the surface, which starts part way through the laser pulse and persists until the end of the pulse.

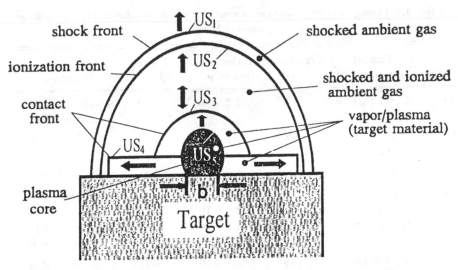

Figure 4.17. A schematic diagram of the discontinuities observed in the blast wave emanating from a copper target irradiated at 248 nm (from Callies *et al.* 1994).

Kools (1993) performed a Monte Carlo simulation of the expansion of laser ablated species into a dilute gas and found that, under certain conditions, self-focusing of the particle flow may occur. In a related experiment the effect of ambient pressure on the morphology of the plume vaporized from barium with doubled Nd:YAG laser radiation (532 nm) was studied by Cappelli *et al.* (1990). They found that the plume develops a 'neck' above the surface. This neck separates a high density, well localized vapor cloud adjacent to the surface from a larger, more diffuse cloud that exists several centimeters from the surface. The origin of this neck is unclear but it may be related to collisional ionization of neutral atoms by electrons heated in the plume by inverse *Bremsstrahlung*.

4.7 LASER-ASSISTED ETCHING

Irradiation of metals with UV laser radiation in the presence of a precursor gas has been shown to lead to efficient chemical etching (Ehrlich *et al.* 1980, 1981, Loper and Tabat 1985, Koren *et al.* 1985, Bäuerle 1986, Ashby 1991). Etching can proceed both by thermally activated and by photolytically driven processes. For gaseous precursors such as Cl_2 or NF_3 irradiated with excimer laser pulses, the primary effect may be the photolytic generation of radicals, which subsequently attack the

Table 4.3. *Etching rates for various metals exposed to UV laser radiation in the presence of gaseous etchants.*

Metal	Etchant (gas, pressure)	Wavelength (nm)	Etching rate (μm pulse^{-1})	Fluence (J cm^{-2})	Reference
Ag	Cl$_2$, 0.1 Torr	337	1.7×10^{-4}	0.12	Sesselmann and Chuang (1985)
Al	Cl$_2$, 0.4 Torr	308	0.1	0.7	Koren *et al.* (1986)
	Cl$_2$, 0.1 Torr		0.03		
Cu	Cl$_2$, 0.4 Torr	308	0.008	0.3	Ritsko *et al.* (1988)
Mo	F$_2$CO	193	0	0.06	Loper and Tabat (1985)
	NF$_3$, 750 Torr		2.2×10^{-5}		
	Cl$_2$, 0.1 Torr	337	3.3×10^{-4}	0.12	Sesselmann and Chuang (1985)
Ti	NF$_3$, 750 Torr	193	2.9×10^{-5}	0.115	Loper and Tabat (1985)
	Br$_2$, 0.1 Torr	248	4×10^{-4}	0.275	Tyndall and Moylan (1990)
	CCl$_3$Br, 0.8 Torr		5.5×10^{-4}		
Ni	Br$_2$, 0.1 Torr	248	2×10^{-4}	0.7	Tyndall and Moylan (1989)

metal surface. This attack leads to the creation of surface compounds, which either spontaneously desorb, or are desorbed during the following laser pulse.

A comprehensive study of etching of Ti with 248 nm KrF laser radiation in the presence of Br$_2$ or CCl$_3$Br (Tyndall and Moylan 1990) showed, however, that the initial step in the etch process may be the reaction between the etchant and the metal surface between laser pulses. This reaction (in the presence of Br$_2$) leads to the formation of TiBr$_2$ and may involve reaction with Br$_2$ as well as with Br. With CCl$_3$Br, both Br and CCl$_3$ radicals are indicated. The large etching rates observed (Table 4.3) show that more than one atomic layer can be removed with each laser pulse, suggesting that the reaction proceeds spontaneously between laser pulses. The laser pulse then acts in part as a mechanism for the desorption of adsorbed surface species. In the case of Ti and Br$_2$ this is probably TiBr$_2$, whereas with Al irradiated in Cl$_2$, AlCl$_3$ is desorbed (Koren *et al.* 1986). Mass spectrometric study of the 355, 351, 248 and 193 nm laser-induced desorption from chlorinated Cu surfaces (Sesselmann *et al.* 1986, Chen *et al.* 1988) has shown that a variety of chlorinated species such as $CuCl_2$, Cu_2Cl, Cu_2Cl_2, $Cu_2Cl_3 \ldots Cu_3Cl_3$ are desorbed. The kinetic energies of some of these heavier molecules can be substantial, suggesting that emission occurs under non-thermal equilibrium conditions (Küper et al. 1992).

High resolution etching is possible through control over the intensity distribution of laser radiation on the metal surface together with optimization of such parameters as the pressure of the etching precursor and any inert buffer gases. Blurring of the etching pattern can result from the diffusion of reactive species as well as thermal diffusion in the target. However, for ideal photo-assisted etching, the laser intensity should be reduced to the point at which the temperature rise in the target is negligible. Under these conditions, spatial resolution can be in the 0.3 μm range (Loper and Tabat 1985).

Etching rates under laser-assisted chemical reaction conditions can be substantial and may reach about 1 μm per pulse dependent on the laser fluence, the pulse repetition rate and the gas pressure. Figure 4.18 shows the strong pressure dependence of the etching rate of Al on Si using Cl_2 gas activated with 308 nm XeCl laser radiation (Koren et al. 1985, 1986). The strong peak in etching rate at a Cl_2 pressure of about 0.2 Torr separates a low pressure etching regime, in which the etching rate is limited by the availability of reactants, from a high pressure regime, in which the presence of higher pressure may inhibit desorption of the products or enhance redeposition.

A variation on laser-assisted chemical etching of metals has been reported by Ehrlich et al. (1981). They show that Zn atoms created by the laser photo-decomposition of $Zn(CH_3)_2$ diffuse into an irradiated Al film. The result is the formation of an inhomogeneous mixed phase Al/Zn alloy that is relatively chemically reactive. This alloy can be subsequently removed by chemical etching in the regions of the Al surface defined by Zn doping. Changes in the surface color of stainless steel surfaces exposed to KrF laser radiation have been discussed by Sugioka et al. (1990).

4.8 PROPERTIES OF IRRADIATED SURFACES

4.8.1 General

Exposure of metal surfaces to excimer laser radiation at intensities exceeding $10^7 \, W \, cm^{-2}$ results in changes in surface morphology,

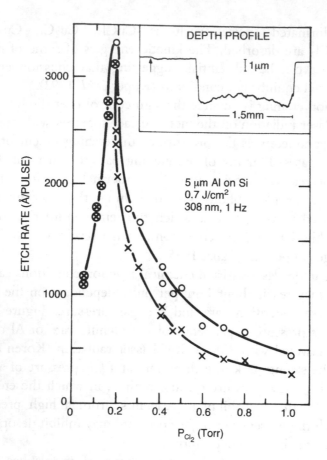

Figure 4.18. Sides (O) and centre (×) etching rates versus the Cl₂ pressure in the cell. The curves are only guides for the eye. In the inset is shown a depth profile of a partially etched film at 0.4 Torr Cl₂ after 15 laser pulses of $0.7\,\mathrm{J\,cm^{-2}}$ at 1 Hz (Koren *et al.* 1985).

hardness and chemistry. A dominant effect is the production of roughening on a microscopic ($<10\,\mu\mathrm{m}$) scale. At low intensity, this roughness can take the form of quasi-periodic structures with one- or two-dimensional symmetry (Ursu *et al.* 1985, Kelly and Rothenberg 1985, Dubowski 1986, van de Riet *et al.* 1993). Higher intensities bring more disordered turbulent morphologies together with the formation of a high concentration of particulates. The composition of this disordered surface is strongly influenced by the ambient atmosphere. Processing in air results in the appearance of oxide and hydroxide particulates (Kinsman and Duley 1988). Under certain conditions, these particulates become entrained in the porous metal substrate to form what is, in effect, a composite material.

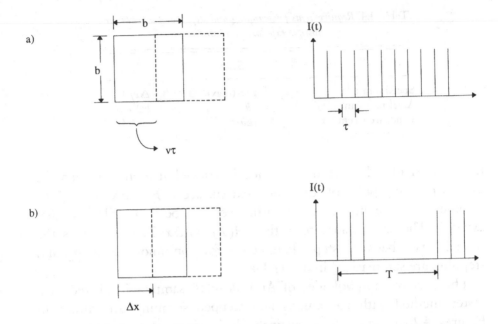

Figure 4.19. (a) The scanning of a square beam across a surface at velocity v showing the position of the beam after a time interval $\Delta t = \tau$, where $\tau = f^{-1}$ and $\Delta t = b/v$. (b) A step scan. The beam is stepped a distance Δx and then pulsed. In each case the time-dependent intensity applied to the workpiece is shown. T is the interval between steps and τ is the interval between pulses.

An ability to change the surface properties of a metal in this way has been shown to yield materials with useful new properties. These involve the scanning or stepping of the excimer beam across the surface or moving the workpiece in a similar manner under the beam. If we consider the excimer beam to be focused onto a square area of side b (Figure 4.19) and to move at speed v while being pulsed repetitively at a rate f, then the total fluence, E', at the surface is

$$E' = \frac{bf}{v} E \ (\mathrm{J\,cm^{-2}}) \tag{45}$$

where E is the fluence per pulse. The first and last 'pixel' in a scan will have lower values of E' than given by equation (45). For example, if a total fluence $E' = 10\,\mathrm{J\,cm^{-2}}$ is required when $v = 1\,\mathrm{cm\,s^{-1}}$, $b = 0.1\,\mathrm{cm}$ and $f = 10^2\,\mathrm{Hz}$, then $E = 1\,\mathrm{J\,cm^{-2}}$. This assumes that the fluence is constant over the focal area. The areal scan rate is $A = vb\ (\mathrm{cm^2\,s^{-1}})$.

An alternate method of irradiation is to step the beam or workpiece a certain distance Δx and then pulse the laser (Figure 4.19(b)). If N is

Table 4.4. *Requirements for scanning and stepping of a beam for multiple exposure of a sample.*

	Scan	Step
Speed (cm s^{-1})	b/τ (max)	$\Delta x/T$
Areal rate (cm^2 s^{-1})	vb	$b\,\Delta x/T$
Exposure (J cm^{-2})	fEb/v	$bNE/\Delta x$

the number of pulses, each of fluence E, applied at each step then the exposure at any point on the irradiated surface is $E' = bNE/\Delta x$. This assumes that the fluence is uniform over the beam profile and that $\Delta x < b$. The areal scan rate is then $A(T) = b\,\Delta x/T$, where T is the time interval between steps. Parameters for continuous scanning and stepping are summarized in Table 4.4.

The surface morphologies of Al 1100-H14 samples irradiated in a raster method with continuous and stepped scanning are shown in Figures 4.20(a) and (b), respectively. It is apparent that microscopic surface morphology can be superimposed on macroscopic structure. This offers the possibility of novel surface reflective and scattering properties (Kinsman 1991, Kinsman and Duley 1993).

4.8.2 Radiative properties

The absorptive and emissive properties of a metal surface can be altered by exposure to excimer laser radiation at intensities $I \geq 10^7\,\mathrm{W\,cm^{-2}}$. Since this exposure involves a change both in surface roughness and in composition, the radiative properties of an irradiated surface will be wavelength-dependent. This effect has been studied by Kinsman (1991) and by Kinsman and Duley (1993). Figure 4.21 shows the wavelength dependence of the reflectivity of several metals irradiated with 248 nm KrF laser radiation at intensities well above the threshold for plasma formation.

The near normal specular reflectance $\rho'(\lambda)$ is related to surface roughness through the expression

$$\rho'(\lambda) = \rho'_F(\lambda) \exp\left(-\frac{(4\pi\sigma)^2}{\lambda^2}\right) \tag{46}$$

where σ is the RMS roughness and the ratio σ/λ is the optical roughness. ρ'_F is the near normal reflectance of a metal surface with no roughness at wavelength λ. This expression is valid when the optical roughness is small (i.e. $\sigma/\lambda < 1$) and therefore should hold in the infrared ($\lambda \geq 5\,\mu m$) for $\sigma \leq 1\,\mu m$. Calculations show that equation (46) has a limited range of validity in describing the reflectance of excimer-laser-irradiated surfaces (Kinsman 1991) since σ can range between 0.2 and $10\,\mu m$.

A dominant effect of irradiation is a reduction in infrared reflectivity. However, the effect is not as large as would be predicted from equation (46). The presence of oxide and/or hydroxide can be seen from spectral regions of low reflectivity. For example, the spectrum of Mg irradiated in air shows spectral features at 3, 6.5 and $11.5\,\mu m$ attributable to $Mg(OH)_2$. Irradiation can also have a profound effect on absorptivity at visible wavelengths. Values of the integrated solar absorptivity, α_S, as well as the total normal emissivity, $\epsilon'(T)$, with $T = 300\,K$ for laser-treated metal surfaces are summarized in Table 4.5. Since both α_S and $\epsilon'(T)$ can be controlled by surface irradiation conditions and ambient atmosphere, there is the possibility of generating selective absorbers and emitters for a variety of applications including spacecraft cooling (Duley and Kinsman 1993).

Specific enhancement of the absorption of metal surfaces by exposure to overlapping excimer laser pulses has been reported by Kinsman and Duley (1989). Some data for Al are shown in Figure 4.22. For heavily processed surfaces (exposure $>10^4\,J\,cm^{-2}$), the absorption coefficient for $10.6\,\mu m$ laser radiation can exceed 0.8. Such samples are heavily oxidized. The reflectivity at $10.6\,\mu m$ is strongly influenced by the orientation of macroscopic surface structure (Figure 4.23), which shows the ratio of reflectivity at angle θ_r to that at the specular angle θ. When the plane of incidence lies parallel to this macroscopic structure, the reflectivity is seen to be primarily specular (Figure 4.23(a)). However, when the orientation of the structure is perpendicular to the plane of incidence, a strong angle-dependent diffuse component is observed (Figure 4.23(b)). Similar effects are seen at other wavelengths (Kinsman 1991).

4.8.3 Effect on absorption

The absorptivity of metallic surfaces containing periodic structures induced by laser radiation has been reviewed by Ursu et al. (1987). They show that, under certain conditions, the presence of periodic structures can lead to near-resonant coupling of incident radiation with a resulting

(a)

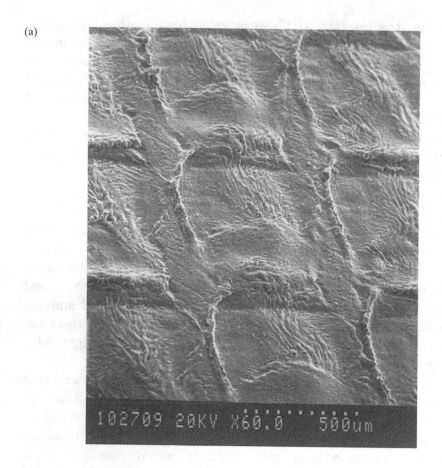

Figure 4.20. SEM micrographs of Al 1100-H14. (a) Continuous scanning with 50 pulses mm^{-1} at $I = 1.9 \times 10^9\,\mathrm{W\,cm^{-2}}$, 248 nm. (b) A step scan with 50 overlapping pulses each with $E = 16\,\mathrm{J\,cm^{-2}}$.

absorptivity close to unity. Such periodic structures can be produced on metals by exposure to overlapping excimer laser pulses (Ursu *et al.* 1985); however, chaotic structures are also frequently observed (Kinsman 1991). Increased absorptivity can be attributed in part to resonant excitation of surface electromagnetic waves. These waves can propagate out of the focal area and can lead to enhanced energy deposition outside the focal spot (Ursu *et al.* 1987).

Experimental studies have shown that the treatment of a metal surface with excimer laser radiation can have a strong effect on the rate of CO_2 laser drilling (Kinsman and Duley 1990), cutting (Kinsman 1991) and welding (Duley *et al.* 1991, Duley and Mao 1994), particularly near

(b)

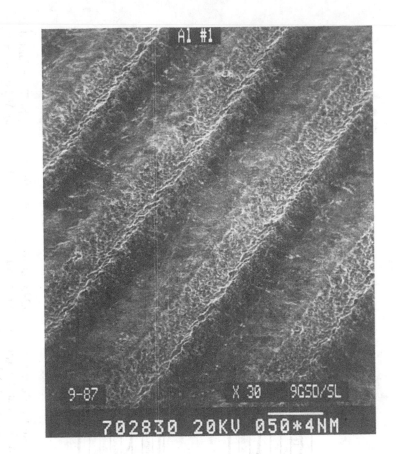

threshold. This occurs as the result of increased coupling of 10.6 μm laser radiation into an irradiated surface. At low CO_2 laser intensity, the effect of pretreatment with 308 nm pulses on welding of 1100 Al has been shown to increase the energy conversion efficiency from several per cent in untreated materials to about 25% in treated material (Duley et al. 1991). Enhancement effects have also been found during simultaneous irradiation with excimer laser pulses during CO_2 laser welding (O'Neill and Steen 1989).

4.8.4 Preparation of metal surfaces for bonding

Surface conditioning by wet chemical or other techniques is customarily used to prepare metals for adhesive bonding. In the case of the adhesive

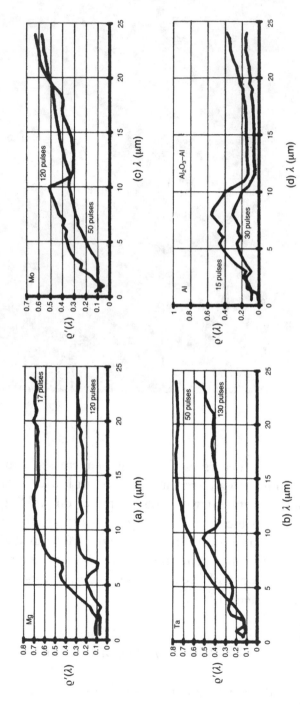

Figure 4.21. $\rho'(\lambda)$ versus wavelength for Mg, Ta, Mo and Al. (a) Mg irradiated at 0.248 µm with $I_0 = 9 \times 10^8\,\mathrm{W\,cm^{-2}}$. (b) Ta irradiated at 0.248 µm with an intensity $I_0 = 8 \times 10^8\,\mathrm{W\,cm^{-2}}$. (c) Mo irradiated at 0.248 µm with an intensity $I_0 = 9 \times 10^8\,\mathrm{W\,cm^{-2}}$. (d) Al 1100-H14 irradiated at 0.308 µm with an intensity $I_0 = 8 \times 10^8\,\mathrm{W\,cm^{-2}}$ (Kinsman and Duley 1993).

Table 4.5. *Room-temperature α_S and ϵ' values for various metals after irradiation with intense KrF excimer laser irradiaion.*

Metal	I_0 ($10^9 \, \text{W cm}^{-2}$)	Pulses	α_S	ϵ'
Al 1100	1.9	50	0.819	0.565
Al 6061 (polished)[a]			0.19	0.042
Al 6061 (unpolished)[a]			0.37	0.042
Cu (electrolytic)	1.2	120	0.813	0.323
Mg	2.2	120	0.928	0.723
Mg	1.5	17	0.891	0.362
Mg[a]			0.27	0.10
Mo	2.2	50	0.916	0.626
Mo	1.0	120	0.920	0.613
Ta	2.2	50	0.834	0.380
Ta	2.2	130	0.871	0.633
Ni (pigmented) Al (anodized)[a]			0.945	0.15

[a] Untreated.

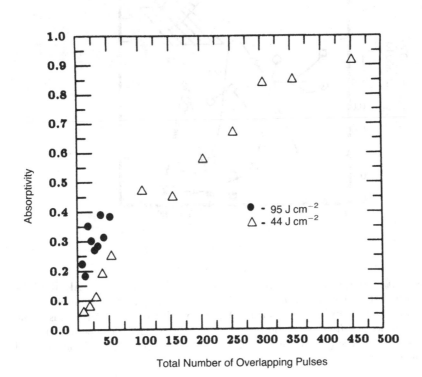

Figure 4.22. Coupling coefficient for 10.6 μm laser radiation on Al versus degree of preprocessing with 308 nm excimer laser pulses (Kinsman and Duley 1989).

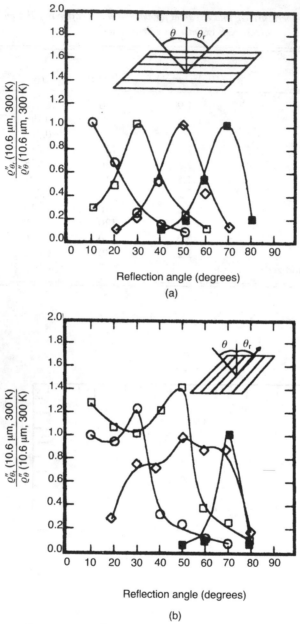

Figure 4.23. $\rho''_{\theta r}$ (10.6 μm)$/\rho''_{\theta}$(10.6 μm) versus the reflection angle for an Al 1100-H14 sheet irradiated at 0.248 μm with $I_0 = 1.9 \times 10^9$ W cm^{-2}. The target was scanned continuously with 50 total overlapping pulses. (a) The direction of the raster scan is in the plane of incidence and (b) the direction of the raster scan is oriented to permit shadowing and masking effects: (O), 10° incidence; (□), 30° incidence; (◇), 50° incidence; and (■), 70° incidence.

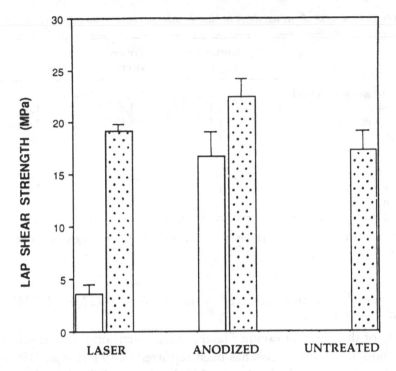

Figure 4.24. A comparison of lap-shear strengths of excimer-laser-treated, anodized and untreated Al 6061 with tie-layer with (dotted columns) and without maleic anhydride. The laser treatment consisted of 13 overlapping pulses each at $I = 6.2 \times 10^8 \, \mathrm{W \, cm^{-2}}$ (Libertucci *et al.* 1993).

bonding of polymers to aluminum, the Al structure is anodized. This form of conditioning promotes adhesion and joint strength through the production of a microfibrous oxide structure. Polymer chains then diffuse into interstices on the oxide layer resulting in enhanced mechanical strength. Since etching with excimer laser radiation in an oxidizing atmosphere yields a structure that is superficially similar to that of anodization, laser treatment has been suggested as a possible alternative to anodizing (Libertucci *et al.* 1993).

The system studied in this work consisted of polypropylene bonded to Al 6061. It was found that, although the wettability of the laser-treated surface was poor and did not vary with number of pulses applied, the correct choice of tie-layer resulted in strong adhesion. Figure 4.24 shows a comparison between the lap-shear strength of bonds to polypropylene made with anodized, untreated and laser-treated Al 6061. The use of a maleic anhydride polypropylene tie-layer was found to yield bonds with lap-shear strengths of $19.1 \pm 0.5 \, \mathrm{MPa}$, which

Table 4.6. *Enhancement of peel test forces in laser-treated galvanneal and galvanized steels.*

	Benchmark (kg cm^{-1})	Treated (kg cm^{-1})	Enhancement
Treatment of galvanneal steel			
Low density	1.1	1.8	1.6
High pulse density	1.1	1.5	1.4
High scan density	1.1	1.8	1.6
Treatment of galvanized steel			
Low density	1.4	3.1	2.2[a]
High pulse density	1.4	4.0[a]	2.8[a]
High scan density	1.4	3.2[a]	2.3[a]

[a] All of the treated temper rolled galvanized steels exhibited some degree of cohesive failure in peel tests.

was comparable to that achieved with anodizing $(22.4 \pm 1.7 \, \text{MPa})$ (Libertucci *et al.* 1993).

A study of the effect of excimer laser pretreatment on the strength of adhesive bonds to coated steels has been reported by Olfert *et al.* (1996). Treatment of temper rolled galvanized and galvanneal sheet steels with 308 nm XeCl laser pulses prior to bonding gave an adhesion enhancement of up to 280% (Table 4.6), although the source of this enhancement is uncertain. Laser enhancement of adhesion of Cu films to sapphire has been discussed by Pedraza *et al.* (1988).

4.8.5 Treatment for corrosion resistance

With the correct choice of laser pulse length and intensity, the surface of a metal can be brought to the melting temperature without significant vaporization. Under these conditions, the metal surface may be smoothed, with the disappearance of surface scratches and eruptions (Bransden *et al.* 1990). These conditions also lead to the appearance of the ripples reported by Ursu *et al.* (1985). In the case of Al, such ripples are spaced by about 3 μm. These effects occur only at low fluences (e.g. 30–50 mJ cm^{-2}).

Talysurf profiles of Cu and stainless steel surfaces before and after irradiation at this fluence level show a dramatic reduction in roughness (Bransden *et al.* 1990). This smoothing, together with the removal of surface inclusions due to explosive vaporization, yielded a dramatic improvement in pitting corrosion in 304 L stainless steel. A similar improvement was observed in the etch resistance of Stellite 6 and the Ni alloy IN100. Other studies of the effect of laser-induced reactions on corrosion resistance in steel have been reported by Okada *et al.* (1992). Cleaning of metals with KrF laser radiation has been discussed by Lu *et al.* (1994).

4.9 REFERENCES

Ashby C. I. H., 1991. *Thin Film Process II*, Academic Press, New York, p. 783.

Bäuerle D., 1986. *Chemical Processing with Lasers*, Springer-Verlag, New York.

Bergmann H. W., Juckenath B. and Lee S. Z., 1988. *DVS Ber.* **113**, 106.

Bransden A. S., Megaw J. H. P. C., Backwill P. H. and Westcott C., 1990. *J. Mod. Opt.* **37**, 813.

Brosda B., Castell-Munoz R. and Kunze H. J., 1990. *J. Phys. D: Appl. Phys.* **23**, 735.

Bütje R., 1990. *Proc. SPIE* **1225**, 196.

Callies G., Schittenhelm H., Berger P., Dansinger F. and Hugel H., 1994. *Proc. SPIE* **2246**, 126.

Cappelli M. A., Paul P. H. and Hanson R. K., 1990. *Appl. Phys. Lett.* **56**, 1715.

Carslaw H. S. and Jaeger J. C., 1976. *Conduction of Heat in Solids*, 3rd Edition, Oxford University Press (Clarendon), London.

Chen L., Liberman V., O'Neill J. A and Osgood R. M. Jr, 1988. *Mater. Res. Soc. Symp. Proc.* **101**, 463.

Chen X. and Mazumder J., 1990. *Appl. Phys. Lett.* **57**, 2178.

Dreyfus R. W., 1991. *J. Appl. Phys.* **69**, 1721.

Dubowski J. J., 1986. *Proc. SPIE* **668**, 97.

Duley W. W., 1976. *CO$_2$ Lasers: Effects and Applications*, Academic Press, New York.

Duley W. W. and Kinsman G., 1993. *Proc. MRS Symp.* **285**, 145.

Duley W. W., Kinsman G. and Mao Y. L., 1991. *Proc. Laser and the Electron Beam*, ed. R. Bakish, Bakish Materials Corp., Englewood, New Jersey, p. 206.

Duley W. W. and Mao Y. L., 1994. *J. Appl. Phys. D: Appl. Phys.* **27**, 1379.

Dyer P. E., 1989. *Appl. Phys. Lett.* **55**, 1630.

Ehrlich D. J., Osgood R. M. Jr and Deutsch T. F., 1980. *IEEE J. Quant. Electron.* **16**, 1233.

Ehrlich D. J., Osgood R. M. Jr and Deutsch T. F., 1981. *Appl. Phys. Lett.* **38**, 399.

Gilgenbach R. M. and Ventzek P. L. G., 1991. *Appl. Phys. Lett.* **58**, 1597.

Guo H. and Lou Q., 1990. *Opt. Commun.* **77**, 381.

Guo H., Lou Q., Chen S. C., Cheung N. H., Wang Z. Y. and Lin P. K., 1993. *Opt. Commun.* **98**, 220.

Harrach R. J., 1977. *J. Appl. Phys.* **48**, 2370.

Herziger G. and Kreutz E. W., 1984. *Laser Processing and Diagnostics*, ed. D. Bauerle, Springer-Verlag, Berlin, p. 90.

Ivri J. and Levin L. A., 1993. *Appl. Phys. Lett.* **62**, 1338.

Jervis T. R., Zocco T. G. and Steele J. H., 1991. *Mater. Res. Soc. Symp. Proc.* **201**, 535.

Juckenath B., Durchholz H., Bergmann H. W. and Dembowski J., 1988. *Proc. SPIE* **1023**, 236.

Katsuragawa H., Minowa T. and Inamura T., 1989. *Nucl. Instrum. Methods* **1343**, 259.

Kawamura Y., Toyoda K. and Kawai M., 1984. *Appl. Phys. Lett.* **45**, 308.
Kelly R. and Dreyfus R. W., 1988. *Surf. Sci.* **198**, 263.
Kelly R. and Rothenberg J., 1985. *Nucl. Instrum. Methods* **137/8**, 755.
Kinsman G., 1991. Ph.D Thesis, York University.
Kinsman G. and Duley W. W., 1988. *Proc. SPIE* **957**, 105.
Kinsman G. and Duley W. W., 1989 *Appl. Phys. Lett.* **54**, 7.
Kinsman G. and Duley W. W. 1990. *Appl. Phys. Lett.* **50**, 996.
Kinsman G. and Duley W. W., 1993. *Appl. Opt.* **32**, 7462.
Kools J. C. S., 1993. *J. Appl. Phys.* **74**, 6401.
Kools J. C. S. and Dieleman J., 1993. *J. Appl. Phys.* **74**, 4163.
Koren G., Ho F. and Ritsko J. J., 1985. *Appl. Phys. Lett.* **46**, 1006.
Koren G., Ho F. and Ritsko J. J., 1986. *Appl. Phys.* **A40**, 13.
Küper S., Brannon K. and Brannon J., 1992. *Laser Ablation of Electronic Materials* ed.
    E. Fogarassy and S. Lazare, Elsevier Science, Amsterdam, p. 213.
Libertucci M., Kinsman G., North T. H. and Duley W. W., 1993 (unpublished work).
Loper G. L. and Tabat M. D., 1985. *Appl. Phys. Lett.* **58**, 3649.
Lu Y. F., Takai M., Komuro S., Shinkowa T. and Aoyagi Y., 1994. *Appl. Phys.* **A59**,
    281.
Mehlman G., Chrisey D. B., Burkhalter P. G., Horwitz J. S. and Newman D. A., 1993.
    J. Appl. Phys. **74**, 53.
Mihailov S., 1992. Ph.D. Thesis, York University.
Moustaizas S. D., Tatarakis M., Kalpcuzos C. and Fotakis C., 1992. *Appl. Phys. Lett.* **60**,
    1939.
Okada N., Katsumura Y. and Ishique K., 1992. *Appl. Phys.* **A55**, 207.
Olfert M., Duley W. W. and North T., 1996. *Proc. NATO ASI* (in press).
Omar M. A., 1975. *Elementary Solid State Physics: Principles and Applications*, Addison-
    Wesley Publishing Co., Reading, Massachusetts.
O'Neill W. and Steen W. M., 1989. *Proc. SPIE* **1031**, 574.
Pedraza A. J., 1987. *J. Metals* Feb. p. 14.
Pedraza A. J., Godbole M. J., Kenik E. A., Lowndes D. H. and Thompson J. R., 1988.
    J. Vac. Sci. Technol. **A6**, 1763.
Poprawe R., Beyer E. and Herziger G., 1984. *Inst. Phys. Conf. Ser.* **72**, 67.
Poprawe R., Beyer E. and Herziger G., 1985. *Proc. 5th Int. Symp. Gas Flow and Chemical
    Lasers*, ed A. S. Kaye and A. C. Walker, Adam Hilger, London.
Puell H., 1970. *Z. Naturforsch.* **A25**, 1807.
Ready J. F., 1971. *Effects of High Power Laser Radiation*, Academic Press, New York.
Ritsko J. J., Ho F. and Hurst J., 1988. *Appl. Phys. Lett.* **53**, 78.
Rosen D. J., Mitteldorf J., Kathandaraman G., Pirri A. N. and Pugh E. R., 1982. *J. Appl.
    Phys.* **53**, 5190.
Rykalin N. N. and Uglov A. A., 1971. *High Temp.* **9**, 522.
Rykalin N., Uglov A., Zuev I. and Kokora A., 1988. *Laser and Electron Beam Material
    Processing*, Mir, Moscow.
Saenger K. L., 1994. *Pulsed Laser Deposition of Thin Films*, ed. D. B. Chrisey and G. K.
    Hubler, J. Wiley, New York, p. 199.
Sappey D. and Gamble T. K., 1992. *J. Appl. Phys.* **72**, 5095.
Sedov L. I., 1959. *Similarity and Dimensional Methods in Mechanics*, Cleaver Hume Press,
    London.
Sesselmann W. and Chuang T. J., 1985. *J. Vac. Sci. Technol.* **133**, 1507.
Sesselmann W., Marinero E. E. and Chuang T. J., 1986. *Appl. Phys.* **A41**, 209.
Singh R. K., Singh A. K., Lee C. B. and Narayan J., 1990. *J. Appl. Phys.* **67**, 3448.
Stafast H. and von Przychowski M., 1989. *Appl. Surf. Sci.* **36**, 150.

Sugioka K., Tashiro H., Murakami H., Takai H. and Toyoda K., 1990. *Jap. J. Appl. Phys.* **29**, L1185.

Sussman M. V., 1972. *Elementary General Thermodynamics*, Addison-Wesley, Reading, Massachusetts.

Tosto S., Lazzaro P. D., Letardi T. and Martelli S., 1988. *Proc. SPIE* **1023**, 208.

Tyndall G. W. and Moylan C. R., 1989. *Mater. Res. Soc. Symp. Proc.* **129**, 327.

Tyndall G. W. and Moylan C. R., 1990. *Appl. Phys.* A**50**, 609.

Ursu I., Mihailescu I. N., Popa A., Prokhorov A. M., Ageev V. P., Gorbonov A. A. and Konov V. I., 1985. *J. Appl. Phys.* **58**, 3909.

Ursu I., Mihailescu I. N., Prokhorov A. M., Tokarev V. N. and Konov V. I., 1987. *J. Appl. Phys.* **61**, 2445.

van de Riet E., Nillesen A. J. C. M. and Dieleman J., 1993. *J. Appl. Phys.* **74**, 2008.

Viswanathan R. and Hussla I., 1986. *J. Opt. Soc. Am.* **133**, 796.

von Gutfeld R. J. and Dreyfus R. W., 1989. *Appl. Phys. Lett.* **54**, 1212.

Wang H., Salzberg A. P. and Weiner B. R., 1991. *Appl. Phys. Lett.* **59**, 935.

Zinovev A. V. and Lugovskoi V. B., 1978. *Sov. Phys. Tech. Phys.* **23**, 900.

Zinovev A. V. and Lugovskoi V. B., 1980. *Sov. Phys. Tech. Phys.* **25**, 953.

CHAPTER 5

Interaction of UV radiation with organic polymers

5.1 ABSORPTION OF UV RADIATION

The interaction of intense UV radiation with organic substrates is complicated by the occurrence of photochemical in addition to thermal and other dynamic (e.g. ponderomotive) effects. The relative importance of photochemical and photothermal effects depends primarily on wavelength but can also depend on pulse duration as well as pulse repetition rate and the ambient atmosphere. In general, photochemical effects can occur whenever the energy of a single photon exceeds the dissociation energy D_0 for a component of the absorber, i.e.

$$\hbar w > D_0 \tag{1}$$

Since photochemistry can occur with single photons, there is no associated intensity threshold for this effect. On the other hand, photothermal effects may exhibit an intensity threshold since the equilibrium or peak temperature reached will be the result of both heat input and loss terms.

At high laser intensities the requirement for the onset of photochemical effects can be expanded to include the possibility of two, three or multiphoton processes. Then

$$n\hbar w > D_0 \tag{2}$$

where n is the order of the non-linear interaction. Since the rate $W^{(n)}$ of an n-photon process

$$W^{(n)} \propto I^n \tag{3}$$

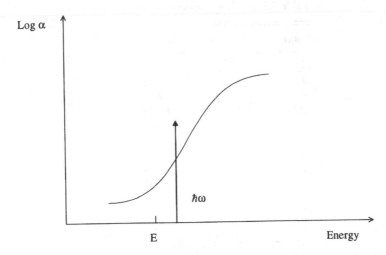

Figure 5.1. A schematic plot of the absorption threshold in a molecular solid. For photon energies $\hbar w < E$ the solid is weakly absorbing. For $\hbar w > E$ strong single-photon excitations dominate whereas two-photon transitions are resonantly enhanced.

where I is intensity, multiphoton-induced photochemical effects will exhibit an intensity threshold. In addition, the relative importance of multiphoton effects will also depend strongly on wavelength and the absorption of the material at this wavelength. This effect is shown schematically in Figure 5.1 for an organic molecular absorber. When the photon energy

$$\hbar w \geq \frac{E}{2} \tag{4}$$

where E is the onset of electronic absorption, then two-photon transitions are possible energetically. If the final state has energy $2\hbar w > D_0$ then a two-photon induced photochemical effect can occur.

When $\hbar w > E$, so that α is large, single-photon processes are strongly allowed and single-photon photochemistry is important, provided that one has sufficient photon energy to break one or more chemical bonds. A summary of bond dissociation energies for several different types of bond is given in Table 5.1.

Even when photochemical effects occur, the quantum efficiency, or yield, need not approach unity. As a result, internal energy can be retained by the absorber. This energy will be disposed of in a variety of ways. For example, it may be promptly reradiated in the form of

Table 5.1. *Bond dissociation energies.*

Bond	D_0 (eV)
H_2	4.48
O_2	5.12
N_2	9.76
CO	11.09
C–C	3.62
C=C	6.40
C≡C	8.44
C–H	4.30
C–N	3.04
C=N	6.40
C≡N	9.27
C=S	4.96

fluorescence, reradiated after some delay as phosphorescence, or retained as internal vibrational energy until it is conducted to the surroundings. In many cases, a combination of these effects occurs with only part of the absorbed energy being reemitted.

The remainder is retained as internal excitation that is eventually degraded to heat. Figure 5.2 shows how this occurs. In this simple example, a photon of energy $\hbar w$ is absorbed leading to the formation of an electronically and vibrationally excited molecule in state A. This extra vibrational energy, E', is shared among other molecular vibrational modes and is eventually lost to the surroundings as heat (Dlott 1990). Emission of a photon at energy $\hbar w_e$ returns the molecule to a vibrationally excited level of the ground state. In this example, the fraction of the original photon energy converted to heat, i.e. the photothermal component is $(E' + E'')/(\hbar w)$.

The quantum efficiency for photodissociation, ϕ, will be wavelength-dependent and need not follow that of the absorption coefficient α, or the cross-section σ, although ϕ is generally expected to increase as α becomes larger. An example of this situation is shown in Figure 5.3 in the photodissociation of formaldehyde (Moorgat *et al.* 1983). It is important to note that, since σ and ϕ do not in general have the same wavelength dependence, the relative proportion of photochemical and photothermal effects will depend on wavelength.

A variety of dissociation channels is frequently available for all but the simplest molecules. For example, in the photodissociation of H_2CO one

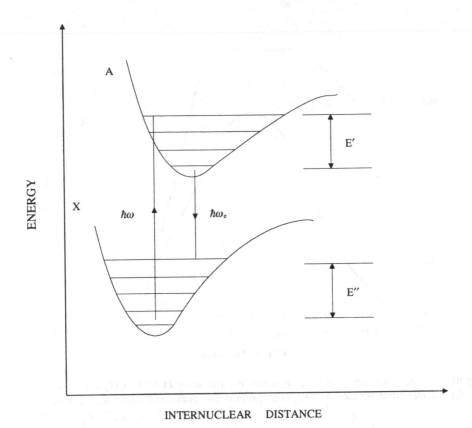

Figure 5.2. Molecular energy levels showing absorption at energy $\hbar w_e$ followed by vibrational deactivation, E', reemission of energy and ground state vibrational deactivation E''.

has the following possible routes:

$$H_2CO + \hbar w \rightarrow H_2 + CO$$

$$\rightarrow H + HCO$$

$$\rightarrow CH + OH$$

with the first two channels dominating. The resulting atoms and molecules can be formed with considerable translational and vibrational energy (see the review by Moore and Weisshaar 1983) after excitation of gaseous H_2CO. On the other hand, H_2CO does not photodissociate when in condensed media (Lewis and Lee 1978). Thus quantum efficiencies derived from measurements on gaseous species may not be applicable to solid state systems.

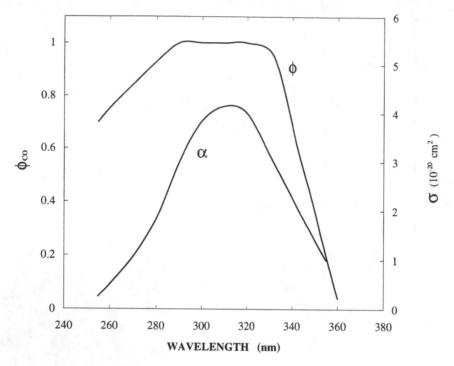

Figure 5.3. The quantum yield ϕ_{CO} for photodecomposition $H_2CO \rightarrow H_2 + CO$ versus the wavelength dependence of the absorption cross-section (Moorgat *et al.* 1983).

The interaction of relatively low intensity UV light with organic solids has been the subject of much investigation, particularly with respect to the photodegradation of polymers (MacCallum and Schoff 1972). Photodegradation occurs when an incident photon is absorbed by a chromophore or covalently unsaturated molecular group (e.g. C=C, C=O, etc.) This electronically excited group may dissociate immediately (on a timescale of picoseconds), undergo reaction with another molecular entity or retain some or all of the initial energy as electronic/vibrational excitation. Direct bond breaking can lead to the evolution of gas and the formation of radicals within the irradiated volume. It is the rapid evolution of gas that is responsible in part for the clean edges associated with UV photoablation phenomena.

A study of UV-induced photolytic decomposition of poly(methyl methacrylate), PMMA, at low intensities (10^{-4}–$10^{-5}\,W\,cm^{-2}$) shows that the primary effect is the random scission of polymer main chain bonds leading to depolymerization (MacCallum and Schoff 1971a,b). The quantum efficiency for scission ϕ_s was found to be strongly temperature dependent with ϕ_s in the range 0.012–0.039 at room

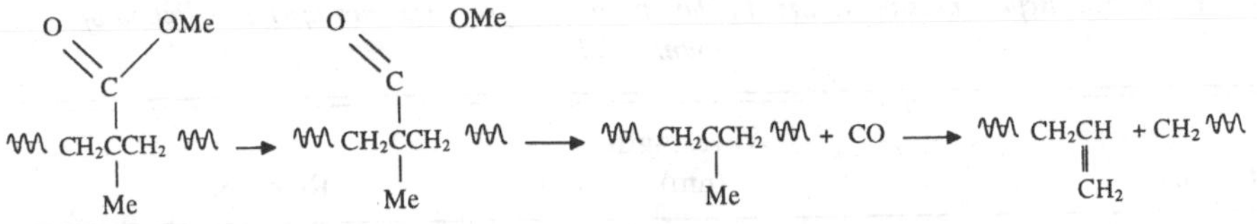

Figure 5.4. Initial stages of the UV-induced photodegradation of PMMA (MacCullum and Schoff 1971a).

temperature rising to 0.2–1.02 at 200°C. This decomposition will evolve CO (Figure 5.4). A mass spectrometric study by Estler and Nogar (1986) confirmed the presence of CO over PMMA irradiated with 266 nm YAG radiation. In addition, the methylmethacrylate monomer was also detected as a gaseous molecule. These observations indicate that, at least for PMMA, the chemical products of UV irradiation at low and high intensities are similar.

5.2 GASEOUS PRODUCTS OF UV LASER ABLATION

Basic data on the processes involved in UV ablation of organic polymers can be obtained from mass spectroscopic measurements of ejection products. A list of some experimental studies on this subject is given in Table 5.2. These experiments were performed at a variety of wavelengths, including the fourth harmonic of the Nd:YAG laser at 266 nm.

One of the earliest studies was of the composition of gases evolving from PET irradiated at 193 nm (Srinivasan and Leigh 1982). As expected, CO, CO_2 and H_2 were major products. However, a wide variety of organic products was also observed. The most important of these products was benzene, followed by toluene and benzaldehyde.

The quantum yield of benzene was found to rise rapidly with applied fluence above a threshold of about $20\,mJ\,cm^{-2}$. At high fluence ($100\,mJ\,cm^{-2}$) the yield saturated. It was postulated that this saturation was due to the onset of a two-photon destruction channel at high laser fluence. A subsequent study by Hansen (1989), performed with 266 nm radiation, confirmed the presence of benzene in the gaseous products. In addition, a strong signal attributable to phenyl-CO^+ was also observed. This molecular ion probably derives from the decomposition of the PET monomer.

Table 5.2. *Representative studies of ablation products from excimer laser irradiation of organic solids.*

Material	Wavelength (nm)	Reference
Polymethylmethacrylate	193	Srinivasan (1993)
		Srinivasan *et al.* (1986b)
		Danielzik *et al.* (1986)
	240	Estler and Nogar (1986)
	248	Srinivasan (1993)
		Srinivasan *et al.* (1986b)
		Larciprete and Stuke (1987)
	266	Estler and Nogar (1986)
Polyimide	266	Hansen (1989)
	308	Tsunekawa *et al.* (1994)
		Taylor *et al.* (1988)
		Otis (1989)
Polyethylene terephthalate	193	Srinivasan and Leigh (1982)
	222	Taylor *et al.* (1988)
	266	Hansen (1989)
	308	Andrew *et al.* (1983)
		Taylor *et al.* (1988)
Polystyrene	193	Feldman *et al.* (1987)
	248	Larciprete and Stuke (1987)
	308	Tsunekawa *et al.* (1994)
Polycarbonate	193	Hansen (1989, 1990)
	248	Hansen (1989, 1990)
	266	Hansen (1989, 1990)
	355	Hansen (1989, 1990)
Polybenzimidazole	266	Hansen (1989 1990)
Poly-α-methylstyrene	266	Hansen (1989, 1990)
Tetrafluoroethylene	266	Estler and Nogar (1986)
Biogenic	248	Singleton *et al.* (1986)

This decomposition would be a primary process, i.e. one occurring as the result of the initial photochemical interaction. Hansen (1989) found that larger molecules such as naphthalene or biphenyl were also abundant in the ejected gas. The presence of these molecules was attributed to a thermally induced structural rearrangement occurring during ablation. The PET monomer was not observed as an ablation product. However, carbonaceous debris are observed in the vicinity of the ablated area (Taylor *et al.* 1988). Hansen (1989) reported that the initial rapid stage of ablative material removal appears to involve the ejection of bare carbon clusters, with C_3 as the dominant species. These may be related to this sooty debris, but the composition of redeposited material is not that of elemental carbon (Hansen and Robitaille 1988).

A wide range of carbon clusters has also been observed in the 308 nm ablation of polyimide (Creasy and Brenna 1988, Otis 1989). Clusters with masses as large as 5000 amu were detected in three primary size ranges; <300 amu, 400–1300 amu; >2000 amu (Otis 1989). Both pure and heteroatom clusters were detected and a peak in the mass distribution near 700 amu (C_{60}) was clearly evident. In this regard, the observed mass spectrum was found to be quite similar to that obtained from laser ablation of pure carbon or graphite (Kroto *et al.* 1985).

The origin of these carbon clusters is uncertain (Hansen 1989, 1990, Otis 1989). It is unlikely that they are direct photodissociation products of the parent compound since they contain many more carbon atoms than any chemical precursor in these systems. It is possible that these clusters are created by collisional recombination of smaller species in the plasma in front of the irradiated surface. One other possibility is that clusters may originate from laser-induced decomposition of much larger ejected particles. These particles would be related to the soot that is redeposited on the surface.

The rate of formation of soot has been shown to be strongly dependent on the composition and pressure of the ambient atmosphere and can be suppressed in the ablation of polyimide by ambient He or H_2 (Küper and Brannon 1992). This may be due to the fact that ablation into a light ambient gas permits a more rapid expansion and limits the time for plume reactions to occur. The symmetry of the debris deposit around an ablated region on polymers such as polyimide and PET has been found to exhibit structure that can be attributed to gas dynamical interactions within the plume (Miotello *et al.* 1992).

There have been several studies of the ablation products of PMMA (Srinivasan *et al.* 1986a,b, Danielzik *et al.* 1986, Estler and Nogar 1986, Larciprete and Stuke 1987). Since PMMA is virtually transparent at 308 nm, most experiments have been performed at shorter wavelengths. Mass spectroscopy shows that the primary ejecta are CO, CO_2, the monomer methylmethacrylate (MMA) and methylformate, HCO_2CH_3. In addition, one also observes a low molecular weight (2500 amu) fraction of PMMA. The relative proportion of MMA is found to be larger under irradiation at 193 nm than at 248 nm whereas methylformate is found only below (but not at) 240 nm.

An experiment by Larciprete and Stuke (1987) using a picosecond UV laser to ionize the ablation products following 248 nm etching of PMMA showed a different range of species in the plume. They found that the monomer was virtually absent but larger molecules were present. These larger molecules did not appear to be fragments of the polymer chain. It was suggested that these species were formed during ablation via fast photochemical or thermal reactions.

Danielzik *et al.* (1986) studied the products of 193 nm ablation of PMMA using conventional quadrupole mass spectrometry. Under these conditions, MMA was found to be an important component of the plume. A fit to the time of flight data was consistent with a Maxwell–Boltzmann velocity distribution at a kinetic temperature that increases with laser fluence. This temperature ranged from about $800 \, K$ at threshold ($60 \, mJ \, cm^{-2}$) to about $3000 \, K$ at $120 \, mJ \, cm^{-2}$. Under these conditions, the emission would appear to be consistent with a thermal equilibrium process. However, the use in these experiments of $100 \, eV$ electrons for impact ionization prior to mass sampling may distort the product distribution. At higher fluence fast non-thermal ablation products were observed.

A study of the ablation products of PMMA using 308 nm radiation (Tsunekawa *et al.* 1994), which is only weakly absorbed by PMMA, showed that the dominant species are the MMA monomer and its dimer. The strong non-linear dependence of this signal from these species is compatible with a multiphoton excitation. Equivalent kinetic temperatures inferred from time of flight measurements were 600–1200 K suggesting a thermal origin for these decomposition products. Similar results have been obtained in the irradiation of polytetrafluoroethylene (PTFE) at 248 and 308 nm (Dickinson *et al.* 1993) and has been attributed to the creation of radicals via an initial photochemical interaction followed by a radical catalyzed unzipping reaction in the polymer which is thermally activated by laser heating.

Laser-induced fluorescence (LIF) studies (Srinivasan *et al.* 1986a) of the plume generated during ablation of PMMA showed the presence of C_2 molecules with high translational energies (about $6 \, eV$) even at incident fluences below the threshold for etching at 248 nm. It was suggested that C_2 is formed via multiphoton excitation and decomposition of PMMA.

Other small molecules such as CH, CN, N_2 and CO can be seen spectroscopically in the plume over PMMA irradiated at 193 nm (Koren and Yeh 1984, Davis et al. 1985). An analysis of emission in the CH band at 431 nm showed a vibration–rotation equilibrium at a temperature of 3200 ± 200 K. This implies that the plasma temperature should be similar to this value just above the interaction region. On the other hand, the translational temperature was estimated to be about 11 000 K implying that the expansion was supersonic. It was suggested that these high excitation and translational temperatures are the result of direct bond breaking by UV photons.

Laser-induced fluorescence of CO (Goodwin and Otis 1989) and C_2 (Dreyfus 1988) over polyimide during laser ablation are characterized by rotational temperatures of about 1150 and 1400–2800 K, respectively. Dreyfus (1992) has shown that the rotational temperature of CN evolving from polyimide under etching at 193 nm depends on laser fluence and rises rapidly through the threshold region to about 1710 ± 140 K at high fluence ($E > 100$ mJ cm^{-2}). This value is close to the estimated surface temperature for thermal decomposition (Mihailov and Duley 1991).

Koren (1988) has measured the plume temperature produced by 248 nm laser ablation of polyimide films using IR spectroscopy. A fit to the IR vibration–rotation band of HCN near 700 cm^{-1} resulted in an excitation temperature of about 2250 ± 150 K. It was suggested that that would be a lower limit to the effective temperature at the ablated surface.

A mass spectroscopic study of 193 nm laser ablation of polystyrene using VUV laser ionization of the resulting products (Feldman et al. 1987) showed that the kinetic energy of product molecules peaked at about 0.7 eV. Time of flight data demonstrated, however, that the velocity distribution was more sharply peaked than predicted by this kinetic energy. Feldman et al. (1987) postulated that this discrepancy can be understood if the ablation products undergo an adiabatic 'nozzle-type' expansion away from the ablation site. This would lower the Maxwell–Boltzmann temperature of the beam while maintaining a high mean velocity. The Maxwell–Boltzmann temperature characterizing the beam velocity distribution was found to be 2350 K, which is close to that measured by Koren (1988) in the plume over polyimide irradiated at 193 nm. An analysis of the yield of low mass ions near threshold (Feldman et al. 1987) concluded that the internal (i.e. electronic, vibrational, and rotational) energy of styrene monomers ejected during ablation cannot exceed 2 eV. However, chemical processing within the plume itself may influence product distributions and presumably the level of internal excitation. For example, Larciprete and Stuke (1987) found evidence for molecular fragments with masses greater than that of the monomer in polystyrene ablation products. These species can be attributed to chemical reprocessing during the expansion phase.

Hansen (1990) found further evidence for a thermally processed product in polycarbonate ablated with 193, 248, 266 and 355 nm laser radiation. This product evolves from the target at a speed of about $3 \times 10^4\,\mathrm{cm\,s^{-1}}$ and is emitted after a hyperthermal plasma of carbon clusters and smaller molecules. The product mass distribution and the kinetic energy of these molecules are compatible with material that has been produced at elevated temperatures although subsequent photochemical processing in the plume is also likely.

The picture that emerges from these studies is then one in which the initial stages of laser ablation lead to the production of small molecules and carbon atom clusters. These leave the interaction region at high (about $10^6\,\mathrm{cm\,s^{-1}}$) speed. They are followed by a slower (10^4–$10^5\,\mathrm{cm\,s^{-1}}$) wave containing species whose chemistry often reflects high temperature thermal equilibrium. The molecules and radicals observed in this wave are often of higher molecular weight than the monomer. The excess internal vibrational energy in these species is likely much less than the energy of ablating photons. This again is consistent with a relatively high density, high temperature processing environment.

Dyer and Srinivasan (1989) proposed a causal relationship between the components of the fast wave and those of the slower wave. By measuring the impact of plume components using a pyroelectric detector, Dyer and Srinivasan were able to show that the velocity V of larger fragments of PMMA was related to that of the faster small molecules v, through the ratio $v/V = (M/m)^{1/2}$ where M and m are large and small masses, respectively. This suggests that the small molecules created in the initial dissociation collisionally accelerate larger fragments. These larger fragments constitute the slow thermal component observed in time of flight measurements.

Later experiments in which the gas dynamic behavior of ejecta was imaged (Ventzek *et al.* 1990, Braren *et al.* 1991, Kelly *et al.* 1992, Srinivasan 1993) showed that the light particles initially emitted from the target form a contact front, which is preceded by a shock wave (Figure 5.5). Heavier particles are retained close to the surface during the initial stages of expansion of the shock and form a Knudsen layer within which chemical equilibration can occur. Material is emitted from this layer over a substantially longer timescale and may be redeposited on the surface.

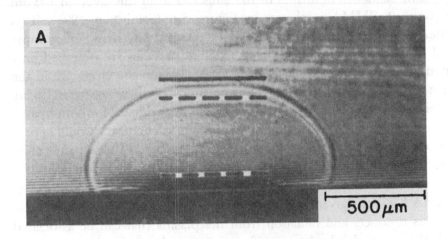

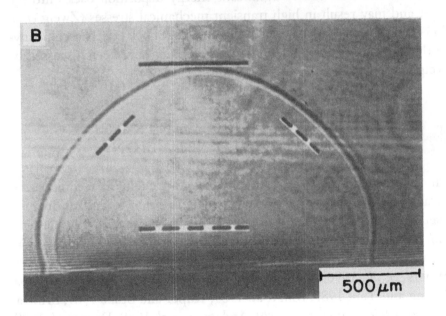

Figure 5.5. PI targets with a thickness of 125 µm were exposed to single laser pulses (308 nm, about 20 ns, pulse diameter 800 µm, 2.3 J cm^{-2}, normal incidence) in air. The ejecta, contact front, and shock wave were photographed by firing parallel to the target surface a second ('probe') laser (596 nm, about 1 ns). The imageable limit of the ejecta and the contact front of the light particles are marked with dashed lines and the shock wave is marked with a solid line. (a) Delay of 190 ns and (b) delay of 410 ns (Kelly *et al.* 1992).

A time sequence of images of the emission from this layer on 193 nm ablation of PMMA is shown in Figure 5.6 (Srinivasan 1993). This material is of high molecular weight (5000 amu) and may be accompanied by particles or liquid droplets. Melted particles are observed only when ablation is carried out at 248 nm. At 193 nm the ejecta are in the form of small solid particles. The dynamics of these ejecta is highly wavelength- and material-dependent and shows evidence for both hydrodynamical and photothermal interactions. Much chemistry would appear to occur in the Knudsen layer so that the fragments collected after ablation may be chemically modified components of the ablated material.

A full discussion of gas dynamic effects in the UV laser ablation of PMMA has been given by Braren *et al.* (1991). The expansion of these gaseous products and radiation from the plasma that can be associated with them can also lead to significant energy deposition back onto the target and may result in high transient mechanical stresses (Zweig *et al.* 1993) and can increase material removal rates. Shock effects accompanying the ablation of polyimide in a water ambient have been discussed by Zweig and Deutsch (1992).

5.3 SURFACE MEASUREMENTS DURING AND PRIOR TO ABLATION

A number of studies have used physical and chemical techniques to characterize the surface of organic polymers prior to and during the onset of ablation (Dyer and Karnakis 1994a,b, Küper *et al.* 1993, Ediger *et al.* 1993, Lazare and Benet 1993, Zweig *et al.* 1993, Mihailov and Duley 1993, Ediger and Pettit 1992, Paraskevopoulos *et al.* 1991, Bahners *et al.* 1990, Singleton *et al.* 1990, Srinivasan *et al.* 1990a, Chuang and Tam 1989, Kim *et al.* 1989, Küper and Stuke 1989, Srinivasan *et al.* 1989, Zyung *et al.* 1989, Anderson *et al.* 1988, Dyer *et al.* 1988, Meyer *et al.* 1988, Mihailov and Duley 1988, Davis and Gower 1987, Dijkkamp *et al.* 1987, Dyer *et al.* 1986, Dyer and Srinivasan 1989). One of the important factors to be addressed is the question of precursor effects prior to the onset of ablation. These effects would include bond breaking, radical generation and trapping/recombination and transient thermal heating. The overall result might be expected to produce a surface layer whose physical and chemical properties were altered from that of pristine material and which might be more susceptible to subsequent ablation.

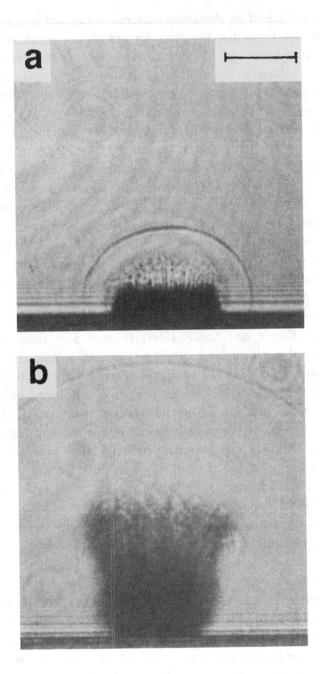

Figure 5.6. Photographs of the blast wave and the ejected material from the surface of a PMMA film on the impact of a single UV (193 nm) laser pulse. Scale bar length 0.5 mm. The time intervals after the start of the laser pulse are (a) 500 ns and (b) 2.9 μs (Srinivasan 1993).

When the surface of an absorbing material is exposed to radiation at a fluence level less than that required for ablation, individual molecular groups are excited and may dissociate. However, the rate at which this occurs is not sufficient to result in a large enough concentration of molecular fragments to initiate the onset of ablation. Some of the energy generated from recombination of these fragments as well as that resulting from non-dissociative photoexcitation is dissipated in the surface layer. The thickness of this layer will be typically α^{-1} cm. As a result of some bond breaking together with these photothermal heating effects, the composition of this layer will be changed. In addition, a low concentration of reactive species will be trapped in this layer. The recombination energy of these fragments may be liberated when the temperature of the surface is subsequently raised by application of additional laser pulses.

Ultraviolet (Küper and Stuke 1989) and infrared (Mihailov and Duley 1988, Küper and Stuke 1989) spectra of irradiated surfaces show spectral changes that can be attributed to a modification of surface composition. UV transmission spectra of PMMA irradiated at $40\,\mathrm{mJ\,cm^{-2}}$ with 248 nm light (Küper and Stuke 1989) (Figure 5.7) show increasing absorption in the 300 nm region. The species responsible for this absorption cannot be identified from this spectrum due to its broadness. However, IR spectra obtained under the same conditions indicate that unsaturated species are present on the irradiated surface. IR absorption attributable to triple or cumulated double bonds is also observed. Surprisingly, these effects can also be produced after a relatively short (about 5 s) exposure to broad band UV radiation from a conventional light source. Methyl formate is known to be one of the products of such irradiation of PMMA (Gupta *et al.* 1980).

A series of measurements by Srinivasan *et al.* (1990a) has demonstrated that irradiated material retained in the surface layer after exposure to multiple pulses at above threshold fluence is easily removed with a solvent. This showed that incubation effects were still present in underlying material even under conditions of significant ablation (Dyer and Karnakis 1994a,b). These irradiated sublayers consisted of depolymerized material. Significantly, the first few pulses were observed to result in a surface that was elevated above the surrounding (unirradiated) area (Figure 5.8). This would be consistent with the partial dissolution of the irradiated surface with evolution of gas so that the irradiated area swelled up from the surface.

Imaging of the polymer surface and plume during ablation has been reported by Srinivasan (1993), Kelly *et al.* (1992), Ventzek *et al.* (1990),

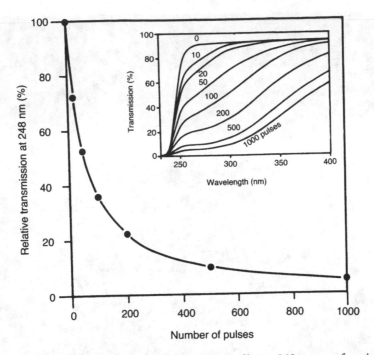

Figure 5.7. Transmission of a 40 μm thick PMMA film at 248 nm as a function of exci-mer laser irradiation dose at 248 nm with a fluence of 40 mJ cm^{-2}. After 1000 pulses, the transmission has dropped to less than 6% of its initial value. Insert: the UV spectrum of the same sample exhibits a broad absorption for wavelengths up to the visible (Küper and Stuke 1989).

Srinivasan *et al.* (1990b, 1989), Kim *et al.* (1989), Simon (1989) and Zyung *et al.* (1989). Observation of PMMA during ablation with exci-mer laser radiation (248 and 308 nm) shows that physical and chemical changes are apparent in the focal area within a few nanoseconds of the start of the ablating pulse. Figure 5.9 shows a series of photographs tracing the development of craters in PMMA during ablation with XeCl laser radiation (Simon 1989). It is apparent that some distortion is present at the bottom of craters during the ablation event. The effect of a reduction in fluence is to increase the time delay from the beginning of the pulse to the onset of visible damage at the focus.

Images of the plume obtained with about 50 ns resolution (Srinivasan 1993, Srinivasan *et al.* 1989) show that the first evidence for ablation is a hemispherical shock wave that expands at about 5×10^4 cm s^{-1} away from the laser focus. Solid material is not observed to leave the surface until the laser pulse is terminated. Emission of ejecta then persists for

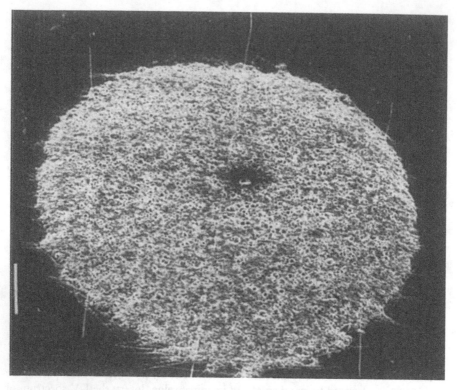

Figure 5.8. A scanning electron micrograph of a PMMA surface after ten pulses
at 248 nm, 1.3 J cm^{-2}. The bar is 100 µm in length (Srinivasan *et al.* 1990a).

several microseconds and can consist of particles and liquid droplets as
well as molecular material.

Kim *et al.* (1989) and Zyung *et al.* (1989) have studied the ablation of
PMMA by 532 nm radiation using 2 ps pulses from a dye laser for imaging. Since PMMA is quite transparent at this wavelength but absorbing
at half this wavelength (266 nm), coupling of incident radiation probably
involves a two-photon process. Although some photochemistry may
result from this primary interaction, overall the dominant mechanism
for material removal would appear to be a thermally induced explosive
decomposition.

Ablation was produced by 100 ps pulses. A surprising result is that
strong self-focusing occurs during the ablation pulse. This self-focusing
leads to the creation of a small diameter filament within the focal area
but of smaller diameter (micrometers) than the laser focus. This
filament persists for about 20 ps. Non-linear absorption induced by the
high laser intensity within this filament results in a rapid rise in

temperature to 500–1000°C. At this point, the material within the filament explosively decomposes by ejecting particles whose diameter is approximately one laser wavelength. These particles act to shield the underlying surface from the trailing edge of the laser pulse. These particles undergo a supersonic expansion into the surrounding medium. The associated blast wave contains about 0.25 of the laser pulse energy above threshold and persists for hundreds of nanoseconds. The region adjacent to the ablation core is intensely heated prior to and during the explosive decomposition phase. This heat is conducted into the surrounding volume over a timescale determined by the thermal diffusivity. The final resulting morphology is shown in Figure 5.10 (Kim *et al.* 1989).

Ablation of PMMA by 160 fs 308 nm pulses (Srinivasan *et al.* 1987) shows efficient etching at fluences much less than those required with 30 ns pulses. Because PMMA is only weakly absorbing at 308 nm, two-photon processes must be important in the initiation of material removal, as was observed at 532 nm (Kim *et al.* 1989). Srinivasan *et al.* (1987) estimate that the two-photon absorption depth in PMMA under the conditions of their experiment was about 5 μm. This is similar to that for one-photon attenuation at 193 nm. The one-photon absorption depth at 308 nm was found to be 0.02 cm.

Imaging studies were not carried out under these conditions, but it seems likely that self-focusing leading to filament generation may play a rôle in etching with femtosecond pulses. However, photochemical effects will be favored at 154 nm (0.5 × 308 nm) over doubled 532 nm radiation. The morphology of laser-etched holes produced with 308 fs pulses suggests that photochemistry dominates. Küper and Stukes (1987) studied the ablation of PMMA with 300 fs 248 nm pulses and obtained similar results.

Measurements of the time-dependent reflectivity of polyimide ablated at 355 nm (Chuang and Tam 1989), 308 nm (Mihailov and Duley 1993), 248 nm (Singleton *et al.* 1990, Paraskevopoulos *et al.* 1991) and 193 nm (Ediger and Pettit 1992, Ediger *et al.* 1993) have elucidated the rôle played by plume material in shielding the surface and chemical changes in the polymer that alter surface reflectivity. Chuang and Tam found that the trailing edge of the laser pulse may be blocked by ejecta. Singleton *et al.* (1990) found that, for fluences $<20 \, \text{mJ cm}^{-2}$, specularly reflected radiation from the irradiated surface has the same temporal profile as that of the laser pulse. They found that reflection ceases to occur after the fluence is increased to $23 \, \text{mJ cm}^{-2}$. This effect was

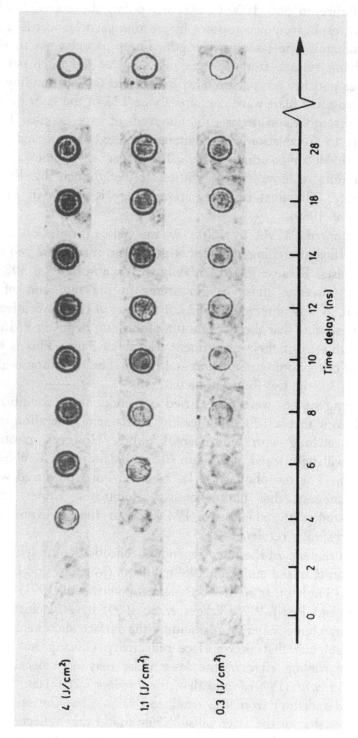

Figure 5.9. The temporal development of craters etched in polyimide with different laser fluences. The final picture in each row shows the crater appearance several seconds after ablation (Simon 1989).

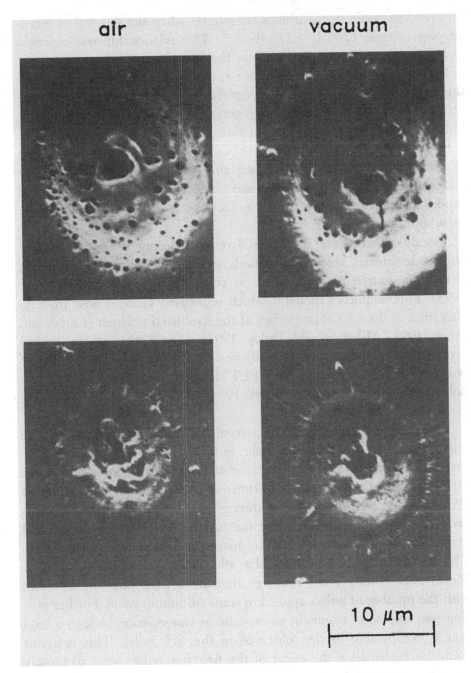

Figure 5.10. Electron micrographs of polymethylmethacrylate samples damaged in air and in vacuum by 230 μJ (top) and 47 μJ (bottom), 50 μm diameter, 93 ps, 0.532 μm laser pulses. The dark cavity in the center is created when PMMA is removed from the surface by explosive thermal decomposition. The melted region is slightly larger in vacuum than in air, because the ablation plume dissipates faster in vacuum than it does in air (Kim *et al.* 1989).

observed to occur even within a laser pulse when the integral of intensity versus time $\int I(t)\,dt \geq 23$ mJ cm^{-2}. This relationship was observed to hold over a wide range of incident fluences (20–150 mJ cm^{-2}).

It seems likely that this behavior can be attributed to a change in the properties of the irradiated surface after the 23 mJ cm^{-2} fluence threshold has been reached. Singleton *et al.* suggested that refractive index changes may produce a reduction in reflected intensity. However, it is more likely that changes in surface morphology result in trapping of incident radiation, thus reducing the specular component. Surface roughening could also reduce the specular term. In this case, one would expect an increase in the diffuse reflectivity at fluences exceeding the threshold for ablation.

A study of diffuse and specular reflectivities during 193 nm ablation of polyimide showed, however, that both decrease, implying that scattering is not the source of a reduction in reflected intensity (Ediger *et al.* 1993). This supports a model in which reflectivity changes arise from an alteration in the optical properties of the irradiated polymer (Ediger and Pettit 1992, Mihailov and Duley 1993). These changes in optical properties in response to UV irradiation can be observed in a variety of polymers including mylar and PET (Lazare and Benet 1993) and polyethylene (Dyer and Karnakis 1994a,b) as well as in polyimide (Srinivasan *et al.* 1990b).

Time-resolved transmission measurements of PMMA and polystyrene during ablation with excimer laser radiation have been reported by Davis and Gower (1987) and Meyer *et al.* (1988). At fluences well above threshold the transmission of 248 nm radiation through a thin film of PMMA decreases dramatically throughout the laser pulse. This is probably due to the absorption at the surface as well as absorption and scattering of incident radiation by material ejected during the pulse. The transmission recovers somewhat after the termination of the pulse. Meyer *et al.* (1988) found, however, that the net transmission decreases with the number of pulses applied to some minimum value. Further pulsing past this point results in an increase in transmission back to a level that is comparable to that observed in the first pulse. This behavior could be explained if the effect of the first few pulses were to modify underlying material so as to cause it to become more absorbing. Eventually, application of a large number of pulses to the same area seems to reverse this process, resulting in enhanced transmission. A similar effect has been observed in the irradiation of polyethylene at 193 nm (Dyer and Karnakis 1994a,b).

A novel experiment by Dijkkamp *et al.* (1987) in which thin PMMA films on Si were irradiated with 248 nm pulses has been used to investigate ablation mechanisms. Because PMMA is relatively transparent at 248 nm ($\alpha \simeq 50\,\mathrm{cm}^{-1}$) whereas Si is strongly absorbing, incident excimer radiation is dissipated as heat in the Si with little direct excitation of the PMMA layer. Under these conditions, decomposition occurs at the PMMA–Si surface by heat transfer from the Si substrate. The resulting reverse side ablation showed the usual physical characteristics of conventional ablative material removal, including high spatial resolution and threshold behavior.

The redeposition of debris generated in the explosive decomposition of various organic polymers has been studied by many groups (Miotello *et al.* 1992, Küper and Brannon 1992, Srinivasan *et al.* 1987, von Gutfeld and Srinivasan 1987, Koren and Oppenheim 1987, Dyer *et al.* 1987, 1988, Taylor *et al.* 1988). These debris are known to be primarily carbonaceous in nature, charged and of dimensions comparable to excimer wavelengths, although agglomeration undoubtedly occurs during redeposition. They are easily removed with CO_2 laser pulses (Koren and Donelon 1988). Debris are redeposited inside and around the periphery of ablated areas (Figure 5.11). It is probable that this debris is the result of reactions involving carbon clusters in the plume ejecta. Alternatively, these particles may be the dehydrogenated products of larger pieces of polymer that are mechanically removed by the ablating gas during etching.

Debris deposited in the area of ablation acts to shield the surface from subsequent etching. The effect of a particle is to cast a shadow on the surface (Taylor *et al.* 1988). This shadow is actually a Fresnel diffraction pattern of the particle. It includes a region of high intensity at its center (Jenkins and White 1957). The resulting image projected on the surface takes the form of a cone, occasionally with a hole at its center (Figure 5.12). The full cone angle, θ_m, is related to the threshold fluence, E_T, and the applied fluence, E, by (Dyer *et al.* 1988)

$$\theta_m = 2\sin^{-1}\left(\frac{E_T(1-R_0)}{E(1-R(\theta_m))}\right) \qquad (5)$$

where R_0 is the reflectivity at normal incidence for the polymer surface whereas $R(\theta_m)$ is the reflectivity at the cone angle $\theta_m/2$. The threshold fluence may be obtained from equation (5) so that a measurement of θ_m as a function of E provides E_T when $R(\theta_m)$ is available. Dyer *et al.*

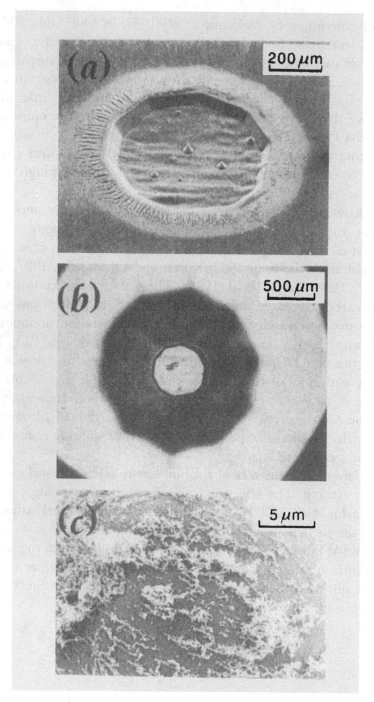

Figure 5.11. (a) A SEM photograph of cones in a 0.56 mm diameter crater formed by
XeCl laser ablation of Kapton at a fluence of $0.09\,\mathrm{J\,cm^{-2}}$, 562 pulses. The crater is an
image of a ten-sided iris used to select the central portion of the laser beam. (b) An optical microscope photograph using top illumination of the symmetrical distribution of

Table 5.3. *A comparison of fluence thresholds obtained from measurement of cone angle* (E_T) *with those obtained using a thermocouple probe* (E'_T) *(from Dyer et al. 1988).*

Polymer	Wavelength (nm)	E_T (mJ cm^{-2})	E'_T (mJ cm^{-2})
Polyimide	308	32 ± 5	45
	193	18 ± 5	45
	157	27 ± 7	36 ± 7
Polyethylene terephthalate	308	97 ± 12	190
	193	14 ± 3	25
	157	19 ± 6	29 ± 6
Polyethylene	157	25 ± 7	67 ± 14
Nylon 66	193	15 ± 3	
	157	33 ± 7	

(1988) used this expression to estimate E_T for a variety of polymers at several laser wavelengths (Table 5.3). Comparison with data obtained from measurements of etch depths shows that E_T estimated from the cone angle is less than these values by up to a factor of two. Dyer *et al.* (1988) suggested that this apparent discrepancy is due to the difficulty of measuring etching depths near threshold and to the high sensitivity required for thermocouple measurements. The spontaneous formation of conical structures may also be an adaptive process on ablating surfaces near threshold (see Chapter 3). Extended areas of periodic and quasi-periodic structure consisting of cones and/or ripples are also frequently observed on irradiated surfaces (Arenholz *et al.* 1991, 1992).

The rôle of ejected material in determining the morphology of etched surfaces has been studied by Mihailov and Duley (1991) using separate 308 nm pulses applied to the surface of polyimide with a variable time delay. The fluence of each pulse was adjusted such that it was less than E_T, but sufficiently large that, when both pulses were overlapped temporally, etching occurred. As the time separation between these pulses τ_p was increased, etching diminished and then ceased at $\tau_p = \tau_{ce}$. A plot of τ_{ce} versus the peak intensity in each beam is

Caption for Figure 5.11 (*continued*)

debris surrounding a 0.56 mm diameter crater formed by XeCl laser ablation of Kapton at a fluence of 0.8 J cm^{-2}, 62 pulses. (c) A SEM photograph of debris lying on the top surface of a cone in a 3.5 mm diameter crater formed by XeCl laser ablation of Kapton, at a fluence of 0.12 J cm^{-2} (Taylor *et al.* 1988).

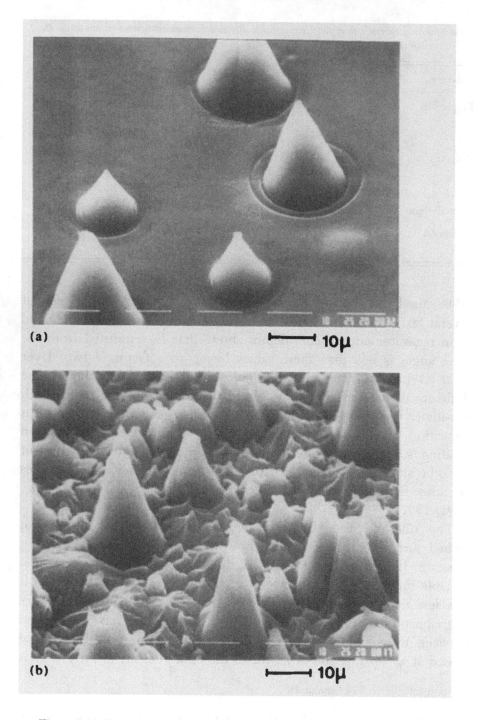

Figure 5.12. Cone structures produced in (a) nylon 66 and (b) PET using 300 pulses each at $0.11\,\mathrm{J\,cm^{-2}}$ from a F_2 laser ($\lambda = 157\,\mathrm{nm}$) (Dyer *et al.* 1988).

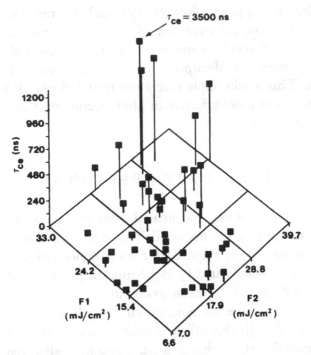

Figure 5.13. The cessation time of ablation etching τ_{ce} versus the incident fluences of two overlapping excimer beams. The initial beam has fluence F_1 and the second beam has fluence F_2 (Mihailov and Duley 1991).

shown in Figure 5.13. It was found that irradiation with a low intensity subthreshold pulse followed by a high intensity subthreshold pulse was more likely to result in etching than when the order of these pulses was reversed.

Other experimental data on the effect of pulse sequencing on the etching of polyimide, PMMA and PET (Srinivasan and Braren 1988b, 1990) showed that the efficiency of etching is a function of each pulse fluence and pulse–pulse separation. In polyimide, shielding of the second pulse by material ejected by the first pulse appears to dominate the overall interaction. Etching with overlapping sub-picosecond pulses at total fluences just above threshold shows that the etching depth per pulse pair is enhanced in highly transparent materials such as teflon (PTFE) but can be reduced in absorbing polymers such as polyimide (Preuss *et al.* 1993).

Mixed wavelength studies in which pulses of different wavelength are applied to polymer surfaces have been reported by Srinivasan and Braren (1990). They found that the etching rate produced by spatially and temporally overlapped pulses of 193 and 308 nm radiation on PMMA is greatly enhanced over that of 193 nm radiation on its own. The mechanism for this enhancement may be the preferential excitation of dissociative states via absorption of 308 nm photons by excited chromophores. This would imply that a resonant 193 plus 308 nm two-photon excitation was more effective in photodecomposition than was a 193 plus 193 nm excitation.

5.4 ETCH DEPTH PER PULSE

In general, the amount of material ablated from an organic polymer per pulse depends strongly on the fluence delivered to the surface. Typically a threshold fluence, E_T, is required for measurable etching to occur. Estimation of E_T is complicated by incubation or precursor effects, and reliable values of E_T may be difficult to obtain.

For fluences $E > E_T$ the incremental etch depth per pulse increases rapidly with E, although this effect is both material- and wavelength-dependent. Generally, the etch depth per pulse in a particular material *decreases* at constant fluence with decreasing wavelength. This effect is surprising in view of the expected higher probability of photochemical reactions at shorter wavelengths.

The absorption coefficient, α, of a polymer at a particular wavelength is also significant since it determines the nature of the primary inter-action process. For example, when $\alpha \geq 10^4 \, \text{cm}^{-1}$ single-photon and resonant two-photon absorptions are important. However, when α is small, as occurs for example in PMMA at 308 nm, two-photon absorption dominates the initial stages of ablative photo-decomposition. It is significant in this regard that PMMA can be efficiently etched with sub-picosecond pulses at 308 nm, but not with nanosecond pulses of the same fluence (Srinivasan *et al.* 1987).

Measurement of etching depths in organic polymers can be carried out in a variety of ways. The simplest involves counting the number of pulses required to penetrate a film of given thickness. Unfortunately, blast wave effects may lead to structural failure and the loss of remaining material before complete ablative penetration can occur. An alternate method involves the use of a profilometer or the microbalance (Lazare and Granier 1988, Küper *et al.* 1993).

Figures 5.14–5.20 summarize etching rate data for polyimide, PMMA and PET at relevant excimer laser wavelengths. These data have been

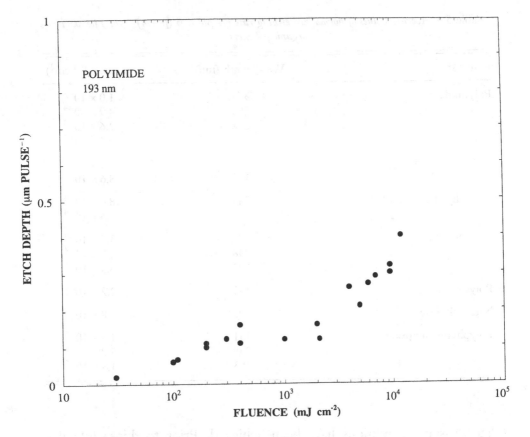

Figure 5.14. Etch depth per pulse for polyimide at 193 nm.

assembled from Srinivasan and Braren (1984), Srinivasan *et al.* (1986b), Taylor *et al.* (1987), Lazare and Granier (1988a,b), Sutcliffe and Srinivasan (1986) and Babu *et al.* (1992). Data points represent selected observations. However, all data given by these groups are not reproduced in these plots.

These data are characterized by a threshold fluence (which is generally poorly defined (Küper *et al.* 1993)) from the plot of etching rate per pulse, X, versus E. Above this threshold, X increases continuously with increasing E although some saturation may occur in PMMA at 248 nm for $E > 4\,\mathrm{J\,cm^{-2}}$ and in polyimide at 308 nm. A summary of average etching rates normalized to unit fluence, β, for fluences above threshold is given in Table 5.4. Since plots of X versus E are in general non-linear, these values must be used with caution.

It is important to note that these values of β are obtained only after incubation effects have been eliminated. In practice, this implies that

Table 5.4. *Average values of normalized etching rate per pulse, β ($\mu m\, mJ^{-1}\, cm^2$) for various organic polymers.*

Polymer	Wavelength (nm)	β ($\mu m\, mJ^{-1}\, cm^2$)
Polyimide	351	1.0×10^{-3}
	308	4.9×10^{-4}
	248	2.6×10^{-4}
	193	0.3×10^{-4}
Polymethylmethacrylate	248	4.0×10^{-4}
	193	8.6×10^{-4}
Polyethylene terephthalate	248	8.0×10^{-4}
	193	5.6×10^{-4}
Polycarbonate	308	3.0×10^{-4}
	248	2.7×10^{-4}
	193	6.3×10^{-5}
Polystyrene	193	2.2×10^{-4}
Nitrocellulose	193	2.8×10^{-4}
Polyphenylquinoxaline	351	1.9×10^{-4}
	248	5.0×10^{-5}
	193	1.5×10^{-5}

'steady-state' conditions have been achieved. Prior to this, while dry etching rates are less than steady-state values, incubated material can be removed by washing with solvent (Lazare and Granier 1988a,b, Srinivasan *et al.* 1990a,b). An example of this effect for PET etched at 193 nm is shown in Figure 5.21.

In a careful series of experiments Küper *et al.* (1993) have shown that etching can be observed even at incident fluences that are much less than E_T. Under these conditions the etching rate per pulse was found to have an exponential dependence on E for ablation at 248, 308 and 351 nm. The behavior at 193 nm was found to be different, with a sharp threshold and a linear dependence of X on E. At low fluences the ablation efficiency, defined as the number of monomers removed per incident photon, was greatest at 193 nm and decreased with increasing wavelength.

The effect of pulse duration on etching rates has been studied by Taylor *et al.* (1987). They found that the threshold fluence for etching was virtually identical for 7 and 300 ns XeCl laser pulses. In addition, the overall etching rate is almost identical for both extreme pulse

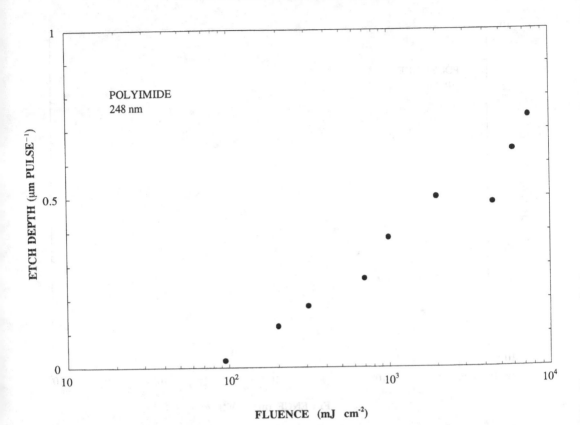

Figure 5.15. Etch depth per pulse for polyimide at 248 nm.

durations at the same fluence, suggesting that pulse intensity is not a factor in determining the magnitude of etching rates. However, comparison with the data of Nikolaus (1986) on etching of polyimide with 5 ps pulses shows a significant reduction in etching rate with picosecond pulses at high fluence. This can be attributed to the onset of plasma absorption at the 10^{10}–$10^{11}\,\mathrm{W\,cm^{-2}}$ intensity level. Near threshold, however, picosecond pulses can result in enhanced etching (Srinivasan et al. 1987) rates. This is possibly due to non-linear interactions and self-focusing effects. Multiphoton absorption has been shown to be effective in the ablative etching of transparent materials such as teflon with femtosecond pulses (Küper and Stuke 1989).

Etching with shorter wavelengths has been reported by Henderson et al. (1985) and by Dyer and Sidhu (1986) using 157 nm radiation from an F_2 laser. At this wavelength the optical path must be evacuated or purged with He or Ar gas. In general no new etching phenomena were

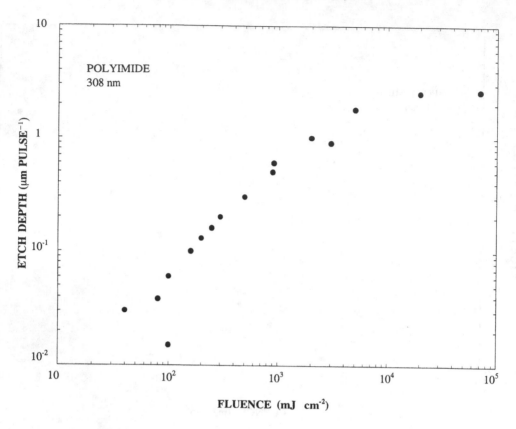

Figure 5.16. Etch depth per pulse for polyimide at 308 nm.

observed at 157 nm. However, because of the shorter wavelength, materials such as polyethylene that are transparent at longer wavelengths could be etched. Ablation of polytetrafluoroethylene (PTFE) with a range of wavelengths between 160 and 184 nm has been reported by Wada *et al.* (1993).

Etching of transparent polymers such as PMMA and PTFE can be facilitated by doping of the polymer with an absorbing compound (Hiraoka *et al.* 1988, Srinivasan and Braren 1988a, Chuang *et al.* 1988, Bolle *et al.* 1990, Davis *et al.* 1992, Egitto and Davis 1992, D'Couto *et al.* 1993). Polyimide has been found to be effective as a dopant in PTFE and results in strong etching at 248 and 308 nm when the concentration exceeds about 0.1 wt% (D'Couto *et al.* 1993). The threshold fluence and limiting etching rate at high fluence were found to decrease with increasing dopant concentration. The ablation mechanism would appear to be primarily photothermal in origin.

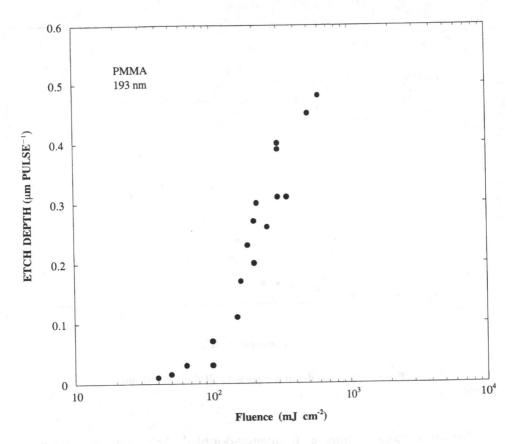

Figure 5.17. Etch depth per pulse for PMMA at 193 nm.

Srinivasan (1991a,b, 1992) has demonstrated that polymers such as polyimide and PET can be etched using tightly focused radiation from a CW UV laser in a scanning geometry if the dwell time, defined as the transit time over a beam diameter, is reduced to $\leq 200\,\mu$s. If this dwell time is taken to be equivalent to the pulse width in a conventional ablation experiment, then effective etching rates of 0.5–1 μm per pulse can be achieved at incident intensities as low as 10–$10^2\,\text{kW cm}^{-2}$. Some melting is observed in PET under these conditions, but the primary material removal mechanism appears to be via photothermal decomposition. This effect can also be used to create conducting links in polyimide (Srinivasan *et al.* 1993). An increase in the efficiency of material removal due to an increase in average temperature has also been observed in a scanning geometry using a high repetition rate excimer laser (Bishop and Dyer 1985). Etching of vias in polyimide using CW 257 nm

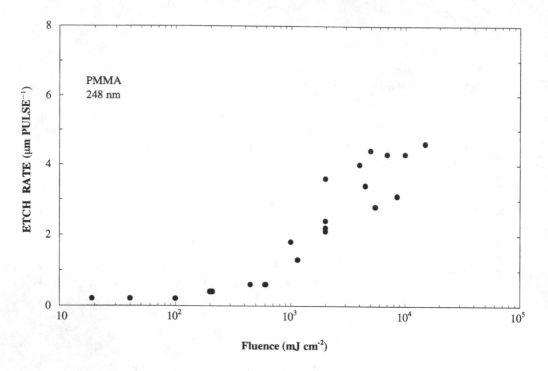

Figure 5.18. Etch depth per pulse for PMMA at 248 nm.

radiation obtained from a frequency-doubled Ar^+ laser has been reported by Treyz *et al.* (1989).

5.5 THEORY OF POLYMER ABLATION

The empirical aspects of polymer etching are becoming better understood, but a complete quantitative physical and chemical description of ablative removal mechanisms has yet to be developed. This is not surprising given the complexity of the problem and the interrelation between photochemical, photothermal and gas dynamic effects. The previous sections have shown that ablative photodecomposition (APD) is a process with specific temporal and spatial characteristics. Strong hysteresis effects associated with incubation are also observed, particularly at or near threshold. Such processes are amenable to empirical elaboration but are difficult to treat theoretically in any quantitative way.

A major problem in developing a quantitative theory of APD involves these hysteresis and attendant incubation effects. For

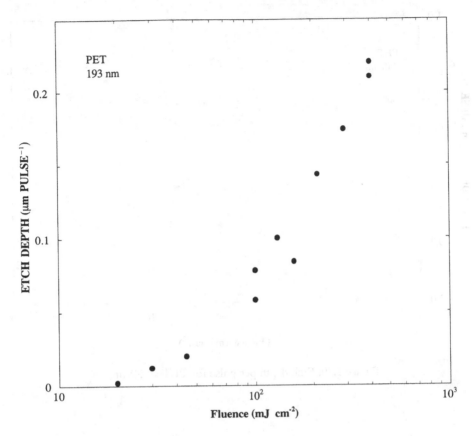

Figure 5.19. Etch depth per pulse for PET at 193 nm.

example, even when material has been cleanly removed by laser etching, the underlying solid will have been exposed to heat and radiation and will have had its physicochemical properties altered. The effect of this alteration on the rate of subsequent etching can be seen in experimental etching rate data but is difficult to quantify because the chemical state of this partially processed material cannot be easily characterized.

There have been a large number of theoretical studies of excimer laser ablation effects in organic polymers (Dyer and Karnakis 1994a,b, Sobehart 1993, Pettit and Sauerbrey 1993, Cain *et al.* 1992a,b, Gai and Voth 1992, Furzikov 1990, Kalontarov 1991, Mahan *et al.* 1988, Jelinek and Srinivasan 1984, Garrison and Srinivasan 1985a,b, Srinivasan *et al.* 1986b, Keyes *et al.* 1985, Mahan *et al.* 1988, Sutcliffe and Srinivasan 1986, Küper and Stuke 1987, Ye and Hunsperger 1989). Emphasis has

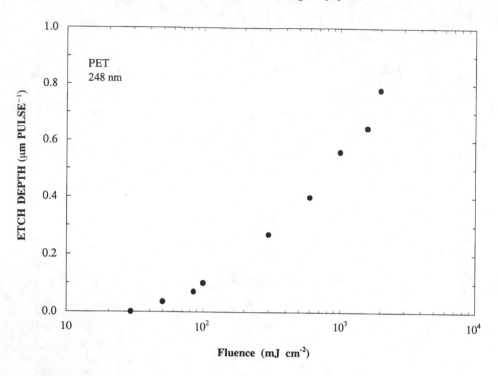

Figure 5.20. Etch depth per pulse for PET at 248 nm.

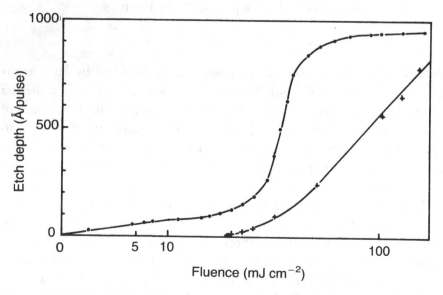

Figure 5.21. PET etch depth per pulse (+) and soluble depth (●) (in acetone) versus fluence at 193 nm plotted on a logarithmic scale (Lazare and Granier 1988a,b).

been placed on calculation of the ablation rate as a function of fluence. A simple analysis based on the use of the Beer–Lambert law leads to the prediction that

$$X = \frac{1}{\alpha} \ln \left(\frac{E}{E_T} \right) \tag{6}$$

where α is the absorption coefficient at the laser wavelength. Equation (6) predicts that X should be linearly related to log E and that the slope of this line will be α^{-1}. In general, neither of these predictions is in agreement with experiment because X is usually found to exhibit a superlinear dependence on E near and above threshold. In addition, the absorption coefficient predicted from equation (6) will not be the same as that measured in conventional absorption spectra at low light levels. Furthermore, equation (6) would apply for continuous as well as pulsed sources whereas APD and ablation are observed only for high intensity pulses of short duration.

The fundamental problem in applying equation (6) arises because it assumes that all material to a depth α^{-1} remains in the beam throughout the laser pulse. In fact, ablation is observed to begin within the first few nanoseconds after the start of the pulse. Furthermore, radiation penetrates to a considerably larger depth than α^{-1} even if it is not of sufficient intensity to yield ablative decomposition. Thus the total exposure of a given layer to radiation following several pulses may be larger than that given by the Beer–Lambert relation.

A correct treatment of APD must then incorporate the presence of a receding ablation surface whose position relative to the front surface of the sample depends on time during the one laser pulse. This problem has been addressed by Keyes *et al.* (1985) and by Lazare and Granier (1988a,b). The geometry of this moving interface is shown in Figure 5.22. The speed of this interface is taken to be

$$v = r[I(x) - I_T] \tag{7}$$

where r is an ablation rate constant with the units $\mu m \, ns^{-1} \, MW^{-1} \, cm^2$ (Lazare and Granier 1988a,b).

$I(x)$ is related to I_0 through

$$I(x, t) = I_0 \exp(-\gamma x(t)) \tag{8}$$

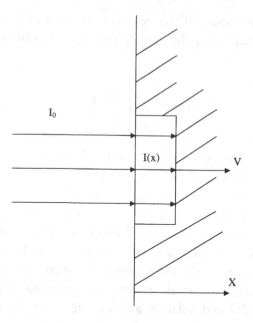

Figure 5.22. A moving interface in laser APD of a polymer. I_0 is the intensity at the front surface and $I(x)$ is the instantaneous intensity at the moving interface at depth x. The interface is receding at a speed v.

where γ is an effective attenuation coefficient for the vapor over the surface. Then, integrating to obtain the etch depth at time t_2,

$$X(t_2) = \int_{t_1}^{t_2} r(I_0(t') \, e^{-\gamma x(t')} - I_T) \, dt' \qquad (8a)$$

where $t_2 - t_1$ is the time interval during the laser pulse when $I(x) > I_T$. Lazare and Granier (1988a,b) showed that equation (8a) can be used to fit experimental etching rate data for the polymer polyphenylquinoxaline (PPQ) using r and γ as variable parameters. They found that γ obtained from this fitting procedure is not equal to the absorption coefficient α measurement at low intensity. For PPQ, r was found to range between $1.25 \times 10^{-3} \, \mu m \, ns^{-1} \, MW^{-1} \, cm^2$ for irradiation at 193 nm and $0.92 \times 10^{-3} \, \mu m \, ns^{-1} \, MW^{-1} \, cm^2$ at 351 nm. Although this theory is successful in incorporating shielding effects due to ablation products, it does not address the mechanisms associated with ablation. One characteristic parameter is the threshold fluence, E_T.

Table 5.5 summarizes some representative threshold fluence data for several materials at different laser wavelengths. The fluence threshold

$$E_T = F_T \hbar w \qquad (9)$$

Table 5.5. *A summary of data for photoetching of polyimide at threshold.*

	Wavelength (nm)				
	157	193	248	308	351
E_T (mJ cm^{-2})	36	45	27	45	120
F_T (10^{16} cm^{-2})	2.9	4.4	3.4	7.0	21
α (10^5 cm^{-1})	1.1	3.7	2.8	0.95	0.28
$\sigma n_c F_T$	1.4	7.1	4.1	2.9	2.6
η_T (eV mode^{-1})	0.10	0.41	0.18	0.11	0.08

where F_T is the photon fluence (photons cm^{-2}) and $\hbar w$ is the energy per photon. It is evident that threshold fluences in the range 50–150 mJ cm^{-2} correspond to photon fluences of 10^{16}–10^{17} photons cm^{-2}.

The significance of these photon thresholds can be seen by examining the product σE_T, where

$$\sigma = \frac{\alpha}{N_c} \tag{10}$$

is the cross-section per chromophore and N_c is the number of chromophores per cm^3 in the polymer. N_c can be obtained from the density of monomer N_m by

$$N_c = n_c N_m \tag{11}$$

where n_c is the number of chromophores per monomer. In Kapton^{TM}, where the density is 1.4 g cm^{-3} and the monomer has an atomic mass of 382 amu, $N_m = 2.3 \times 10^{21}$ cm^{-3}. In regions of strong absorption $\sigma \simeq 10^{-16}$–10^{-17} cm^2.

Since the absorption coefficient

$$\alpha = \sigma N_c \tag{12}$$

$$\sigma n_c = \alpha/N_m \tag{13}$$

For polyimide

$$\sigma n_c = 4.35 \times 10^{-22} \alpha \tag{14}$$

The product $\sigma n_c F_T$ then gives the number of photons absorbed per monomer at threshold. This quantity is given for polyimide in Table 5.5. It is apparent that $\sigma n_c F_T$ is greater than unity at all wavelengths. Since there are three aromatic rings per monomer in polyimide, these data suggest that, with the exception of excitation at 157 nm, the threshold for ablation corresponds to the absorption of at least one photon by each ring. Since the energy required to break an internal ring bond is about 10 eV (Table 5.1) and photon energies are all less than this value, it is unlikely that excitation leads directly to $-C{=}C$ bond breaking and the liberation of carbon atoms. Some energy redistribution must therefore occur prior to dissociation.

Some idea of this redistribution can be obtained from a simple model in which electronic energy is assumed to be shared on a rapid timescale (picoseconds) between the $(3N - 6)$ vibrational modes of the monomer. In polyimide, whose monomer has the formula $C_{22}N_2O_5H_{10}$, there are 39 atoms and 111 such modes. The average energy per vibrational mode at threshold is then

$$\eta_T = \frac{\sigma n_c F_T \hbar w}{(3N - 6)} = \frac{\sigma n_c F_T \hbar w}{111} \tag{15}$$

where $\hbar w$ is the incident photon energy. An estimate of η_T at various wavelengths is given in Table 5.5. Values range from 0.08 eV per mode at 351 nm to 0.41 eV per mode at 193 nm. A vibrational energy $\eta = 0.1$ eV per mode would correspond to an internal vibrational temperature $T \simeq 1160$ K.

An equilibrium non-statistical model predicts a dissociation rate

$$R_T \simeq \nu_v \exp\left[-D/(kT)\right] \tag{16}$$

where ν_v is the vibrational frequency (s^{-1}) for the bond with dissociation energy D. With $\nu_v \simeq 3 \times 10^{13} \, s^{-1}$ and $D = 3.5$ eV as representative values, one obtains $R_T = 0.02 \, s^{-1}$ when $kT = 0.1$ eV and $R_T = 4.8 \times 10^9 \, s^{-1}$ when $kT = 0.4$ eV. Under these conditions the product $R_T \tau_p$ provides an approximate estimate of the probability of dissociation during a laser pulse of duration τ_p. With $\tau_p = 10$ ns it is evident that $R_T \tau_p$ ranges from $\ll 1$ when $kT = 0.1$ eV to >1 when $kT = 0.4$ eV. Thus this highly simplified analysis predicts that redistribution of electronic excitation to available vibrational modes of the monomer may under certain conditions result in a high probability for dissociation of specific bonds during the laser pulse even at threshold fluence. As E exceeds E_T, η would increase proportionately, leading to a higher effective dissociation rate.

In reality, the redistribution of energy within a monomer unit following excitation occurs statistically. The resulting dissociation rate, θ, for the decomposition

$$MX^* \rightarrow M + X \tag{17}$$

is given as follows (Eyring et al. 1980):

$$\theta = \frac{\nu_v (j - m + s - 1)! j!}{(j + s - 1)!(j - m)!} \tag{18}$$

where

$$j = \frac{\hbar w}{\hbar \nu_v} \tag{19}$$

is the number of vibrational quanta (assumed to be an integer number) produced by the initial excitation $\hbar w$. Also

$$m = \frac{D}{\hbar \nu_v} \tag{20}$$

is the number of vibrational quanta corresponding to dissociation. In this approximation all vibrational modes are taken to have the same vibrational frequency and all bonds the same dissociation energy, D. The total number of modes is $s = 3N - 6$, where N is the number of atoms in the molecule. In this formalism, dissociation will occur if $\theta \tau_p \simeq 1$.

As an example, we take $\nu_v = 3 \times 10^{13} \text{ s}^{-1}$ and $\hbar \nu_v = 0.125 \text{ eV}$ so that, with $D = 3 \text{ eV}$ and $\hbar w = 5 \text{ eV}$, $j = 40$ and $m = 28$. Then, with $s = 111$, as is appropriate for the monomer of polyimide, $\theta \simeq 10^{-7} \text{ s}^{-1}$. A reduction in D or an increase in $\hbar w$ would increase θ appreciably. However, θ would have to increase to about 10^8 s^{-1} for significant dissociation to occur during the 10 ns excimer pulse length.

These estimates of dissociation rates have all neglected the presence of other energy loss mechanisms that may result in the deactivation of monomers before dissociation can occur. Such channels can involve radiative decay (i.e. luminescence) or conduction to the surrounding medium. Given the short duration of excimer laser pulses and the low thermal diffusivity of these organic polymers, conductive losses will be minimal. Radiative losses may be significant in some instances but polymers such as polyimide are not usually strongly luminescent.

The existence of other loss channels, each with an associated time-scale, implies that some intensity threshold exists for ablative etching. The rôle of laser intensity in etching has been discussed by several groups (Sutcliffe and Srinivasan 1986, Ye and Hunsperger 1989, Single-ton *et al.* 1990) and is somewhat controversial. There is little doubt, however, that a threshold intensity must exist for ablation since no abla-tion is produced by conventional low intensity light sources at fluence levels comparable to those obtained with excimer laser irradiation.

An estimate of the threshold intensity can be obtained from a model that assumes that the onset of ablation is accompanied by the establish-ment of a critical level of radicals or molecular fragments during the laser pulse. Figure 5.23(a) shows this process in schematic form.

If the density of radicals generated by photodissociation of monomeric units is n cm^{-3}, then the rate equation for the radical concentration is

$$\frac{dn}{dt} = \frac{Q(N_c - n)\sigma I}{\hbar w} - kn^2 \tag{21}$$

where N_c is the chromophore density and Q is the quantum efficiency for photodissociation at the photon energy. $\sigma = \alpha N_c$ is the absorption cross-section per chromophore and k is the rate constant for radical recombination. This will be given by

$$k = \Sigma v \tag{22}$$

where Σ is the cross-section for collision and v is a diffusion velocity.

If a dynamic equilibrium is assumed then $dn/dt = 0$ and

$$I = \frac{kn^2 \hbar w}{Q(N_c - n)\sigma} \tag{23}$$

The threshold intensity for ablation will then correspond to the estab-lishment of some critical radical density n'. Then

$$I_T = \frac{kn'^2 \hbar w}{Q(N_c - n')\sigma} \tag{24}$$

(a)

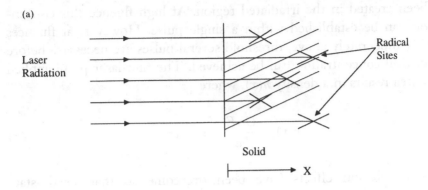

(b)

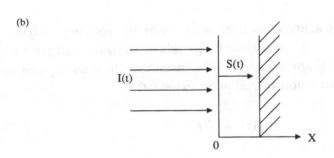

Figure 5.23. (a) The interface region at the surface of an organic polymer exposed to UV laser radiation. (b) The moving interface region when ablation is occurring. $S(t)$ is the instantaneous position of this interface at time t measured from its initial position at $t = 0$.

Values for many of the quantities in equation (24) are unknown. Some approximate values, typical of a polymer, might be $\Sigma = 10^{-16}\,\mathrm{cm}^2$, $v = 10^4\,\mathrm{cm\,s^{-1}}$, $\hbar w = 5\,\mathrm{eV} = 8 \times 10^{-19}\,\mathrm{J}$, $Q = 0.1$, $\sigma = 2 \times 10^{-17}\,\mathrm{cm}^2$, $N_\mathrm{c} = 10^{22}\,\mathrm{cm}^{-3}$ and $n' = 0.1 N_\mathrm{c} = 10^{21}\,\mathrm{cm}^{-3}$. Then $I_\mathrm{T} \simeq 4 \times 10^7\,\mathrm{W\,cm^{-2}}$, which is somewhat higher than values estimated by Sutcliffe and Srinivasan (1986) from etching rate data for PMMA and polyimide.

To find the dependence of etch depth on fluence in this model we follow the analysis of Mahan *et al.* (1988). In this approximation ablation is taken to occur when a certain threshold concentration n' of radicals has been created in the irradiated region. At high fluence this concentration can be established within a single pulse. However, at fluences near, or moderately above, threshold several pulses are necessary before this critical concentration can be achieved. The first laser pulse applied to an area results in a density $n(x)$, where

$$n(x) = \frac{Q\alpha E}{\hbar w} \exp(-\alpha x) \tag{25}$$

When incubation effects have been overcome so that steady-state ablation is occurring, $n(x,t)$ can be obtained by solving the following equation:

$$\frac{dn(x,t)}{dt} = \frac{Q\alpha I(t)}{\hbar w} \exp[-\alpha(x - s(t))] \tag{26}$$

where $I(t)$ is the instantaneous laser intensity at the receding polymer surface. The position of this surface is $s(t)$ relative to the initial position of the front surface (Figure 5.23(b)). The instantaneous intensity can be expressed in terms of a normalized pulse shape $i(t)$

$$I(t) = Ei(t) \tag{27}$$

where

$$\int_{-t_r}^{t_r} i(t') \, dt' = 1 \tag{28}$$

and $2t_r$ is the pulse repetition period.

The position of the ablating surface $s(t)$ (where $n(x,t) = n'$) can be obtained from the equation

$$e^{\alpha s(t)} = 1 + \frac{Q\alpha E}{\hbar w n'} \int_{-t_r}^{t} i(t') \, e^{\alpha s(t')} \, dt' \tag{29}$$

yielding the solution

$$s(t) = \frac{QE}{\hbar w n'} \int_{-t_r}^{t} i(t') \, dt' \tag{30}$$

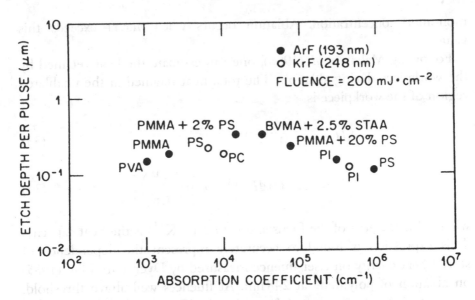

Figure 5.24. Etch depth per pulse versus the absorption coefficient α at the laser frequency. The data show little dependence upon absorption coefficient; the latter span three decades of values. All points were measured for the same fluence (Mahan *et al.* 1988).

If $t = t_r$, i.e. the position of s is calculated at $t = t_r$ then the integral in equation (30) becomes unity and

$$X \equiv s(t_r) = \frac{QE}{\hbar w n'} \tag{31}$$

so that the etch depth per pulse, X, is predicted to be linearly related to E and independent of the absorption coefficient, α. These predictions appear to be well substantiated at low laser fluence (Figure 5.24) for a range of materials irradiated at 193 or 248 nm. Data for polyimide at 308 nm (Figure 5.16) also show a linear dependence on E over a fluence range extending to $6 \, J \, cm^{-2}$. These data can be used to obtain an estimate of the critical density of bond breaking n' since

$$n' = \frac{QE}{X \hbar w} \tag{32}$$

Using the data given in Figure 5.24, one obtains $EX^{-1} \simeq 1.67 \times 10^4 \, J \, cm^{-3}$ and thus $n'Q^{-1} \simeq 1.6 \times 10^{22}$ and $2.1 \times 10^{22} \, cm^{-3}$ at 193 and 248 nm, respectively. With $Q \simeq 0.1$, this would imply that the critical density of broken bonds (or radical pairs) would be about 10% of the monomer concentration. Ablation occurs when $n(x,t)$ exceeds this value.

Following Mahan *et al.* (1988), one can estimate the heat retained by the workpiece during ablation. The total heat retained in the unablated region of the workpiece is

$$\Delta H = \frac{AC}{\alpha} \Delta T(X) \tag{33}$$

$$= (1 - Q)AE_T \left[1 - \exp\left(\frac{-E}{E_T}\right) \right] \tag{34}$$

where A is the area of the focus and C ($\mathrm{J\,cm^{-3}\,K^{-1}}$) is the heat capacity. This expression is found to accurately represent the dependence of stored heat energy on laser fluence measured by Dyer and Sidhu (1985) on ablation of polyimide at 248 nm. At fluences well above threshold, equation (34) shows that $\Delta H \rightarrow (1 - Q)AE_T \simeq AE_T$ if Q is small. Sutcliffe and Srinivasan (1986) have shown that the thermal effect is constant above threshold but increases linearly up to the threshold. This is in agreement with the above result and implies that the concentration of radical pairs can increase up to a limiting value. Solutions in excess of this value are accompanied by ablation.

A more general solution for the thermal heating component would include provision for the heat released by the recombination of radical pairs below the threshold for ablation. Equation (34) then becomes

$$\Delta H = \frac{A}{\alpha} C \, \Delta T(X) + \frac{A}{\alpha} n' D \tag{35}$$

with $n' \simeq 10^{21} \, \mathrm{cm^{-3}}$ and $D = 3.5 \, \mathrm{eV}$, $n'D \simeq 5.6 \times 10^2 \, \mathrm{J\,cm^{-3}}$. Since $C = 1.55 \, \mathrm{J\,cm^{-3}\,K^{-1}}$ for polyimide (Goredetsky *et al.* 1985) and $\Delta T \simeq 800\text{--}2200 \, \mathrm{K}$, $C \, \Delta T(X) \simeq 3 \times 10^3 \, \mathrm{J\,cm^{-3}}$. Thus the heating effect attributed to radical recombination is somewhat smaller than that due to direct excitation of lattice modes.

Some insight into the retention of energy by ablation products can be obtained from the ratio

$$\xi = (E - E_T)/X \tag{36}$$

where ξ ($\mathrm{J\,cm^{-3}}$) is the energy density dissipated in producing ablation. Some data for polyimide etching at 193, 248 and 308 nm are given in Table 5.6.

Table 5.6. *Energy per unit volume of ablated material ξ ($J\,cm^{-3}$) for polyimide as a function of etch depth per pulse, X (Srinivasan et al. 1987).*

Wavelength (nm)	E_T (mJ cm^{-2})	E (mJ cm^{-2})	X (μm)	ξ (J cm^{-3})
193	45	60	0.035	4.3×10^3
		1000	0.13	7.3×10^4
		2000	0.17	1.2×10^5
248	27	1000	0.35	2.9×10^4
		2000	0.52	3.8×10^4
308	45	200	0.14	1.1×10^4
		1000	0.55	1.8×10^4
		2000	1.15	1.7×10^4

Surprisingly, ξ is largest for 193 nm ablation and actually *decreases* (at constant E) with increasing wavelength. This suggests that ablation at 193 nm is relatively inefficient compared with ablation at other wavelengths because of enhanced energy storage (see, however, Küper *et al.* 1993). This energy storage could be in the form of internal (vibronic) or external (kinetic) terms. There does not appear to have been a systematic experimental study of such energy storage (or of another organic polymer) at three or more excimer wavelengths. However, data that are available (e.g. Srinivasan *et al.* 1987) show that neither the velocity distribution of C_2 emitted during APD nor the average translational energy depend strongly on laser wavelength or fluence. However, it is possible that the bulk of stored energy resides as internal energy in larger ejected fragments. Since the radiative lifetime for radiative infrared emission is typically about 1–10 ms, such vibrationally excited fragments will not deactivate during the ablation event.

Since ablation of polymers involves at least some combination of photochemical and photothermal effects, it has been suggested that the etch depth per pulse can be written (Srinivasan *et al.* 1986b)

$$X = X_{\mathrm{pc}} + X_{\mathrm{th}} \tag{37}$$

$$= \frac{1}{\alpha} \ln\left(\frac{E}{E_T}\right) + A_1 \exp\left(\frac{-\epsilon_1}{kT}\right) \tag{38}$$

where the last term is a thermally activated reaction rate with an activation energy ϵ_1 and a simple Beer's law dependence has been assumed for the photon penetration depth. If the excess energy $E - E_T$ is taken to be converted to heat within the layer extending to a depth α^{-1} then

$$E - E_T = \rho C X \, \Delta T \tag{39}$$

where ρ is the density and C is the heat capacity ($J\,g^{-1}\,K^{-1}$). Using the approximation from Beer's law that

$$\Delta T = \alpha \frac{E - E_T}{\rho C} \left[\ln \left(\frac{E}{E_T} \right) \right]^{-1} \tag{40}$$

$$= T - T_0 \tag{41}$$

where T_0 is an ambient temperature and assuming $T \gg T_0$, equation (38) can be rewritten

$$X = \frac{1}{\alpha} \ln \left(\frac{E}{E_T} \right) + A_1 \exp \left[\frac{-\epsilon_1 \rho C}{k \alpha (E - E_T)} \ln \left(\frac{E}{E_T} \right) \right] \tag{42}$$

This equation yields a good fit to empirical etching rates in a number of polymers, particularly at high fluence. Figure 5.25 shows a comparison with etching rate data for polyimide at 308 nm (Babu *et al.* 1992).

Cain *et al.* (1992a,b) have adopted a pure photothermal description of the ablation process in which absorbed photon energy is converted to heat which then drives a thermal degradation wave into the polymer. The position of this surface is given by the solution to the equation

$$\frac{ds}{dt} = k_0 \exp \left[-\epsilon / (kT) \right] \tag{43}$$

where k_0 is an Arrhénius factor, ϵ is the activation energy and k is Boltzmann's constant. The temperature of this receding surface is

$$T = \frac{\alpha E \, e^{-\alpha s}}{\rho C} \tag{44}$$

The temperature evolution within the irradiated material in response to this heating can be obtained from equations (4.14) and (4.15) (Ready 1971). Since this result does not account for the removal of material

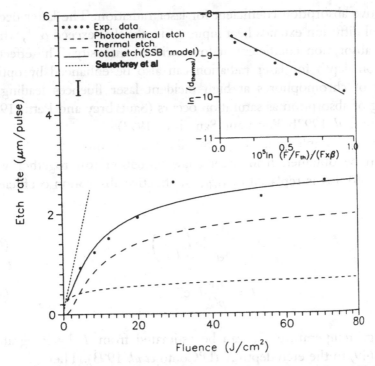

Figure 5.25. The etching rate for polyimide at 308 nm, 25 ns pulse width. A typical error bar for the data is shown by the vertical line. Inset: a plot of $\ln(E/E_{th})/(E\beta)$ versus $\ln d_{thermal}$ (Babu *et al.* 1992).

once it achieves the decomposition temperature, it overestimates the etch depth because energy stored within material at the front surface can continue to heat the interior of the solid past the time when it would have been removed. A moving etching front can be included by solving the coupled equations (Cain *et al.* 1992a)

$$\frac{dX'}{dt'} = \exp\left(-\frac{\epsilon}{kT(x')}\right) \tag{45}$$

$$T(x', t' + dt') = T(x', t') + \delta T(x') \tag{46}$$

$$\delta T(x') = \kappa' \frac{\delta^2 T(x')}{(\delta x')^2} \delta t' \tag{47}$$

where $x' = \alpha s$, $t' = \alpha k_0 t$, $X' = \alpha S$ and $\kappa' = \alpha \kappa k_0^{-1}$. The effect is to limit the etching rate at long times following the laser pulse while decreasing the effective absorption coefficient for laser radiation. The latter occurs as thermal diffusion extends heat input to depths that exceed α^{-1}, where α is the absorption coefficient at low incident intensity. The effective penetration depth for laser radiation can also be enhanced by optical pumping of chromophores at high incident laser fluences leading to bleaching of absorption as saturation occurs (Sauerbrey and Pettit 1989, 1990, Cain *et al.* 1992b, Pettit and Sauerbrey 1993).

A relatively simple estimate of the fluence dependence of the etch depth can be obtained from Beer's law (equation (6)) together with equation (43) if α is replaced by α_{eff}, an effective absorption coefficient. Then

$$X = \frac{1}{\alpha_{\text{eff}}} \ln\left(\frac{E}{E_T}\right) \tag{48}$$

$$T = \frac{\alpha_{\text{eff}}}{\rho C} E \, e^{-\alpha_{\text{eff}} s} \tag{49}$$

An average temperature T_{av} can be estimated from T by integrating equation (49) to the etch depth X (D'Couto *et al.* 1993). Then

$$T_{\text{av}} = \frac{\alpha_{\text{eff}} E}{\rho C X} \int_0^X e^{-\alpha_{\text{eff}} s} \, ds \tag{50}$$

$$= \frac{\alpha_{\text{eff}}}{\rho C} \frac{1}{\ln(E/E_T)} (E - E_T) \tag{51}$$

and we have

$$\frac{ds}{dt} = k_0 \exp\left[-\epsilon/(kT_{\text{av}})\right] \tag{52}$$

or

$$\ln\left(\frac{ds}{dt}\right) = \ln k_0 - \frac{\epsilon}{kT_{\text{av}}} \tag{53}$$

$$= \ln k_0 - \frac{\epsilon \rho C}{k\alpha_{\text{eff}}} \frac{\ln(E/E_T)}{(E - E_T)} \tag{54}$$

Some numerical values for the parameters in equation (54) derived from fits to the dependence of etching rate on fluence in polyimide and

Table 5.7. *Activation energy and pre-exponential Arrhenius factor for polyimide and polyimide-doped PTFE derived from etching rate (D'Couto et al. 1993).*

Dopant level (wt%)	E_T (J cm^{-2})	$\epsilon \rho C/(k\alpha_{eff})$ (J cm^{-2})	k_0 (µm/pulse)	λ (nm)	Pulse length (ns)
100	0.03	0.10	0.68	248	16
5	0.08	0.05	1.55		
1	0.10	0.06	1.97		
0.2	0.74	0.73	3.56		
100	0.05	0.53	1.34	308	16, 25
5	0.21	0.89	4.04		
1	0.37	0.46	4.11		
0.2	0.85	0.62	7.72		

PTFE doped with polyimide (D'Couto *et al.* 1993) are given in Table 5.7. These fits are to the overall etch depth per pulse, i.e. $ds/dt \equiv X$ per pulse and yield a good simulation of empirical etching rate data.

The activation energy ϵ for decomposition of polyimide has been estimated to be 8.3×10^{-19} J (Cain *et al.* 1992b). With $\rho C = 1.55$ J cm^{-3} K^{-1} for polyimide, the data in Table 5.7 yield $\alpha_{eff} \simeq 9.3 \times 10^5$ and 1.8×10^5 cm^{-1} at 248 and 308 nm, respectively. These values are *larger* than those measured at low incident intensities (Table 5.5). In addition, $T_{av} \geq 10^4$ K (equation (51)), which is considerably higher than the surface decomposition temperature (1560–1730 K) for polyimide implied by threshold ablation experiments (Mihailov and Duley 1991) and direct temperature measurement (Brunco *et al.* 1992). Thus values of ϵ and α_{eff} derived from fitting equation (54) to experimental etching rate data should not be given undamental significance.

To summarize these models and their predictions, Figure 5.26, adapted from Sobehart (1993), shows temperature, laser intensity and density across the ablation front in a polymer such as polyimide during ablation. In Figure 5.26, the etching front is the boundary between regions 3 and 4, and is moving to the right at a speed ds/dt. Region 4 is the volume directly ahead of the etching front. Material in this region has been exposed to a precursor radiative flux and is at elevated temperature. It may also exhibit bleaching due to saturated absorption, particularly near the regions 3–4 interface. The temperature, T_A, inside region 4 is elevated and approaches the decomposition temperature for

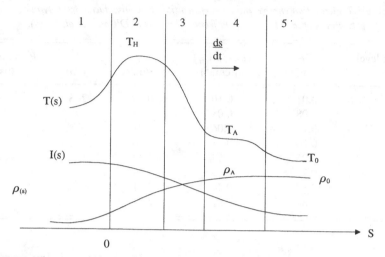

Figure 5.26. The spatial dependence $T(s)$, $I(s)$ and $\rho(s)$ across the ablation front in an absorbing polymer such as polyimide (after Sobehart 1993). ρ_0 is the density of the bulk polymer and T_0 is the ambient temperature.

the polymer whereas the density inside this region $\rho_A < \rho_0$ due to precursor effects such as the loss of volatile components.

The interface between regions 4 and 5 is less well defined spatially and represents a region where incubation effects may be important and where the integrated fluence $E \leq E_T$. Farther inside region 5, the density and temperature reduce to those of the ambient medium and the laser intensity drops to zero.

Region 3 occurs directly above the decomposing surface of the polymer. It is a region of strong heating and high thermal gradients where the density is close to that of the solid polymer. Energy deposition in this region continues due to absorption of incident laser radiation although the absorption coefficient may be much less than that of the polymer itself. This heating occurs back into region 2, where an equilibrium is established between excitation due to absorption of laser radiation and deexcitation due to gas ejection, radiative cooling and electron–ion recombination. The equilibrium is achieved at T_H, which may be considerably higher than the decomposition temperature, T_A. The gas in this region is ejected with a velocity

$$v \simeq \left(\frac{\gamma k T_H}{M} \right)^{1/2} \tag{55}$$

where M is the average mass of a decomposed fragment and $\gamma \simeq 5/3$. Expansion occurs into the dilute ambient gas (region 1).

5.6 LITHOGRAPHY WITH EXCIMER LASERS

Photolithography with UV radiation is a standard technique in the fabrication of microelectronic devices. UV light offers the advantage of increased resolution but suffers from the disadvantage that incoherent light sources are generally of low intensity. In addition, the emission of UV light from a conventional light source is often accompanied by large amounts of light at visible and IR wavelengths. This excess light leads to heating with attendant distortion of the substrate. Another disadvantage is that the low intensity available from conventional light sources results in long exposure times and therefore long wafer processing times.

The use of lasers as photolithographic sources to expose resists has been known for some time (see, for example, Becker *et al.* 1978). The low laser power from CW laser sources means that scanning must be used with a corresponding increase in processing time. Furthermore, the high coherence of CW laser sources such as the Ar^+ ion laser limits image quality because of speckle (Upatnieks and Lewis 1973).

The availability of high average power excimer lasers eliminates many of these problems. For example, the low spatial and temporal coherence of the excimer output serve to reduce speckle while the high intensity enables exposures to be made over timescales of the order of 10^{-8} s. Such short exposure times virtually eliminate problems associated with vibration, while wavelengths as short as 157 nm (from the F_2 laser) promise improved spatial resolution.

The interaction of high intensity excimer laser radiation with organic polymers involves a combination of thermal and photochemical effects that results in surface etching at fluences exceeding a few tens of millijoules per cm^2. The ablative nature of this interaction yields well defined etch structures with sharp boundaries that show little evidence for thermal heating of the surrounding region. This offers the possibility of high resolution direct photoetching in the fabrication of semiconductor devices using excimer laser radiation (Jain 1986, Rothschild and Ehrlich 1988).

The possibility of utilizing excimer laser radiation in self-developing photoetching applications was noted by Srinivasan and Mayne-Banton

Table 5.8. *A comparison between excimer laser and other techniques in direct etching of nitrocellulose films (Geis et al. 1983).*

Exposure technique	Sensitivity ($eV\ nm^{-3}$)	Resolution (nm)	Residue (nm)
ArF laser 193 nm 10 ns pulse 20 mJ cm^{-2}	24	300	<3
Hg lamp ≥200 nm 0.8 mW cm^{-2}	600		Yes
Ar$^+$	190	100	<3
H$_2^+$ 100 keV	580		Yes

(1982). In this experiment, pulses from an ArF laser were used to irradiate polyethylene terephthalate (PET) films at fluences of 50–400 mJ cm^{-2} per pulse. The resulting etching rate was found to be linearly dependent on the total dose above threshold. Etch depths per pulse were of the order of α^{-1} where α is the absorption coefficient of PET at 193 nm. An example of the resolution attainable can be seen in Figure 5.27.

High resolution photoetching in nitrocellulose using 193 nm laser radiation was subsequently reported by Geis *et al.* (1983). The threshold for photoetching in nitrocellulose was found to be about 20 mJ cm^{-2}, somewhat less than that observed in etching of PET films. Excimer laser photoetching of nitrocellulose was compared with Ar$^+$ and H$_2^+$ ion beam etching (Table 5.8) and was shown to exhibit comparable resolution. Because of the ablative nature of the interaction, little non-volatile residue remained following etching with excimer radiation.

The use of the excimer laser as a source for photolithography was initially reported in 1982 (Kawamura *et al.* 1982a,b, Jain *et al.* 1982a,b). Kawamura *et al.* (1982a) irradiated PMMA photoresist with pulses from a KrF laser at fluence levels between 0.09 and 2.7 J cm^{-2}. The resist was subsequently developed to reveal etching to a depth ≤1 μm. Since fluences in this range also result in direct etching of PMMA it is surprising that no mention of direct etching appears in the paper by Kawamura *et al.* (1982a). Jain *et al.* (1982a) used a contact mask to print

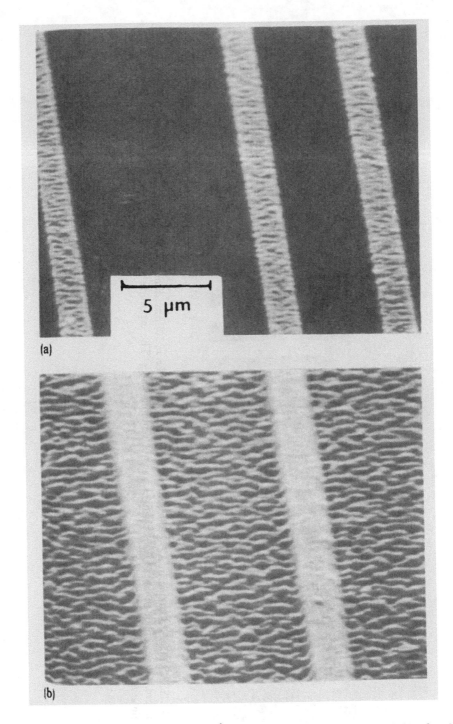

Figure 5.27. PET film with lead (2000 Å) lines. (a) Before irradiation and (b) after 30 pulses of 193 nm radiation (110 mJ cm^{-2}) for exposure in air (Srinivasan and Mayne-Banton 1982).

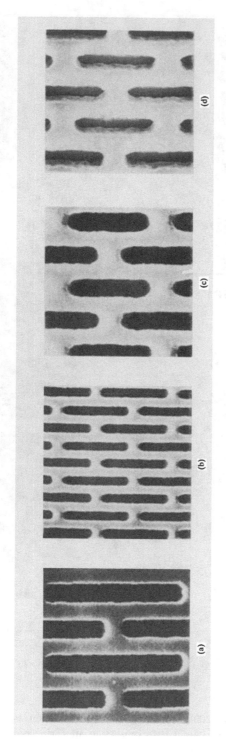

Figure 5.28. SEM images obtained in AZ-2400 photoresist with a XeCl laser at 308 nm: dimensions of lines and spaces are (a) 2 μm, (b) 1 μm, and (c) 0.5 μm. (d) An oblique view of (c) showing a trace of standing waves. The exposures were made with two 10 ns wide, 50 mJ cm^{-2} laser pulses (Jain et $al.$ 1982b).

roughly 0.5 µm wide lines and spaces in various photoresists using pulses from XeCl, KrF and KrCl lasers. Generally fluence levels were considerably less than that utilized by Kawamura *et al.* (1982a). Typical exposures consisted of two pulses each of 10 ns duration and with about 50 mJ cm^{-2}.

Kawamura *et al.* (1982b) later used an ArF laser to expose PMMA photoresist. Lines of ≤0.5 µm width separated by about 2 µm were produced with a fluence of 0.93 J cm^{-2} irradiating a chromium photomask. Figure 5.28 shows a selection of images of lines and spaces produced using contact lithography with pulses from an XeCl laser (Jain *et al.* 1982b). Here the photoresist was AZ-2400 and the fluence level was about 50 mJ cm^{-2}. This study and the work of Kawamura *et al.* (1982a,b) demonstrated for the first time the feasibility of excimer laser photolithography at a resolution of about 0.5 µm.

Contact lithography using an F$_2$ excimer laser oscillating at 157 nm has been shown to be capable of generating features as small as 0.15 µm (Craighead *et al.* 1983, White *et al.* 1984). Limits to the size of features that could be reproduced were determined by the availability of suitable masks. The masks used were fabricated on CaF$_2$ substrates with electron beam lithography. A three-level resist consisted of a polyimide layer coated with 25 nm of Ge, which was then etched in a two-step process by a combination of photon and anisotropic ion etching.

The resist used in excimer laser lithography by Craighead *et al.* (1983) was a copolymer of methyl methacrylate (MMA) and methacrylic acid (MAA). This resist was found to exhibit higher sensitivity and to permit deeper etching than PMMA while retaining the capability of reproducing high resolution features.

Etching of diamond-like carbon photoresist with 193 nm ArF laser radiation has been discussed by Rothschild *et al.* (1986) and Rothschild and Ehrlich (1987). Etching of lines with widths as small as 0.13 µm was reported, which is close to the ultimate line width of an imaging system ($\lambda/(4\mathrm{NA})$, where NA is the numerical aperture of the focusing system). It is notable that this is less than the laser wavelength. Etching occurs by a combination of direct vaporization and photoconversion to a more strongly absorbing graphitic phase. The overall efficiency defined as the ratio of the number of carbon atoms removed per laser photon was found to be as high as 0.12.

The use of masks in lithography has a number of disadvantages. Most notable is the occurrence of defects in the mask due to mechanical damage during placement and removal from the wafer. Projection

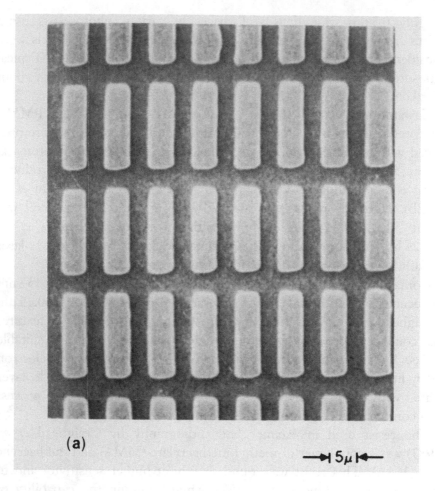

(a)

$\longrightarrow$ | 5μ |$\longleftarrow$

Figure 5.29. Images obtained using a XeCl excimer laser, $\lambda = 308$ nm, on a commercial 1 : 1 scanning projection printer. The full-wafer exposure time in 1 μm thick positive photoresist was 15 s (Jain and Kerth 1984).

lithography eliminates these problems while permitting full wafer exposure with superimposed laser pulses. The use of an excimer laser for projection lithography has been discussed by Jain and Kerth (1984) and Latta *et al.* (1984). Figure 5.29 shows images obtained using an XeCl laser in a commercial 1:1 scanning projection printer using this technique. The laser was operated at 100 Hz and it was estimated that each point on the wafer was exposed to about 20 overlapping pulses, each with fluence of 50–100 mJ cm^{-2}. Features as small as 1 μm were accurately reproduced with little evidence for speckle. Wafer exposure times were 15 s at an average XeCl laser power of 13 W. Extension of

(b)

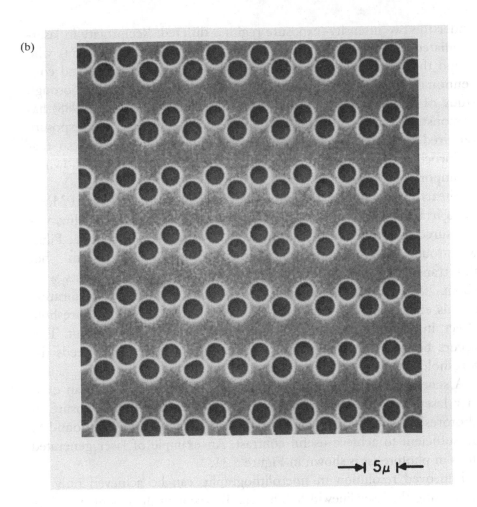

$\longrightarrow$ 5µ $\longleftarrow$

projection photolithography to 193 nm was reported by Latta *et al.*
(1984).

Reciprocity is an important consideration when using excimer laser
sources in lithography. Photoresists such as PMMA that are effectively
etched with low intensity UV light and long exposures are also easily
etched with excimer laser pulses on a timescale that is reduced by a
factor of at least 10^8. In conventional photolithography the resist is
exposed to light of intensity about $1\,\mathrm{mW\,cm^{-2}}$ for several minutes. The
integrated fluence would then be about $0.1\text{--}1\,\mathrm{J\,cm^{-2}}$. This is comparable to the integrated fluence utilized in excimer photolithography,
where the intensity is typically $5\text{--}10\,\mathrm{MW\,cm^{-2}}$ but where the pulse
length is about $10^{-8}\,\mathrm{s}$ and several pulses may be applied to the same
area. Reciprocity failure would be important if the exposure produced

under the two intensity–exposure regimes differed. Reciprocity in laser-irradiated photoresists has been studied by Rice and Jain (1984) who found that, from a lithographic point of view, excimer lasers and conventional light sources behave in a similar way. However, a thorough study of the fluence dependence of reciprocity by Abe *et al.* (1988) has demonstrated that photothermal effects may decrease the exposure required to achieve a given level of reaction at high fluence so that reciprocity is not conserved. Incubation effects have also been shown to be important (Preuss *et al.* 1992).

Sheats (1984) studied the photobleaching of acridine doped PMMA films irradiated with a KrF laser. The transmittance of these films was measured as a function of the intensity, I, of a bleaching pulse. Films were found to have low transmittance ($\leq 1\%$) for $I \leq 2.6\,\mathrm{MW\,cm^{-2}}$ but this transmittance rose rapidly for I above this threshold. Films were about 70% transmitting when $I \leq 4\,\mathrm{MW\,cm^{-2}}$. A possible application of this effect was discussed by Sheats (1984). By utilizing the threshold effect, in principle an image can be formed at the diffraction limit. This occurs because only that part of an irradiating beam that exceeds the threshold intensity will be registered.

A schematic diagram showing a pattern generator based on an excimer laser source is shown in Figure 5.30 (Austin 1988). Exposure of photoresist to about $50\,\mathrm{mJ\,cm^{-2}}$ of 308 nm laser radiation was found to be sufficient to achieve useful contrast. An example of laser generated lines in photoresist is shown in Figure 5.31.

Improved resolution in microlithography can be achieved only by decreasing the laser linewidth. This can be accomplished using an injection locked oscillator–amplifier system (Rückle *et al.* 1988) whereby the laser linewidth is reduced from about 100 to about $0.4\,\mathrm{cm^{-1}}$. The output of such a system can be tuned over the gain-bandwidth of the laser amplifier. This is typically ± 0.2 nm and allows the laser output to be accurately matched to the optical projection system. With a 248 nm laser source the spatial resolution can be 0.5 µm or better (Austin 1988). This configuration has been incorporated in commercial wafer stepper systems (Austin 1988). An example of UV microlithography using a line-narrowed excimer laser source is shown in Figure 5.32. A general discussion of factors affecting spatial resolution can be found in the review by Rothschild and Ehrlich (1988). Specific considerations related to the use of 248 nm excimer laser radiation have been discussed by Hibbs (1991). Bachmann (1989) has reported on the use of KrF lasers for via formation in a production environment.

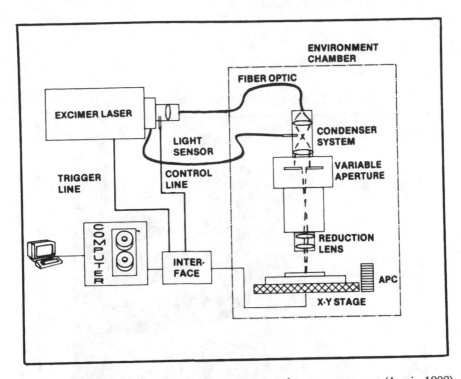

Figure 5.30. The scheme of an excimer laser-equipped pattern generator (Austin 1988).

Periodic etching structures consisting of an array of lines can be created on polymers using an interferometer to project an interference pattern such that the fluence within the interference peaks exceeds the threshold fluence for ablation (Phillips *et al.* 1991, 1992, Phillips and Sauerbrey 1993). The periodicity is half the period of the grating and can be smaller than the laser wavelength whereas the width of the lines produced depends on the size of the region within a fringe for which $E > E_T$. The spatial dependence of fluence $E(x)$, where x is a coordinate transverse to a set of fringes, is

$$E(x) = 2E_{av} \cos^2 (2\pi x/d) \tag{56}$$

where E_{AV} is the average fluence in the interference pattern and d is the grating spacing. This yields a line etch width, W

$$W = \left(\frac{2d}{\pi}\right) \cos^{-1} \left(\frac{E_T}{2E_{AV}}\right)^{1/2} \tag{57}$$

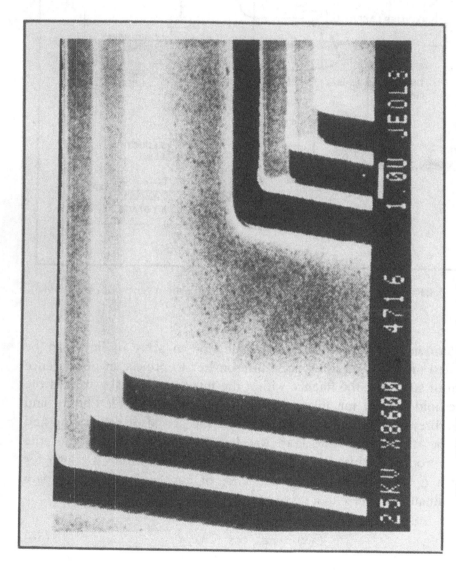

Figure 5.31. The excimer laser print shows 0.2 μm resist lines (Austin 1988).

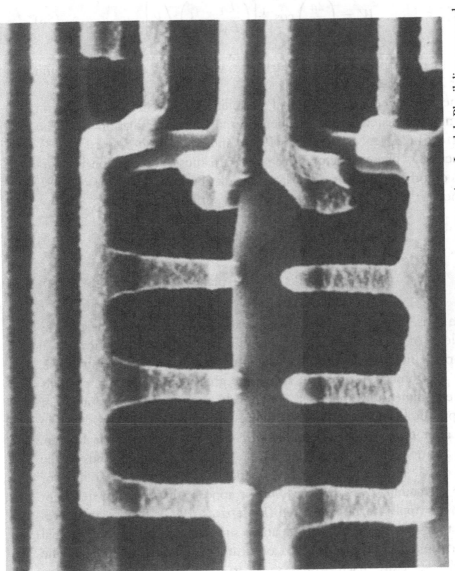

Figure 5.32. Three-level 0.5 μm features obtained by a prototype stepper using a Lambda Physik line-narrowed excimer laser. Courtesy of AT&T Bell Labs, Murray Hill, USA (Rückle *et al.* 1988).

in the limit of perfect fringe visibility $V = (I_{max} - I_{min})/(I_{max} + I_{min})$. Modification of equation (57) to include $V < 1$ gives (Phillips and Sauerbrey 1993)

$$W = \left(\frac{2d}{\pi}\right) \cos^{-1}\left[\left(\frac{E_T}{E_{AV}} - 1\right)\bigg/ V\right] \tag{58}$$

This effect has been utilized to etch lines with widths as small as 300 nm using 248 nm laser radiation.

The spontaneous formation of periodic line and dot structures on polymers irradiated with low fluence excimer laser radiation has been reported by Dyer and Farley (1990) and Bolle and Lazare (1993), respectively. These structures are similar to those observed on the surfaces of other materials after laser irradiation (van Driel *et al.* 1982, Emmory *et al.* 1973). Line structures have periods Λ given by

$$\Lambda = \lambda/(n \pm \sin\theta) \qquad \bar{k} \,\|\, \text{plane of incidence} \tag{59}$$

$$= \lambda/(n - \sin^2\theta)^{1/2} \qquad \bar{k} \perp \text{plane of incidence} \tag{60}$$

where λ is the laser wavelength, n is the real refractive index of the sample or of the surrounding medium, θ is the angle of incidence ($\theta = 0$ at normal incidence) and $\bar{k}$ is the grating vector of the line structure. They are observed over an area that greatly exceeds any spatially coherent dimension and over a relatively narrow fluence range. For example, Bolle and Lazare (1993) found that ripples can be produced in PET after exposure to 10^3 pulses with $3 \leq E \leq 5\,\mathrm{mJ\,cm^{-2}}$ at 193 nm. The ablation threshold under these conditions was found to be $17\,\mathrm{mJ\,cm^{-2}}$. Dyer and Farley (1990) found a wider fluence range for ripple formation, $E_T \leq E \leq 3E_T$, but it appears that ablation need not occur for ripples to be generated (Bolle *et al.* 1992). It appears, therefore, that this structure accompanies surface melting or the partial degradation that occurs prior to or near the ablation threshold. The periodic structure itself would seem to arise from the interference between the incident laser beam and waves scattered from surface defects. The mechanism by which this interference wave couples into the polymer to produce ripple and dot-like structures is uncertain at present (Niino *et al.* 1990, Dyer and Farley 1993, Heitz *et al.* 1994, Hiraoka and Sendova 1994).

5.7 EXCIMER INTERACTION WITH HDPE SURFACES

Clear polyethylene is quite transparent to 308 nm excimer radiation, but the addition of a dye or filler, as, for example, in commercial samples of black high density polyethylene (HDPE), yields a material that absorbs strongly at UV wavelengths. Under these conditions excimer irradiation leads to a change in surface roughness, together with evidence for the loss of particles. These features can be seen in Figure 5.33. In addition, the visual appearance of irradiated samples was altered, showing a whitening due presumably to the removal of surface pigment in the irradiated area. Under some conditions, a fine scale surface roughness can be observed.

An increase in surface roughness can also be inferred from IR spectra of irradiated samples. Spectra of HDPE obtained in reflection show the characteristic features of CH_2 groups at 2930, 2855, 1490 and 750 cm^{-1}. In irradiated samples, the amplitude of CH_2 stretching vibrations at 2930 and 2855 cm^{-1} increases with total fluence, saturating at an exposure level of about 20 J cm^{-2}. At higher total fluence, surface roughness decreases slightly due to increased surface melting. There is no evidence from these spectra for any chemical change (e.g. oxidation) in surfaces irradiated under these conditions. Chromatographic analysis of material taken from the surface area shows no change in molecular size in irradiated or untreated material. The effect of excimer laser irradiation then appears to be primarily one of particle heating and ejection. The mechanism for ejection may involve heating to the point at which absorbed gas is explosively removed, leading to particle ejection. This effect is expected to be largest for small particles.

5.8 ABLATION OF COMPOSITE MATERIALS

There have been few studies of the effect of excimer laser radiation on composite materials containing organic polymers (Mohiuddin *et al.* 1996) despite the wide use of these materials in industrial applications. One such material is sheet molding compound (SMC), which is commonly used in automotive components. This material contains a mixture of $CaCO_3$ particles and glass fibers in an epoxy matrix.

The structure and composition of SMC surfaces after irradiation with excimer laser pulses depend on laser wavelength, fluence and the

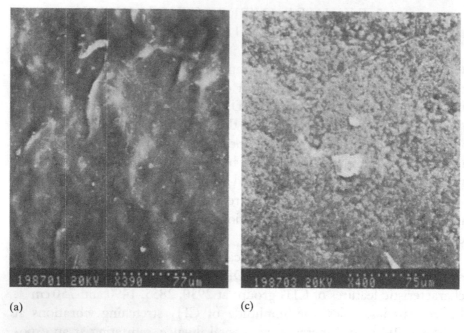

(a)   (c)

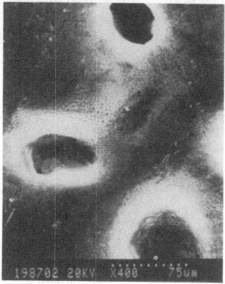

Figure 5.33. The surface of HDPE after
exposure to 308 nm pulses from a XeCl
laser (Duley *et al.* 1992).

(b)

number of laser pulses. An SEM picture of surface morphology in SMC
following irradiation at 248 nm is shown in Figure 5.34. Removal of the
organic matrix and filler material is evident, but glass fibers remain on
the surface. The fluence threshold for single pulse damage is typically
about 350 mJ cm^{-2} at 248 nm and about 400 mJ cm^{-2} at 308 nm. FTIR

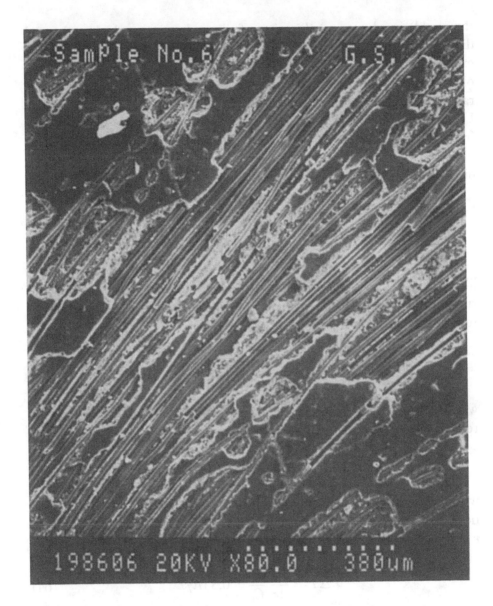

Figure 5.34. The surface structure of SMC after etching with KrF laser radiation.

spectra of irradiated surfaces show an apparent increase in the concentration of CH_n surface functional groups and a decrease in absorption attributable to $CaCO_3$ particles. The increase in the signal due to CH_n groups can be attributed to an increase in surface area.

By analogy with excimer laser interactions with other organic polymers, the primary interaction at modest fluences (several times threshold) involves a combination of photochemical and photothermal dissociation. Photochemical interactions undoubtedly occur during excimer laser irradiation of SMC but this is complicated by the presence of inorganic filler. Glass fibers are opaque both to 248 and to 308 nm radiation, and shield matrix material. They do not appear to melt under UV irradiation, but do fracture and separate from the surface. This fracture is probably due to the intense pressure pulse associated with photoablation of the matrix. Excimer etching of glass fibers has been discussed by Sigrist and Tittel (1991).

A dominant effect in laser etching of SMC surfaces involves the laser-induced thermal decomposition of $CaCO_3$ particles. The reaction

$$CaCO_3 \rightarrow CaO + CO_2 \tag{61}$$

liberates CO_2 gas that may blow the remaining particle out of the surrounding matrix if the particle mass is not too large. The temperature rise achieved in a particle of volume V is

$$\Delta T = \frac{E_{abs}}{C_v V} \tag{62}$$

where C_v is the heat capacity, $2.16 \, J \, cm^{-3} {}^\circ C^{-1}$ for $CaCO_3$. E_{abs} can be related to incident fluence E by

$$E_{abs} = Q\pi r^2 E \tag{63}$$

where r is the particle radius and Q is the efficiency factor for absorption. At excimer wavelengths, Q will be about unity for particles with $r \geq 0.5 \, \mu m$. Then, under this assumption,

$$\Delta T = \frac{3Q}{4r} \frac{E}{C_v} = 0.35 E/r \tag{64}$$

The temperature rise required for decomposition of $CaCO_3$ is $\Delta T_d = 875 °C$. Thus the relation between the critical fluence for decomposition E_d and r is

$$E_d = 2520r \tag{65}$$

where r is in centimeters and E_d is in $J \, cm^{-2}$. Under irradiation at fluence E, particles with radii less than $r = E/2520$ will be heated to T_d

Table 5.9. *The fraction, f, of CaCO$_3$ particles (by volume) remaining on the SMC surface after exposure to a single excimer laser pulse of fluence E.*

E (J cm^{-2})	f
0.2	0.8
0.5	0.62
0.75	0.53
1.0	0.45
1.5	0.3

and will be ejected from the surface due to a gas-jet effect involving CO_2. Thus only the largest $CaCO_3$ particles will remain in heavily irradiated samples. This is in agreement with IR data that show that average $CaCO_3$ particle size increases with irradiation.

Table 5.9 lists the calculated fraction, f, by volume of $CaCO_3$ particles remaining on the surface of SMC after exposure to single excimer laser pulses of fluence E. Severe depletion of $CaCO_3$ is apparent at fluences exceeding $1 \, \text{J cm}^{-2}$. This suggests that excimer laser treatment may be potentially useful in the alteration of SMC surfaces prior to adhesive bonding. Machining of a variety of aerospace composite materials has been the subject of an extensive study reported by Proudley and Key (1989). Etching of polyurethane coatings on metal wires has been discussed by Brannon *et al.* (1991).

5.9 ABLATION IN BIOGENIC SYSTEMS

The possibility of ablative decomposition of organic materials using UV laser radiation has suggested a number of applications for UV lasers in surgery and medicine. In particular, the excellent control over etch depth that can be achieved with excimer laser radiation is attractive for applications in which precise control of material removal is important. Such applications occur in such diverse areas as the removal of plaque from arteries, changing the curvature of the cornea by selective ablation, in neurosurgery and in treatment of the skin. It is likely that many other applications will be discovered as UV laser technology matures and further information is obtained on the effect of intense UV laser radiation on biogenic tissues. In addition, methods of delivering this radiation to the desired location via optic fibers or wavelengths capable of reliable operation at high intensity must also be developed. Laser irradiation of biogenic tissues, in keeping with irradiation of non-biogenic polymers, is also the source of many chemical species that are potentially harmful. Removal, or deactivation, of these species is an important consideration in the development of safe surgical procedures.

With one notable exception, most of the studies carried out to date have been fundamental in nature with limited clinical application. The goal of these studies has been to delineate the capabilities and limitations of UV laser therapy in a variety of potential applications.

The exception to this has been the development and clinical evaluation of UV laser photorefractive keratectomy (PRK) for the reduction of myopia through flattening of the corneal surface of the eye (Trokel *et al.* 1983, Krueger and Trokel 1985, Marshall *et al.* 1985, 1986, Puliafito *et al.* 1985). An extensive database for PRK obtained on the basis of clinical studies now exists on this procedure (Gartry and Kerr 1992, Sher *et al.* 1992).

Pure liquid water is highly transparent at most excimer laser wavelengths (Chapter 2) and biological materials have a high water content, but the presence of a wide variety of complex biopolymers, organic molecules and ions in biogenic material yields a medium that is highly attenuating at UV laser wavelengths. Attenuation of incident laser radiation in these materials arises from a combination of absorption and scattering and therefore the application of a Beer–Lambert formalism to describe this attenuation may not always be appropriate.

The scattering component in the attenuation coefficient for radiation of wavelength λ will depend on refractive index variations in the propagation medium as well as the size of irregularities in this medium. In general, scattering will be most important when the parameter $2\pi r(n - n_\mathrm{m})/\lambda \simeq 1$, where r is a characteristic dimension of the irregularity (e.g. a particle or fiber), n is the refractive index of the material constituting the irregularity and n_m is the refractive index of the surrounding medium. Propagation in a medium with both absorption and scattering can be described in approximate form using the Kubelka–Munk formulation (see Kortum (1969) for a review of this approximation).

The geometry is as shown in Figure 5.35. I_0 is the incident light intensity and $I(x)$ is the intensity in the forward direction at position x inside the sample. $\mathcal{J}_0$ is the back scattered intensity outside the sample and $\mathcal{J}(x)$ is the back scattered intensity at x. I_D is the transmitted

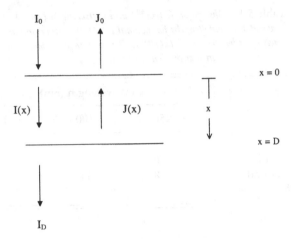

Figure 5.35. The geometry for variables, in the Kubelka–Munk description of the propagation of radiation in an absorbing/scattering medium.

intensity through the sample. The growth of $I(x)$ and $J(x)$ is described by the two equations

$$\frac{dI}{dx} = - SI - KI + SJ \tag{66}$$

$$\frac{dS}{dx} = SJ + KJ - SI \tag{67}$$

where S is the scattering coefficient (cm^{-1}) and K is the absorption coefficient (cm^{-1}). Kubelka (1948) gave the following expressions for S and K derived for the planar slab geometry of Figure 5.35:

$$S = \frac{1}{D}\left[\left(\frac{1 + R^2 - T^2}{2R}\right)^2 - 1\right]^{-1/2} \coth^{-1} x \tag{68}$$

$$X = \frac{1 - R^2 + T^2}{\left[\left(\dfrac{1 + R^2 - T^2}{2R}\right)^2 - 1\right]^{1/2}} \tag{69}$$

$$K = S\{[(1 + R^2 - T^2)/(2R)] - 1\} \tag{70}$$

where

$$R = J_0/I_0 \tag{71}$$

$$T = I_D/I_0 \tag{72}$$

Table 5.10. *Absorption K (cm⁻¹) and scattering S (cm⁻¹) at several UV wavelengths for normal aorta and atheroma as measured by Prince et al. (1986). These data were obtained at low incident light intensity.*

	Wavelength (nm)		
	250	300	350
K (aorta)	155	80	65
K (atheroma)	130	55	50
S (aorta)	8	11	8
S (atheroma)			18

T is then the transmittance and R is the remittance. Additional discussion of the Kubelka–Munk formalism in this context can be found in Welch (1984).

In practice, separation of the absorption and scattering terms from the observation of transmittance and remittance (i.e. integrated diffuse reflectance) is difficult and involves a number of assumptions concerning particle size and sample geometry. A careful empirical study of the wavelength dependence of K and S for normal aortic material and atheromas shows that absorption dominates at UV wavelength, in both these materials (Prince *et al.* 1986). Some of their data are summarized in Table 5.10. Typical values for K lie in the range 50–$200 \, \mathrm{cm}^{-1}$. This is a significant attenuation, but it is far from the absorptivity ($K = 10^4$–$10^6 \, \mathrm{cm}^{-1}$) that pertains to non-biogenic polymers such as polyimide or PMMA at UV wavelengths. This is consistent with the heterogeneous nature of biogenic material and the presence of a large water component in biological tissues.

For a liquid containing a density N (cm^{-3}) of chromophores in solution the overall absorption coefficient

$$K(\lambda) = K_{\mathrm{liq}}(\lambda) + K_{\mathrm{chromo}}(\lambda) \tag{73}$$

where

$$K_{\mathrm{chromo}}(\lambda) \cong \sigma N \tag{74}$$

and σ is the absorption cross-section per chromophore. In spectral regions where absorption is dominated by the presence of chromophores

Table 5.11. *Absorption coefficients measured by Cross et al. (1988) for normal aortic and atheroma material at 248, 308 and 532 nm. Data are given both for irradiation in air and for irradiation in saline solution.*

| Wavelength (nm) | Tissue | Attenuation coefficient (cm^{-1}) | | | |
| | | Photoacoustic | | Etching rate | |
		Saline	Air	Saline	Air
248	Normal	445 ± 60	450 ± 45	7400	7000
	Atheroma	420 ± 60	420 ± 42	9500	9000
308	Normal	300 ± 45	320 ± 40	26 000	15 000
	Atheroma	285 ± 40	310 ± 45	28 000	16 000
532	Normal	80 ± 13	100 ± 17		
	Athroma	70 ± 12	80 ± 14		

and not by the solvent, then $K_{liq}(\lambda)$ can be neglected and

$$K(\lambda) \simeq \sigma N \tag{75}$$

This would appear to be the situation for biological materials at UV wavelengths. Typical values for σ are $10^{-16} \leq \sigma \leq 10^{-18}$ cm^2 so that $K(\lambda) = 10^2$ cm^{-1} implies $N \simeq 10^{18}$–10^{20} cm^{-3}. This should be compared with the density of water molecules in liquid water, $N(H_2O) = 3.3 \times 10^{22}$ cm^{-3}.

Absorption coefficients for excimer laser radiation in normal aortic tissue and in atheromas under high intensity (etching) conditions have been measured by Cross *et al.* (1987, 1988). The absorption coefficient α (cm^{-1}) is derived using the Beer–Lambert formalism, which ignores scattering, i.e.

$$I(x) = I_0 \exp(-\alpha x) \tag{76}$$

It can be related to K and S under certain conditions, which may or may not be appropriate for biological tissue (Kortum 1969). When this approximation can be assumed

$$\alpha \cong [2K(2K + 4S)]^{1/2} \tag{77}$$

Cross *et al.* (1988) estimated absorption coefficients for aortic material and atheroma using both photoacoustic and etch depth versus fluence data. Their results (Table 5.11) yield values of α that are larger than

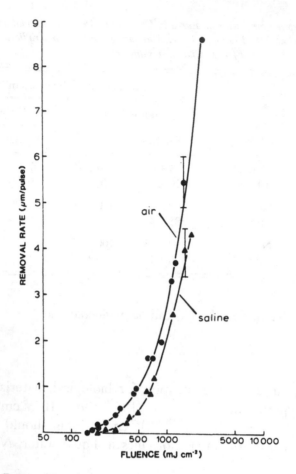

Figure 5.36. Removal rate (μm per pulse) versus fluence for normal aortic material irradiated with 248 nm KrF laser pulses in air and in saline solution (Cross *et al.* 1987).

those for K measured by Prince *et al.* (1986) at low light intensity. This is particularly true when α is estimated from etching rate data. However, the fluence dependence of etching rate in aortic material is highly non-linear when plotted semi-logarithmically (Figure 5.36) so that estimates of α obtained from the expression

$$X = \frac{1}{\alpha} \ln \left(\frac{E}{E_T} \right) \qquad (78)$$

where X is etching depth, E is fluence and E_T is the fluence threshold, are likely to be inaccurate. Nevertheless, despite the uncertainties in estimating α, it is apparent that the effective absorption coefficient is

Table 5.12. *Etch depth per pulse for ablation of several types of biological tissue with excimer laser pulses.*

Wavelength (nm)	Tissue	Etch depth per pulse (µm)	Fluence (mJ cm^{-2})	Reference
308	Aorta	20	8000	Grundfest *et al.* (1985)
351	Aorta	0.5	5000	Murphy-Chutorian *et al.* (1986)
248	Aorta	2.1	850	Cross *et al.* (1987)
	Atheroma	1.3		
248	Skin (guinea pig)	6	1500	Lane *et al.* (1985)
193	Skin (guinea pig)	1.6	1800	Lane *et al.* (1985)
248	Skin (human)	3.4	900	Lane *et al.* (1985)
193	Skin (human)	1.5	1500	Lane *et al.* (1985)
193	Retina (bovine)	20	1200	Lewis *et al.* (1992)
248	Cornea	6.2	1000	Srinivasan and Sutcliffe (1987)
193	Cornea	1	1000	Trokel *et al.* (1983)
193	Cornea (bovine)	0.05	93	Puliafito *et al.* (1985)
193	Cornea (human)	0.025	125	Puliafito *et al.* (1985)

higher than that which would be predicted from measurements of K and S at sub-threshold light intensity. For example, using $\alpha = 440 \, \text{cm}^{-1}$ and $E_T = 155 \, \text{mJ cm}^{-2}$ (Cross *et al.* 1988), the predicted etch depth per pulse for aortic material in air from equation (78) is $X = 38.6 \, \mu\text{m}$ at $E = 1000 \, \text{mJ cm}^{-2}$. This is about ten times larger than observed (Figure 5.36).

Comparable experimental values of X are obtained for other biological tissues under similar conditions (Table 5.12). In general, these values are significantly larger (in some instances by as much as an order of magnitude) than those obtained at similar fluences in non-biogenic organic polymers such as polyimide. They are less, however, than values obtained for laser etching of organic polymeric foams (Duley and Allan 1989), for which etch depths at $E = 1300 \, \text{mJ cm}^{-2}$ were found to be typically 150 µm per pulse.

Some of this difference probably arises from the heterogeneous nature of biological material, with stratified organic material separated by voids or liquid filled interstices. A simple model that represents this geometry in part is shown in Figure 5.37. In Figure 5.37, organic material with density ρ_1 (g cm^{-3}) is interleaved with a separate medium (for example air or liquid water) with density ρ_2. x_i is the thickness of

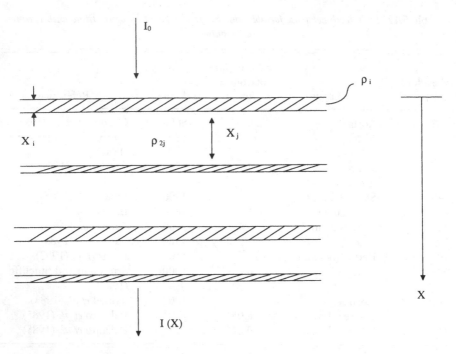

Figure 5.37. The model for biogenic tissue.

the ith slab of organic material and x_j is the thickness of the intervening medium. If α_1 and α_2 are absorption coefficients of the organic and intervening media, respectively, then

$$I(x) = I_0 \exp\left[-\left(\sum_i \alpha_1 x_i + \sum_j \alpha_2 x_j \right) \right] \tag{79}$$

in the absence of scattering, i.e. in the Beer–Lambert approximation, and at low incident intensity, I_0.

If $x_j \to 0$ then a homogeneous organic medium with α_1 is obtained. The etch depth in the Beer–Lambert approximation would then be given by equation (78) with $\alpha = \alpha_1$. The threshold fluence for this removal would be $E_T \equiv E_T(i)$.

For the heterogeneous medium shown in Figure 5.37, there will be at least two threshold fluences, one for removal of the organic material and one for removal of the intervening material. If $\alpha_1 \gg \alpha_2$ then it is likely that $E_T(1) \ll E_T(2)$. Pulses with $E_T(1) < E < E_T(2)$ will then lead to ablation of the organic component without disrupting the intervening medium. This, then, would have the effect of creating a series of contained explosions of organic material along the focus of the laser beam. Material removal in this case would be inhibited as subsurface ablation products would be prevented from escaping from the irradiation volume. At higher fluences, or with repetitive pulsing, the mechanical forces associated with these subsurface explosions would rupture these intervening layers and allow the removal of a larger quantity of material per pulse. The creation of surface waves during excimer laser corneal ablation has been described by Bor et al. (1993). The etch depth per pulse would then be superlinear on an x versus $\ln(E/E_{th})$ plot (Figure 5.36). In the limit in which $\alpha_2 = 0$, as is the case for organic foams such as foamed polystyrene, the effective $E_{th} \equiv E_{th}(1)$. Then the etch depth per pulse will be approximately

$$X(E) = \frac{\rho_1}{\rho} X_1(E) \tag{80}$$

where ρ is the overall density of the foamed material and $X_1(E)$ is the etch depth of the organic material alone at a fluence E. For example, if $X_1 = 1\,\mu m$ per pulse and $\rho_1 = 2\,g\,cm^{-3}$ and $\rho = 0.02\,g\,cm^{-3}$, as is appropriate for a low density foam, then $X = 100\,\mu m$ per pulse.

The threshold fluence for ablation of several biological tissues is summarized in Table 5.13. The large variability in these data is due not only to the different techniques used to estimate E_T, but also to the variability in the composition and properties of tissue samples. Incubation effects are also significant in this regard since the total number of pulses applied to the sample is an important variable near threshold (Murphy-Chutorian et al. 1986).

Comparison of E_T with values obtained when irradiating non-biogenic organic polymers (Table 5.3) shows, in general, that the threshold for ablation in biological tissues is considerably higher. This may be due, in part, to the shielding effect of the intervening relatively transparent material, as represented schematically in Figure 5.37.

The morphology of laser-ablated tissue has been described by several groups (Grundfest et al. 1985, Murphy-Chutorian et al. 1986, Srinivasan 1986, Lane et al. 1985, Lewis et al. 1992, Sinbawy et al. 1991, Marshall et al. 1985, 1986, Puliafito et al. 1985). Comparative studies using conventional mechanical techniques have also been performed (Marshall et al. 1986). The sharp edges characteristic of ablation in non-biogenic organic material are also observed when etching biological samples. It is

Table 5.13. *Representative values of fluence threshold for ablation of biological tissues at excimer laser wavelengths.*

Wavelength (nm)	Tissue	E_T (mJ cm^{-2})	Reference
350	Aorta	4200	Singleton et al. (1986)
308	Aorta	1600	Singleton et al. (1986)
248	Aorta	450	Singleton et al. (1986)
193	Aorta	400	Singleton et al. (1986)
308	Aorta (saline)	310	Cross et al. (1988)
	Aorta (air)	290	
	Atheroma (saline)	370	
	Atheroma (air)	330	
248	Aorta (saline)	170	Cross et al. (1988)
	Aorta (air)	150	
	Atheroma (saline)	280	
	Atheroma (air)	210	
193	Skin (human)	100	Lane et al. (1985)
248	Cornea	300	Srinivasan and Sutcliffe (1987)
248	Cornea (human)	58	Puliafito et al. (1985)
	Cornea (bovine)	71	
193	Cornea (human)	46	Puliafito et al. (1985)
		20	
308	Aorta	1800 (7 ns)	Taylor et al. (1987)
		2200 (300 ns)	

this property, together with precise control over ablation depths, that makes excimer laser ablation potentially attractive for a number of procedures.

An example of the morphology of a laser-excised section in an aortic wall is shown in Figure 5.38. This sequence of sections shows the effect of multiple pulse irradiation at 351 nm to obtain a deeper cut. Evidence for delamination of tissue layers is clearly evident, particularly in the upper figure, which corresponds to a small etch depth near threshold. This delamination may arise from gas evolution from deeper tissue layers. There is also some evidence for density changes in the material underlying the irradiated area.

At shorter laser wavelengths, photochemical effects are of increasing importance and thermal effects may be minimized. The result is often a cleaner cut. An example of such a cut in aortic material achieved using 193 nm radiation is shown in Figure 5.39. In this case, there is little evidence for delamination or peripheral damage.

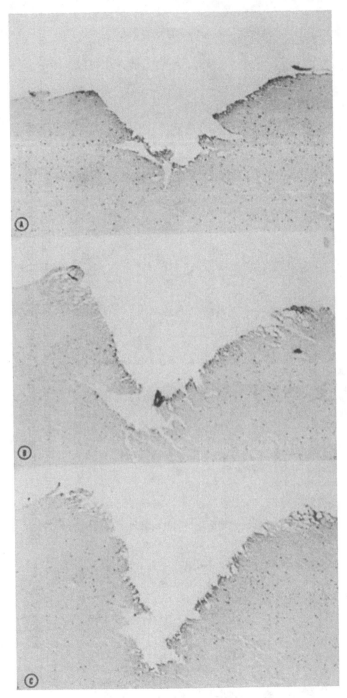

Figure 5.38. Photomicrographs of aortic wall showing the effects of an increasing number of pulses per sequence at constant pulse rate (10 Hz) and energy per pulse (50 mJ per pulse). Top: 150 pulses; middle: 600 pulses; bottom: 1200 pulses. The laser wavelength was 351 nm. (Reprinted by permission from Murphy-Chutorian *et al.* 1986.)

Figure 5.39. (Left) The cross-section of the luminal side of the aortic wall produced by laser radiation at 193 nm; pulse duration 14 ns and fluence $0.25\,\mathrm{J\,cm^{-2}}$. (Right) At 532 nm, pulse duration 5 ns and fluence $1.0\,\mathrm{J\,cm^{-2}}$ (from Srinivasan 1986).

The chemical products of laser ablation of biological tissues in air or in solution are expected to consist of a variety of small molecules, most of which will be volatile. In addition, photochemical products could include radicals and radical ions that will react to form a variety of larger species. A preliminary study of the composition of these products was reported by Singleton *et al.* (1986). These products were found to consist of methane, ethylene and low molecular weight unsaturated hydrocarbons. Acetaldehyde, CH_3CHO was also detected. Molecules such as H_2 and CO were also produced. The total yield of these volatile products was found to be linearly related to laser fluence, from threshold up to fluences $E \simeq 3.2\,J\,cm^{-2}$. The derived quantum yield of all products was about 3×10^{-4}, suggesting that other vaporizing species may have been present but not collected.

5.10 REFERENCES

Abe T., Arikado T. and Takigawa T., 1988. *J. Appl. Phys.* **63**, 1236.

Anderson S. G., McMeyer H., Atanasoska L. and Weaver J. H., 1988. *J. Vac. Sci. Technol.* A6, 38.

Andrew J. E., Dyer P. E., Forster D. and Key P. H., 1983. *Appl. Phys. Lett.* **43**, 717.

Arenholz E., Svorcik V., Kefer T., Heitz J. and Bauerle D., 1991. *Appl. Phys.* A53, 692.

Arenholz E., Wagner M., Heitz J. and Bauerle D., 1992. *Appl. Phys.* A55, 119.

Austin L., 1988. *Microelectron Manufact. Testing*, Oct. 2.

Babu S. V., D'Conto G. C. and Egitto F. G., 1992. *J. Appl. Phys.* **72**, 692.

Bachmann F., 1989. *Chemtronics* **4**, 149.

Bahners T., Knittel D., Hillenkamp F., Bahr U., Benndorf C. and Schollmeyer E., 1990. *J. Appl. Phys.* **68**, 1854.

Becker R. A., Sopori B. L. and Chang W. S. C., 1978. *Appl. Opt.* **17**, 1069.

Bishop G. J. and Dyer P. E., 1985. *Appl. Phys. Lett.* **47**, 1229.

Bolle M., Luther K., Troe J., Ihlemann J. and Gerhardt H., 1990. *Appl. Surf. Sci.* **46**, 279.

Bolle M. and Lazare S., 1993. *J. Appl. Phys.* **73**, 3516.

Bolle M., Lazare S., Le Blanc M. and Wilmes A., 1992. *Appl. Phys. Lett.* **60**, 674.

Bor Z., Hopp B., Racz B., Szabo G., Marton Z., Ratkay I., Mohay J., Suveges I. and Fust A., 1993. *Opt. Eng.* **32**, 2481.

Brannon J. H., Tam A. C. and Kurth R. H., 1991. *J. Appl. Phys.* **70**, 3881.

Braren B., Casey K. G. and Kelly R., 1991. *Nucl. Instrum. Methods* B58, 463.

Brunco D. P., Thompson M. O., Otis C. E. and Goodwin P. M., 1992. *J. Appl. Phys.* **72**, 4344.

Cain S. R., Burns F. C. and Otis C. E., 1992a. *J. Appl. Phys.* **71**, 1415.

Cain S. R., Burns F. C., Otis C. E. and Braren B., 1992b. *J. Appl. Phys.* **72**, 5172.

Chuang T. J., Modl A. and Hiraoka H., 1988. *Appl. Phys.* A45, 277.

Chuang M. C. and Tam A. C., 1989. *J. Appl. Phys.* **65**, 2591.

D'Couto G. C., Babu S. V., Egitto F. D. and Davis C. R., 1993. *J. Appl. Phys.* **74**, 5972.

Craighead H. G., White J. C., Howard R. E., Jackel L. D., Behringer R. E., Sweeney J. E. and Epworth R. W., 1983. *J. Vac. Sci. Technol.* B1, 1186.

Creasy W. R. and Brenna J. T., 1988. *Chem. Phys.* 126, 453.

Cross F. W., Al-Dhakir R. K. and Dyer P. E., 1988. *J. Appl. Phys.* 64, 2194.

Cross F. W., Al-Dhakir R. K., Dyer P. E. and MacRobert A. J., 1987. *Appl. Phys. Lett.* 50, 1019.

Danielzik B., Fabricins N., Röwenkamp M. and von der Linde D., 1986. *Appl. Phys. Lett.* 48, 212.

Davis C. R., Egitto F. D. and Buchwalter S. L., 1992. *Appl. Phys.* B54, 227.

Davis G. M. and Gower M. C., 1987. *J. Appl. Phys.* 61, 2090.

Davis G. M., Gower M. C., Fotakis C., Efthimiopoulos T. and Argyrakis P., 1985. *Appl. Phys.* A36, 27.

Dickinson J. T., Shin J. J., Jiang W. and Norton M. G., 1993. *J. Appl. Phys.* 74, 4729.

Dijkkamp D., Gozdz A. S. and Venkatesan T., 1987. *Phys. Rev. Lett.* 58, 2142.

Dlott D. D., 1990. *J. Opt. Soc. Am.* B7, 1638.

Dreyfus R. W., 1988. *Mater. Res. Soc. Symp. Proc.* 117, 11.

Dreyfus R. W., 1992. *Appl. Phys.* A55, 335.

Duley W. W. and Allan G., 1989. *Appl. Phys. Lett.* 55, 1701.

Duley W. W., Ogmen M., Steel T. and Mihailov S., 1992. *J. Laser Applications* 4, 22.

Dyer P. E. and Farley R. F., 1990. *Appl. Phys. Lett.* 57, 765.

Dyer P. E. and Farley R. F., 1993. *J. Appl. Phys.* 74, 1442.

Dyer P. E., Jenkins S. D. and Sidhu J., 1986. *Appl. Phys. Lett.* 49, 453.

Dyer P. E., Jenkins S. D. and Sidhu J., 1987. *Appl. Phys. Lett.* 49, 453.

Dyer P. E., Jenkins S. D. and Sidhu J., 1988. *Appl. Phys. Lett.* 52, 1880.

Dyer P. E. and Karnakis D. M., 1994a. *Appl. Phys.* A59, 275.

Dyer P. E. and Karnakis D. M., 1994b. *Appl. Phys. Lett.* 64, 1344.

Dyer P. E. and Sidhu J., 1985. *J. Appl. Phys.* 57, 1420.

Dyer P. E. and Sidhu J., 1986. *J. Opt. Soc. Am.* B3, 792.

Dyer P. E. and Srinivasan R., 1989. *J. Appl. Phys.* 66, 2608.

Ediger M. N. and Pettit G. H., 1992. *J. Appl. Phys.* 71, 3510.

Ediger M. N., Pettit G. H. and Sauerbrey R., 1993. *J. Appl. Phys.* 74, 1344.

Egitto F. D. and Davis C. R., 1992. *Appl. Phys.* B55, 488.

Emmony D. C., Howson R. P. and Willis L. J., 1973. *Appl. Phys. Lett.* 23, 598.

Estler R. C. and Nogar N. S., 1986. *Appl. Phys. Lett.* 49, 1175.

Eyring H., Lin S. H. and Lin S. M., 1980. *Basic Chemical Kinetics*, John Wiley, New York.

Feldman, D., Kutzner J., Laukemper J., MacRobert S. and Welge K. H., 1987. *Appl. Phys.* B44, 81.

Furzikov N. P., 1990. *Appl. Phys. Lett.* 56, 1638.

Gai H. and Voth G. A., 1992. *J. Appl. Phys.* 71, 1415.

Garrison B. J. and Srinivasan R., 1985a. *J. Appl. Phys.* 57, 2909.

Garrison B. J. and Srinivasan R., 1985b. *J. Vac. Sci. Technol.* A3, 746.

Gartry D. S. and Kerr M. C., 1992. *Ophthalmology* 99, 1209.

Geis M., Randall J. N., Deutsch T. F., Degraff P. D., Krohn K. E. and Stern L. A., 1983. *Appl. Phys. Lett.* 43, 74.

Goodwin P. and Otis C., 1989. *Appl. Phys. Lett.* 55, 2286.

Goredetsky G., Kazayake T. G., Melcher R. L. and Srinivasan R., 1985. *Appl. Phys. Lett.* 46, 828.

Grundfest W. S., Litvack I. F., Goldenberg T., Sherman T., Morgenstern I., Carroll R., Fishbein M., Forrester J., Margitan J. and McDermid S., 1985. *Am. J. Surgery* 150, 220.

Gupta A., Liang R., Tsay F. D. and Moacanin J., 1980. *Macromolecules* **13**, 1696.

Hansen S. G., 1989. *J. Appl. Phys.* **66**, 1411.

Hansen S. G., 1990. *J. Appl. Phys.* **68**, 1878.

Hansen S. G. and Robitaille T. E., 1988. *J. Appl. Phys.* **62**, 1394.

Heitz J., Arenholz E., Bauerle D., Sauerbrey R. and Phillips H. M., 1994. *Appl. Phys.* **A59**, 289.

Henderson D., White J. C., Craighead H. G. and Adesida I., 1985. *Appl. Phys. Lett.* **46**, 900.

Hibbs M. S., 1991. *J. Electrochem. Soc.* **138**, 199.

Hiraoka H., Chuang T. J. and Masuhara K., 1988. *J. Vac. Sci. Technol.* B6, 463.

Hiraoka H. T. and Sandova M., 1994. *Appl. Phys. Lett.* **64**, 563.

Jain K., 1986. *Proc. SPIE* **710**, 35.

Jain K. and Kerth R. T., 1984. *Appl. Opt.* **23**, 648.

Jain K., Willson C. G. and Lin B. J., 1982a. *Appl. Phys.* **B28**, 206.

Jain K., Willson C. G. and Lin B. J., 1982b. *IBM J. Res. Develop.* **26**, 151.

Jelinek H. G. and Srinivasan R., 1984. *J. Phys. Chem.* **88**, 3048.

Jenkins F. A. and White H. E., 1957. *Fundamentals of Optics*, 3rd Edition, McGraw-Hill, New York.

Kalontarov L. I., 1991. *Phil. Mag. Lett.* **63**, 289.

Kawamura Y., Toyoda K. and Namba S., 1982a. *Appl. Phys. Lett.* **40**, 374.

Kawamura Y., Toyoda K. and Namba S., 1982b. *J. Appl. Phys.* **53**, 648.

Kelly R., Miotello A., Braren B. and Otis C. E., 1992. *Appl. Phys. Lett.* **60**, 2980.

Keyes T., Clark R. H. and Isner J. M., 1985. *J. Phys. Chem.* **89**, 4194.

Kim H., Postlewaite J. C., Zyang T. and Dlott D. D., 1989. *Appl. Phys. Lett.* **54**, 227.

Koren G., 1988. *Appl. Phys.* **B46**, 149.

Koren G. and Donelon J. J., 1988. *Appl. Phys.* **B45**, 45.

Koren G. and Oppenheim U. P., 1987. *Appl. Phys.* **B42**, 41.

Koren G. and Yeh J. T. C., 1984. *J. Appl. Phys.* **56**, 2120.

Kortum G., 1969. *Reflectance Spectroscopy Principles, Methods, Applications*, Springer-Verlag, New York.

Kroto H. W., Heath J. R., O'Brien S. C., Curl R. F. and Smalley R. E., 1985. *Nature* **318**, 162.

Krueger R. R. and Trokel S., 1985. *Arch. Ophthalmol.* **103**, 1741.

Kubelka P., 1948. *J. Opt. Soc. Am.* **38**, 448.

Küper S. and Brannon J., 1992. *Appl. Phys. Lett.* **60**, 1633.

Küper S., Brannon J. and Brannon K., 1993. *Appl. Phys.* **A56**, 43.

Küper S. and Stuke M., 1987. *Appl. Phys.* **B44**, 199.

Küper S. and Stuke M., 1989. *Appl. Phys.* **A49**, 211.

Lane R. J., Linsker R., Wynne J. J., Torres A. and Geronemus R. G., 1985. *Arch. Dermatol.* **121**, 609.

Larciprete R. and Stuke M., 1987. *Appl. Phys.* **B42**, 181.

Latta M., Moore R., Rice S. and Jain K., 1984. *J. Appl. Phys.* **56**, 586.

Lazare S. and Benet P., 1993. *J. Appl. Phys.* **74**, 4953.

Lazare S. and Granier V., 1988a. *J. Appl. Phys.* **63**, 2110.

Lazare S. and Granier V., 1988b. *Appl. Phys. Lett.* **53**, 862.

Lewis A., Palanker D., Herno I., Peer J. and Zauberman H., 1992. *Inv. Ophthalmol. Vis. Sci.* **33**, 2377.

Lewis R. S. and Lee E. K. C., 1978. *J. Phys. Chem.* **82**, 249.

MacCallum J. R. and Schoff C. K., 1971a. *Trans. Faraday Soc.* **68**, 2372.

MacCallum J. R. and Schoff C. K., 1971b. *Trans. Faraday Soc.* **67**, 2383.

MacCallum J. R. and Schoff C. K., 1972. *J. Chem. Soc. Faraday Trans.* I**68**, 499.

Mahan G. D., Cole H. S., Lin Y. S. and Phillipp H. R., 1988. *Appl. Phys. Lett.* **53**, 2377.
Marshall J., Trokel S., Rothery S. and Kruger R. R., 1986. *Br. J. Ophthalmol.* **70**, 482.
Marshall J., Trokel S., Rothery S. and Schubert H., 1985. *Ophthalmol.* **92**, 749.
Meyer J., Kutzner J., Feldmann D. and Welge K. H., 1988. *Appl. Phys.* B**45**, 7.
Mihailov S. and Duley W. W., 1988. *Proc. SPIE* **957**, 111.
Mihailov S. and Duley W. W., 1991. *J. Appl. Phys.* **69**, 4092.
Mihailov S. and Duley W. W., 1993. *J. Appl. Phys.* **73**, 2510.
Miotello A., Kelly R., Braren B. and Otis C. E., 1992. *Appl. Phys. Lett.* **61**, 2784.
Mohiuddin G., Duley W. W. and Uddin N., 1996. *Proc. ICALEO '95* ed. J. Mazumder,
    A. Matsunawa and C. Magnusson, Laser Institute of America, Orlando, in press.
Moore C. B. and Weisshaar J. C., 1983. *Annu. Rev. Phys. Chem.* **34**, 525.
Moorgat G. K., Seller W. and Warneck P., 1983. *J. Chem. Phys.* **78**, 1185.
Murphy-Chutorian D., Selzer P. M., Kosek J., Quay S. C., Profitt D. and Ginsburg R.,
    1986. *Am. Heart J.* **112**, 739.
Niino H., Shimoyama M. and Yabe A., 1990. *Appl. Phys. Lett.* **57**, 2368.
Nikolaus B., 1986. *Conf. Lasers Electro. Optics* (San Francisco: Optical Society of
    America) p. 88.
Otis C. E., 1989. *Appl. Phys.* B**49**, 455.
Paraskevopoulos G., Singleton D. L., Irwin R. S. and Taylor R. S., 1991. *J. Appl. Phys.*
    70, 1938.
Pettit G. H. and Sauerbrey R., 1993. *Appl. Phys.* A**56**, 51.
Phillips H. M., Callahan D. L., Sauerbrey R., Szabo G. and Bor Z., 1991. *Appl. Phys.
    Lett.* **58**, 2761.
Phillips H. M., Callahan D. L., Sauerbrey R., Szabo G. and Bor Z., 1992. *Appl. Phys.*
    A**54**, 158.
Phillips H. M. and Sauerbrey R., 1993. *Opt. Eng.* **32**, 2424.
Preuss S., Langowski H. C., Damm T. and Stuke M., 1992. *Appl. Phys.* A**54**, 360.
Preuss S., Späth M., Zhang Y. and Stuke M., 1993. *Appl. Phys. Lett.* **62**, 3049.
Prince M. R., Deutsch T. E., Mathews-Roth M. M., Margolis R., Parrish J. A. and
    Oseroff A. R., 1986. *J. Clin. Invest.* **78**, 295.
Puliafito C. A., Steinert R. F., Deutsch T. J., Hillenkamp F., Dehm E. J. and Adler
    C. M., 1985. *Ophthalmology* **92**, 741.
Proudley G. M. and Key P. H., 1989. *Proc. SPIE* **1132**, 111.
Ready J. F., 1971. *Effects of High Power Laser Radiation*, Academic Press, New York.
Rice S. and Jain K., 1984. *IEEE. Trans Electron. Dev.* **31**, 1.
Rothschild M., Arnone C. and Ehrlich D. J., 1986. *J. Vac. Sci. Technol.* B**4**, 310.
Rothschild M. and Ehrlich D. J., 1987. *J. Vac. Sci. Technol.* B**5**, 389.
Rothschild M. and Ehrlich D. J., 1988. *J. Van Sci. Technol.* B**6**, 1.
Rückle B., Lokai P., Rosenkranz H., Nikolaus B., Kablert H. J., Burghardt B., Basting
    D. and Mückenheim W., 1988. *Lambda Physik Report* No. 1, p. 1. Feb.
Sauerbrey R. and Pettit G. H., 1989. *Appl. Phys. Lett.* **55**, 421.
Sauerbrey R. and Pettit G. H., 1990. *Appl. Phys. Lett.* **58**, 793.
Sheats J. R., 1984. *Appl. Phys. Lett.* **44**, 1016.
Sher N. A., *et al.* 1992. *Arch. Ophthalmol.* **110**, 935.
Sigrist M. W. and Tittel F. K., 1991. *Appl. Phys.* A**52**, 418.
Simon P., 1989. *Appl. Phys.* B**48**, 253.
Sinbawy A., McDonnell P. J. and Moreira H., 1991. *Arch. Ophthalmol.* **109**, 1531.
Singleton D. L., Paraskevopoulos G., Jolly G. S., Irwin R. S., McKenney D. J., Nip
    W. S., Farrell E. M. and Higginson L. A. J., 1986. *Appl. Phys. Lett.* **48**, 878.
Singleton D. L., Paraskevopoulos G. and Taylor R. S., 1990. *Appl. Phys.* B**50**, 227.
Sobehart J. R., 1993. *J. Appl. Phys.* **74**, 2830.

Srinivasan R., 1986. *Science* **234**, 559.

Srinivasan R., 1991a. *Appl. Phys. Lett.* **58**, 2895.

Srinivasan R., 1991b. *J. Appl. Phys.* **70**, 7588.

Srinivasan R., 1992. *Appl. Phys.* A**55**, 269.

Srinivasan R., 1993. *J. Appl. Phys.* **73**, 2743.

Srinivasan R. and Braren B., 1984. *J. Polym. Sci.* **22**, 2601.

Srinivasan R. and Braren B., 1988a. *Appl. Phys.* A**45**, 289.

Srinivasan R. and Braren B., 1988b. *Appl. Phys. Lett.* **53**, 1233.

Srinivasan R. and Braren B., 1990. *J. Appl. Phys.* **68**, 1837.

Srinivasan R., Braren B. and Casey K. G., 1990a. *J. Appl. Phys.* **68**, 1842.

Srinivasan R., Braren B. and Dreyfus R. W., 1987. *J. Appl. Phys.* **61**, 372.

Srinivasan R., Braren B., Dreyfus R. W., Hadil L. and Seeger D. E., 1986a. *J. Opt. Soc. Am.* B**3** 785.

Srinivasan R., Casey K. G. and Braren B., 1989. *Chemtronics* **4**, 153.

Srinivasan R., Casey K. G., Braren B. and Yeh M., 1990b. *J. Appl. Phys.* **67**, 1604.

Srinivasan R., Hall R. R. and Allbee D. C., 1993. *Appl. Phys. Lett.* **63**, 3382.

Srinivasan R. and Leigh W. J., 1982. *J. Am. Chem.* **104**, 6784.

Srinivasan R. and Mayne-Banton V., 1982. *Appl. Phys. Lett.* **41**, 570.

Srinivasan R., Smrtic M. A. and Babu V. S., 1986b. *J. Appl. Phys.* **59**, 3861.

Srinivasan R. and Sutcliffe E., 1987. *Am. J. Ophthalmol.* **103**, 470.

Sutcliffe E. and Srinivasan R., 1986. *J. Appl. Phys.* **60**, 3315.

Taylor R. S., Singleton D. L. and Paraskevopoulos G., 1987. *Appl. Phys. Lett.* **50**, 1779.

Taylor R. S., Singleton D. L., Paraskevopoulos G. and Irwin R. W., 1988. *J. Appl. Phys.* **64**, 2815.

Treyz G. V., Scarmozzino R. and Osgood R. M., 1989. *Appl. Phys. Lett.* **55**, 346.

Trokel, S., Srinivasan R. and Braren B., 1983. *Am. J. Ophthalmol.* **96**, 710.

Tsunekawa M., Nishio S. and Sato H., 1994. *J. Appl. Phys.* **76**, 5598.

Upatnieks J. and Lewis R. W., 1973. *Appl. Opt.* **12**, 2161.

van Driel H. M., Spie J. E. and Young J. F., 1982. *Phys. Rev. Lett.* **49**, 1955.

Ventzek P. G., Gilgenback R. M., Sell J. A. and Heffelfinger D. M., 1990. *J. Appl. Phys.* **68**, 965.

von Gutfeld R. J. and Srinivasan R., 1987. *Appl. Phys. Lett.* **51**, 15.

Wada S., Tashiro H., Toyoda K., Niino H. and Yabe A., 1993. *Appl. Phys. Lett.* **62**, 211.

Welch A. J., 1984. *IEEE J. Quant. Electron.* **20**, 1471.

White J. C., Craighead H. G., Howard R. E., Jackel L. D., Behringer R. E., Epworth R. W., Henderson D. and Sweeney J. E., 1984. *Appl. Phys. Lett.* **44**, 22.

Ye Y. and Hunsperger R. G., 1989. *Appl. Phys. Lett.* **54**, 1662.

Zweig A. D. and Deutsch T. F., 1992. *Appl. Phys.* B**54**, 76.

Zweig A. D., Venugopalan V. and Deutsch T. F., 1993. *J. Appl. Phys.* **74**, 4181.

Zyung T., Kim H., Postlewaite J. C. and Dlott D. D., 1989. *J. Appl. Phys.* **65**, 4548.

CHAPTER 6

Interactions and material removal in inorganic insulators

6.1 INTRODUCTION

The response of inorganic insulating materials to intense UV radiative fluxes is complex and involves both photophysical and photochemical interactions. The first effect, noticeable at low exposure, involves the production of defect centers or radiative interactions with existing centers or impurities. Such changes often result in an increase in optical absorption at the laser wavelength as well as at other wavelengths. This can have a profound effect on the quality of transmissive optical components such as windows and lenses. With optical fibers, defect formation limits the fluence that can be transmitted.

Higher exposure to radiative fluxes results in changes in composition and density as the sputtering threshold is approached. Defects have been found to play a significant rôle in the initiation of ablation under irradiation with photons of energies less than that of the optical bandgap. The generation of electron–hole pairs via two-photon absorption is also an important process at high laser intensities and would appear to be the initiator of ablative decomposition of transparent insulators.

This chapter begins with a review of the formation and properties of dominant defects in several wide bandgap insulators. The relation between these defects and the coupling of UV laser radiation leading to ablation is then discussed. The chapter concludes with a summary of dry and laser-assisted etching processes and rates in a variety of insulating solids.

6.2 DEFECT FORMATION

6.2.1 Silicon dioxide

A number of intrinsic defect centers have been identified in SiO_2 and in a-SiO$_2$. These defects are based on oxygen vacancies (Griscom 1984). The fundamental defect, called an E$'$ center, consists of an unpaired spin localized on a silicon bonded to three oxygen atoms. This paramagnetic center appears in SiO$_2$ after exposure to ionizing radiation or after heavy-particle impact and is commonly produced at concentrations of 10^{14}–10^{15} per cm^3.

The peroxy radical center consists of an unpaired spin associated with two adjacent oxygens, which may arise when an O$_2$ molecule is trapped by an E$'$ center. The precursor to this center may also be a bridging peroxy link (Griscom and Friebele 1982).

The non-bridging oxygen hole center (NBOHC) involves an unpaired spin in a π orbital normal to the Si–O axis. It can be created by elimination of H from OH bonded to the silicon atom. The NBOHC center is readily observed in γ-irradiated quartz with high hydroxyl content (Nagasawa et al. 1986). The structure of these defects has been discussed in some detail by Elliott (1990).

Since the bandgap of a-SiO$_2$ is about 9 eV, fused silica is relatively transparent at 193 nm (6.4 eV) and longer wavelengths. It is therefore surprising to find that excimer laser radiation in this range is efficient in producing E$'$ centers (Stathis and Kastner 1984a,b, Arai et al. 1988, Tsai et al. 1988). Production of the E$'$ centers customarily occurs when an oxygen atom is displaced from a lattice site as part of an energetic interaction event involving high energy particles or soft X-rays (Triplet et al. 1987). It is unexpected, therefore, that photons with energies less than the bandgap energy would be effective in the production of these centers.

Figure 6.1 (Arai et al. 1988) shows the dependence of spin density attributed to the E$'$ center on irradiation time for ArF and KrF laser excitation and for excitation with a low pressure Hg lamp emitting at 184.9 nm (6.7 eV). It is apparent that spin densities are greatly enhanced after exposure to ArF and KrF laser radiation. The concentration of E$'$ spins has been found to vary as the square of the laser fluence, which implies a quadratic dependence on laser intensity, assuming that pulse shape and duration are not fluence dependent (Arai et al. 1988, Rothschild et al. 1989).

This quadratic dependence suggests that E$'$ centers are created by two-photon absorption. Since two-photon excitation at 193 or 248 nm (but not at 308 nm) results in final states with energies that exceed the bandgap energy, two-photon absorption can result in the creation of excitons. The formation of E$'$ centers can then be associated with the

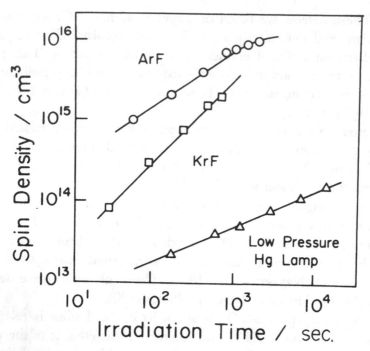

Figure 6.1. The density of the E′ center induced by ArF, KrF excimer lasers, and a low-pressure Hg lamp as a function of the irradiation time. The ArF laser was operated with 45 mJ/(pulse cm^{-2}) at 30 Hz and the KrF laser with 180 mJ/(pulse cm^{-2}). The data for the Hg lamp were taken by M. Kiwaki, M. Kumeda and T. Shimizu at Kanazawa University on the same sample. The spin density obtained by them was multiplied here by a factor of two to correct it for our calibration. The irradiation intensity of the 110 W lamp at the sample was estimated to be 6 mW cm^{-2} at 184.9 nm and 27 mW cm^{-2} at 253.7 nm (Arai *et al.* 1988).

trapping of such excitons at a lattice site followed by non-radiative recombination. The energy released on recombination would be comparable to the bandgap energy (i.e. 8 eV) and this energy would appear to be sufficient to create the E′ defect. Rothschild *et al.* (1989) found that E′ center formation with 193 nm excimer laser radiation was greatly enhanced by an increase in sample temperature. They suggested that this enhancement was due to an increase in the rate of non-radiative relaxation at higher temperature. Non-radiative relaxation leading to exciton luminescence has been observed under similar excitation conditions at low temperatures in MgO (Duley 1984). The excitons formed are likely to involve strongly bound electron–hole pairs with high mobility (Tsai *et al.* 1988).

The observation that the E′ concentration does not appear to saturate even at high levels of irradiation (Tsai *et al.* 1988) is consistent with a

Table 6.1. *The calculated density of E' centers created per pulse at 193 and 248 nm with* $E = 10\,J\,m^{-2}$ *and* $\tau = 23\,ns$.

λ (nm)	$\alpha^{(2)}$ (m MW^{-1})	$\eta^{(2)}$	N_p (cm^{-3})	Reference
193	2×10^{-1}	7.5×10^{-4}	3.2×10^{11}	Rothschild *et al.* (1989)
248	4.5×10^{-3}	2.6×10^{-3}	3.2×10^{10}	Arai *et al.* (1988)[a]

[a] $\eta^{(2)}$ has been corrected to reflect the quantum efficiency per pair of photons absorbed.

model in which defects are created by exciton trapping at lattice sites, rather than at previously existing defect sites. In this description the concentration $N(t)$ of E' centers increases as follows:

$$\frac{dN}{dt} = \alpha^{(1)} F \tau f \eta^{(1)} + \alpha^{(2)} F^2 \tau h \nu f \eta^{(2)} \qquad (1)$$

where $\alpha^{(1)}$ and $\alpha^{(2)}$ are absorption coefficients for defect creation via one- and two-photon processes, respectively. F is the photon flux (number of photons per cm^2 s^{-1}), assumed to be constant over the laser pulse length τ, f is the laser repetition frequency and $\eta^{(1)}$ and $\eta^{(2)}$ are quantum efficiencies for defect creation in one- and two-photon processes, respectively. Equation (1) does not include destruction of the E' center by photo-ionization or by other means. There is also some evidence that E' centers may relax to another defect type after excitation under certain conditions (Leclerc *et al.* 1991a,b); however, equation (1) should still provide an approximate estimate of $N(t)$ during and shortly after a laser pulse.

The number of defect centers created per laser pulse, per unit volume is

$$N_p = \frac{\alpha^{(2)}}{2h\nu} I^2 \eta^{(2)} \tau \qquad (2)$$

$$= \frac{\alpha^{(2)}}{2h\nu} \frac{E^2}{\tau} \eta^{(2)} \qquad (3)$$

where E is the fluence (J cm^{-2}). Some numerical values are given in Table 6.1.

The E' center has been identified with an absorption band at 5.8 eV (Griscom 1988). This absorption band is readily observed in samples exposed to γ radiation and appears under irradiation with excimer laser light under conditions that yield an increase in the E' spin concentration (Arai *et al.* 1988, Leclerc *et al.* 1991a,b). Other optical absorption bands as well as luminescence features also occur in a-SiO$_2$ under these conditions (Stathis and Kastner 1984a, Arai *et al.* 1988, Tohmon *et al.* 1989) but are less well correlated with specific defect centers. A notable example is the absorption band at 4.8 eV, which can be produced by irradiation at 7.9 eV (Stathis and Kastner 1984a). It is also present in unirradiated dehydrated silica glass when its strength may decrease with irradiation at 6.4 and 5.0 eV (Arai *et al.* 1988). However, Devine *et al.* (1986) found that the 4.8 eV absorption is induced by irradiation at 5.0 eV and suggested that it arises as a transition of the NBOHC defect. A 1.9 eV luminescence band is associated with the 4.8 eV center but is not directly attributable to this defect. Tohmon *et al.* (1989) concluded that the defect responsible for the 4.8 eV feature is uncertain but that energy transfer between the 4.8 eV absorber and another defect is responsible for the 1.9 eV luminescence. It is apparent that the final defect state of a-SiO$_2$ irradiated at excimer laser wavelengths is a complex function of the initial composition and level of hydration as well as the thermal and radiative histories of the material. An interaction between defects that is probably temperature-dependent is also indicated by these data. Unfortunately, it is the occurrence of these defects with absorption bands in the 4.5–6.5 eV range that effectively limits the transmission of high intensity UV laser radiation by fused quartz.

Excimer laser irradiation of a-SiO$_2$ has also been shown to lead to structural (Fiori and Devine 1986, Rothschild *et al.* 1989) and chemical changes (Fiori and Devine 1984). Fiori and Devine (1984) have irradiated laser-grown a-SiO$_2$ films with 5.0 eV photons at a fluence of 10^{-1} J cm^{-2} and found that oxygen is desorbed when these samples are heated to about 760 K in vacuum. They concluded that irradiation at 248 nm results in a highly defected dense structure which is deficient in oxygen (SiO$_x$). The oxygen resulting when these defects are formed can be liberated from the thin film sample when this is thermally activated. Fiori and Devine (1984) suggested that this oxygen would ordinarily diffuse in thick samples to recombine with defects.

Significant increases in density have been observed in heavily irradiated 100 nm thick SiO$_2$ films (Fiori and Devine 1986). Figure 6.2 shows the dependence of oxide thickness and refractive index (at 632.8 nm) on total laser fluence at 248 nm at doses of up to 1.7×10^4 J cm^{-2}. Reversible compaction was observed at doses of up to

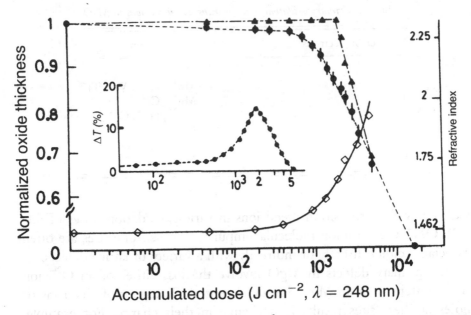

Figure 6.2. (●), Observed compaction of a 1000 Å thick oxide as a function of the UV accumulated dose; (▲), the same sample after thermal annealing at 950°C in vacuum (about 10^{-6} Torr) for 60 min. The inset shows the difference between these two results. (◇), Evolution of the refractive index during the laser irradiation (Fiori and Devine 1986).

2×10^3 J cm^{-2}, and, for doses less than this amount, the original sample thickness and refractive index could be recovered by heating to about 1220 K in vacuum. For higher doses irreversible compaction as well as etching was observed.

The refractive index of a-SiO$_2$ during irradiation was observed to be strongly dependent on total dose at exposures exceeding about 10^3 J cm^{-2}. This effect can be understood as due to volumetric changes. Fiori and Devine (1986) conclude that laser-induced compaction, which can reach about 36% at the limit of the reversible range, produces an essentially continuous range of polymorphs with densities between that of vitreous silica ($\rho = 2.2 \times 10^3$ kg m^{-3}) and $\rho = 3 \times 10^3$ kg m^{-3}.

6.2.2 Magnesium oxide

MgO is an ionic solid with a face-centered cubic lattice. Mg^{2+} and O^{2-} ions are located at the corners of two interpenetrating cubic sublattices. Common substitutional impurities include other alkaline earth ions (e.g.

Table 6.2. *Primary oxidation states of transition metal ions in MgO.*

Number of electrons	Ion
2	V^{3+}
3	Mn^{4+}, Cr^{3+}, V^{2+}, Ti^{+}
4	Mn^{3+}, Cr^{2+}
5	Mn^{2+}, Fe^{3+}, Cr^{+}
6	Fe^{2+}
7	Ni^{3+}, Co^{2+}, Fe^{+}
8	Ni^{2+}, Co^{+}

Ca^{2+}) as well as transition metal ions in various oxidation states (Table 6.2). OH^{-} is a common molecular impurity. These impurities are often associated either directly or indirectly with a variety of lattice defects.

The primary defects in MgO involve the loss either of an O^{2-} ion (F^{2+} center) or of an Mg^{2+} ion (V^{2-} center). Trapping of electrons or holes at these sites results in a change in their charge. For example, trapping of a single hole at the V^{2-} center yields the V^{-} center associated with an absorption band at 2.3 eV (Henderson and Wertz 1977). The F^{+} center with an absorption band near 5 eV is produced by trapping of an electron at the F^{2+} defect. Neutral aggregate centers may be formed when anion and cation vacancy defects occur together.

The association of vacancies with substitutional or interstitial impurities is common in MgO. For example, the V_{OH} center is formed when OH^{-} replaces an O^{2-} ion adjacent to a V^{-} center. Al^{3+} is often associated with a cation vacancy to form a V_{Al} center. The structure of some common defects is shown schematically in Figure 6.3.

Modifications of these defects appear at the surface of MgO crystals and in powders. Such sites are denoted by the subscript 's'. For example, a cation vacancy at the surface would be designated V_s^{2-}. The energies of absorption and luminescence bands for surface sites differ from those in the bulk material (Tench 1972).

F and F^{+} centers are abundant only in samples that have been irradiated with energetic particles (Henderson and Wertz 1977) or thermo-chemically reduced (Ballesteros *et al.* 1984). However, they are also observed in mechanically deformed (Wertz *et al.* 1962) or shocked material (Gager *et al.* 1964). In this way they are analogous to the E′ center in a-SiO$_2$. F and F^{+} centers are not generally produced by absorption of photons with energies less than the bandgap energy (7.7 eV) (Kuusmann and Feldbakh 1981).

Table 6.3. *Primary luminescence bands of common transition metal ions in crystalline MgO.*

Ion	Luminescence (nm)
Cr^{3+}	698.1, $\simeq$720
Mn^{4+}	654.5, $\simeq$672
V^{2+}	870.6
Ni^{2+}	$\simeq$476, $\simeq$810, 1240

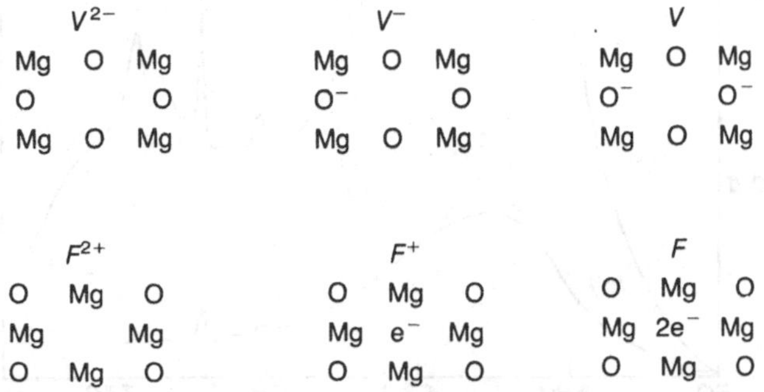

Figure 6.3. Schematic structures of some defect centers in MgO. Unless indicated otherwise explicitly, the charge is $Mg \equiv Mg^{2+}$, $O \equiv O^{2-}$.

The V^-, V and V aggregate centers are common in MgO exposed to ionizing radiation. All absorb near 2.3 eV and are thus not likely to be strong absorbers at excimer laser wavelengths (Agullo-Lopez *et al.* 1982). The absorption of photons within the 2.3 eV band may liberate holes that participate in reactions with a variety of defect and impurity sites.

The absorption of the transition metal ions present as substitutional or interstitial impurities in MgO is of the d → d type and is therefore weak. Nevertheless, transition metal ions can act as strong luminophores (Henderson and Wertz 1977) and also participate as hole and electron traps in a variety of thermochemical processes. Some characteristic wavelengths for luminescence bands due to transition metal ions in MgO are listed in Table 6.3.

Although MgO is nominally transparent to 193, 248 and 308 nm excimer laser radiation, a variety of multiphoton effects have been identified which lead to efficient coupling of UV radiation at high ($10^7 -$ 10^8 W cm^{-2}) intensities (Duley 1984, Dunphy and Duley 1988, 1990, Dickinson *et al.* 1993, Webb *et al.* 1993). A significant effect that occurs at intensities that are just below those that lead to surface etching is the

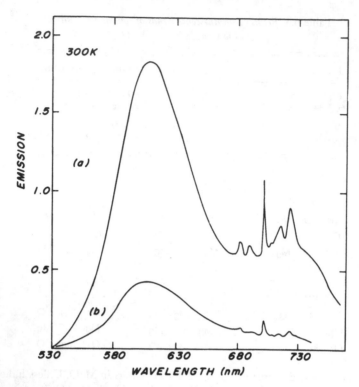

Figure 6.4. (a) Emission from MgO excited with $4 \times 10^8\,\mathrm{W\,cm^{-2}}$ at 308 nm at 300 K. (b) As above with a 0.62 filter in the incident beam (Duley 1984).

appearance of bright luminescence due to transition metal impurities such as Cr^{3+} (Figure 6.4). This luminescence exhibits an I^n dependence on excimer laser intensity, where n is generally 2, 4, 6 or greater (Dunphy and Duley 1988). The fact that only even values of n are generally observed suggests that luminescence is created when excitons deactivate at these impurity centers, since two photons are required to produce each exciton. When luminescence with $n > 2$ is observed, this may indicate that the density of excitons generated at the laser focus may be sufficiently large that bi- or tri-excitons can be created (Duley 1984). Luminescence with a multiphoton dependence on incident laser intensity has been observed from several transition metal ions in MgO (Dunphy 1991). In many cases, n varies with incident laser intensity and may decrease to $n \leq 1$ at laser intensities that are close to the threshold for surface etching. This saturation probably arises from the long radiative lifetime of d $\rightarrow$ d transitions and may not reflect a reduction in the rate of generation of excitons (Dunphy and Duley 1990).

Because the F and F^+ centers absorb in a broad band near 5 eV, they can be excited directly with 248 nm laser radiation. The absorption of two photons by F centers will lead to photo-ionization with the creation of electrons in the conduction band. Trapping of electrons by defect and impurity sites can also lead to the production of luminescence, again with a two-photon dependence on incident laser intensity. In addition, the interaction of laser radiation with these electrons can result in strong plasma heating (Dickinson *et al.* 1993). Transfer of this heat to the lattice produces a large thermal gradient that leads to thermal cracking (Webb *et al.* 1992, 1993) and the formation of additional defects. Dickenson *et al.* (1993) have shown that the presence of surface defects is necessary for plasma formation. When surface breakdown occurs, macroscopic particles of MgO are ejected together with atomic and ionic species. This effect is consistent with a buildup of mechanical strain and deformation in the irradiated MgO surface as multiple defects are created.

Intrinsic luminescence of F and F^+ centers in thermochemically reduced MgO irradiated at 248 nm has been reported by Rosenblatt *et al.* (1989). They found that F center luminescence is favored over F^+ center luminescence at high laser intensities. Both luminescence bands exhibit a long (about 1 s) decay time, which is incompatible with the allowed nature of the electronic transitions involved. This can be attributed to ionization of F and F^+ centers followed by electron trapping at defects such as H^-. Thermal release of electrons from these traps must be the rate limiting step in the luminescence process (Rosenblatt *et al.* 1989).

Another process that may accompany the onset of ablation in MgO involves the trapping of excitons at V centers. Since the V center has the configuration O^- vac O^- (Figure 6.3), trapping of an additional electron–hole pair would yield the configuration O^0 vac O^{2-}. Because O^0 has a radius of 0.060 nm whereas the radius of the O^{2-} ion is 0.14 nm, neutral O can migrate from the vacancy to reside as an interstitial atom, or to form O_2. This is similar to the sputtering mechanism proposed by Itoh and Nakayama (1982), and Rothenberg and Kelly (1984).

The vacancy aggregate, F^{2+}, V^{2-}, formed as the result of the elimination of oxygen from the V center would continue to act as an efficient exciton trap. Trapping of additional excitons would result in the migration of further oxygen atoms away from the defect, yielding an Mg-rich environment. With an excess of electrons available at the FV aggregate,

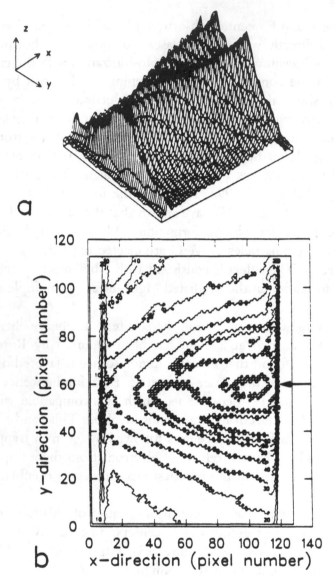

Figure 6.5. (a) The spatial distribution of luminescence excited from an Al : MgO crystal with 308 nm excimer laser radiation. The beam is incident from the right-hand side of the sample. (b) Isophotes of luminescence. The arrow shows the direction of laser impact (Dunphy 1991).

the effect would be to reduce Mg^{2+} and produce metallic magnesium. Dickinson *et al.* (1993) have observed that islands of metallic Mg are present on the surface of MgO irradiated at laser intensities near the ablation threshold.

With a range of impurities and defects that interact with each other by exchange of electrons and holes and are excited either directly by photon absorption, or indirectly by excitons, the nature of the response of MgO to intense UV radiation is expected to be complex. An example of this complexity is shown in Figure 6.5, in which the spatial dependence of the luminescence from an Al:MgO sample excited at one edge by 308 nm excimer laser radiation (Dunphy 1991) is plotted. In addition to 0.005% Al, this sample had trace amounts of Cr^{3+}. It had a scratch about 3 mm from the laser focus and transverse to the optic axis, which appears as a sharp linear emission in Figure 6.5.

Figure 6.6(a) shows the spatial distribution of emission within the G band (510–540 nm) of a CCD camera looking at the flat surface transverse to the optic axis. This emission is observed to be enhanced at points off the optic axis and in fact is reduced in amplitude where the beam is most intense. The G band emission is, however, enhanced at the position of the scratch and at the back surface of the sample. Luminescence in the R band ($\gtrsim 600$ nm) samples emission by Cr^{3+} centers and is localized to the most intense region of the focused beam (Figure 6.6(b)). It is weakly enhanced at the scratch and is also enhanced at the back surface of the sample.

Spectroscopic analysis (Dunphy 1991) shows that emission occurs at about 520 nm off the optic axis and at about 750 nm along the axis. Both spectra show a multiphoton dependence on 308 nm intensity. At threshold, the 520 nm emission has an I^6 dependence whereas the 750 nm on axis emission shows an $I^{\geq 2}$ dependence at threshold. This complicated behavior suggests that the defect centers responsible for 520 nm emission are destroyed in the intense radiative field on the optic axis. The high order dependence for emission off the optic axis implies that excitons can diffuse efficiently from the localized region where they are created.

Enhanced emission at the far end of the sample as well as at the scratch line (Figure 6.6) probably relates to a higher density of defects in regions of mechanical deformation (Wertz *et al.* 1962). The 5.7 eV absorption band, attributed to coordinatively unsaturated O^{2-} ions (Colluccia *et al.* 1978), is also observed in the spectrum of scratched MgO crystals (MacLean and Duley 1984).

6.2.3 Alumina

The bandgap energy in Al_2O_3 is about 7.3 eV (Liu *et al.* 1978). Intrinsic hole centers such as V^{2+}, V^- and V_{OH}^- are observed as in MgO, with

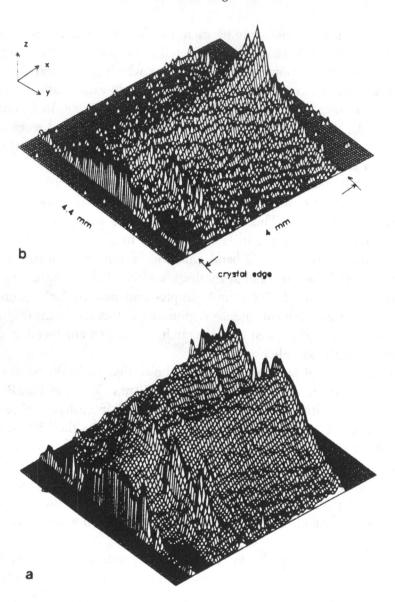

Figure 6.6. (a) As Figure 6.5, G video band intensity. (b) As Figure 6.5, R video band intensity (Dunphy 1991).

absorption bands near 3 eV (Lee and Crawford 1977). Cation vacancies such as F^{2+}, F^+ and F^0 centers are observed only in Al_2O_3 irradiated with energetic particles. These centers do not appear under photon irradiation.

In neutron-bombarded Al_2O_3 strong absorption bands appear at 6.1, 5.4, 4.8 and 4.1 eV (Lee and Crawford 1977). The relative strength of these features can be influenced by exposure to UV radiation. In particular, irradiation in the 6.1 eV absorption band is observed to weaken this feature while enhancing absorption at 4.8 eV. On this basis Lee and Crawford (1977) suggested that the 6.1 eV band is to be associated with the F center, whereas the 4.8 eV band is due to the F^+ center.

Luminescence at 3.75 eV can be produced by irradiation within the 4.8 eV band whereas irradiation in the 4.1 eV band results in luminescence at 2.45 eV. Another emission band at 2.25 eV can be excited by absorption at either 4.1 or 2.7 eV (Lee and Crawford 1977). This behavior in response to radiation suggests (by analogy with MgO) that a complex interaction may exist between intrinsic defects in Al_2O_3. In addition, transition metal ions are often found at the parts per million level in Al_2O_3 (McClure 1959).

Since F centers are produced in Al_2O_3 only by irradiation with energetic particles, they are unlikely to be present at high levels in unirradiated material. The initial coupling of UV laser radiation with $h\nu < 7.3$ eV into Al_2O_3 is therefore likely to occur through two-photon absorption, or via absorption in impurities. As observed with MgO, two-photon absorption can result in the establishment of a population of Wannier-type excitons. Trapping of these excitons will liberate about 7.3 eV at specific lattice sites. This energy may be sufficient to result in the creation of F center defects.

Rothenberg and Kelly (1984) found that Al_2O_3 irradiated at 193 nm emits strongly at about 3.02 eV on excitation at 248 nm. Emission at 3.75 eV is associated with a transition of the F^+ center (Lee and Crawford 1977) and would be expected to be enhanced if these defects were being produced by 248 nm laser radiation. Rothenberg and Kelly (1984) found, however, that the concentration of F^+ centers, as evidenced by emission at 3.75 eV, remained constant and concluded that these defects were not produced by exposure to intense UV pulses. This would not preclude an increase in F center concentration under conditions of exposure to multiple overlapping laser pulses, which would result in stress cracking, as observed for example in irradiated MgO (Webb *et al.* 1992, 1993). Rothenberg and Kelly (1984) found significant thermal stress-related exfoliation in Al_2O_3 irradiated with 532 nm Nd:YAG pulses, but did not observe this at 266 nm (Nd:YAG fourth harmonic), 248 or 193 nm. The rôle of defects in the initiation of ablation in Al_2O_3 is then unclear at present.

6.3 LASER SPUTTERING

6.3.1 General

Inorganic insulators are efficiently sputtered by UV laser radiation even
when the photon energy does not exceed the bandgap energy. Sputtering also occurs at laser fluences that are significantly lower than those
that would be required to produce fast thermal vaporization. As a result,
the description of UV laser sputtering of inorganic insulators has centered on the rôle played by electronic excitation in the sputtering
mechanism; most specifically the formation and trapping of defects and
electron–hole pairs. In this regard, many similarities exist between sputtering of wide bandgap insulators with short wavelength laser radiation,
and that of semiconductors excited at longer wavelengths. The rôle of
surface defects in the initiation of ablation is probably also similar in
these two different classes of materials.

For several important wide bandgap oxides the bandgap energy
$E_g \simeq 7\text{–}9\,\text{eV}$, and so direct band–band excitation is possible only with
F_2 laser radiation. Experiments on F_2 laser ablation ($h\nu = 8\,\text{eV}$) of
quartz (Herman *et al.* 1992) show that ablation occurs at a threshold
fluence of $0.62\,\text{J}\,\text{cm}^{-2}$ (Figure 6.7). Well defined smooth etch regions
are created when ablation is carried out at higher fluence (Figure 6.8).
The etching rate was about $0.023\,\mu\text{m}\,\text{J}\,\text{cm}^{-2}$ per pulse above threshold.
Similar clean ablation has been reported when SiO_2 is etched by a range
of UV wavelengths generated by anti-Stokes stimulated Raman scattering of 266 nm quadrupled Nd:YAG radiation in high pressure H_2 gas
(Sugioka *et al.* 1993). In this case, wavelengths extended to about
130 nm, well within the range of direct band–band excitation in SiO_2.
The threshold fluence was estimated to be about $1.17\,\text{J}\,\text{cm}^{-2}$. Since a
range of wavelengths was present in this experiment, both direct excitation and defect processes were probably involved in ablation, with the
shortest wavelength photons generating electron–hole pairs.

Surprisingly, materials such as SiO_2, MgO and Al_2O_3 can also be
ablated with photons having energies as low as $4\,\text{eV}$ ($\lambda = 308\,\text{nm}$)
although quite high fluences are often required (Ihlemann *et al.* 1992).
In spectral regions such as this, in which these materials are weakly
absorbing, the absorption is often determined primarily by the presence
of impurity species and defects. Such centers undoubtedly help in the
coupling of incident radiation into the material and may initiate damage
that leads to subsequent ablation. However, two-photon absorption can

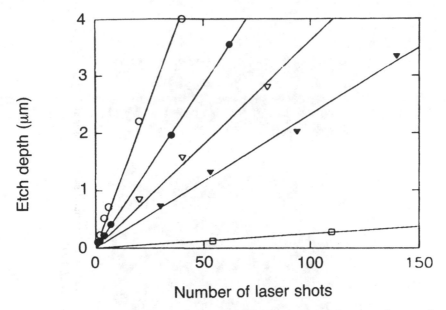

Figure 6.7. Etch depth in quartz plotted against the number of 157 nm laser pulses. Note the linear relationship over the range of fluences 0.68 ($\square$) to 4.9 J cm^{-2} ($\bigcirc$) (Herman *et al.* 1992).

also become significant at high incident fluences and can result in the establishment of a high density of electron–hole pairs in the region of highest excitation (Dunphy and Duley 1988). Since this is often at the surface of the sample, which is also the site of high defect concentration, two-photon absorption can lead directly to enhanced sputtering by electron–hole trapping at these defect sites. Possible mechanisms for sputtering initiated by electron–hole trapping at defect sites have been discussed in some detail by Kelly *et al.* (1985), Itoh (1987), Rothenberg and Kelly (1984), Hattori *et al.* (1992), Ichige *et al.* (1988), Georgiev and Singh (1992), and Singh and Itoh (1990).

The complexity of the interaction between intense UV laser radiation and wide bandgap oxides is well illustrated by the experiments reported by Ihlemann *et al.* (1992). They found that high aspect ratio holes can be drilled in fused silica with 308 nm laser pulses even though this material is quite transparent to light of this wavelength. Polished samples show a fluence threshold of about 20 J cm^{-2} for ablation. However, this is accompanied by damage at the front and back (exit point) of the sample. Ablation is facilitated by the presence of this damage and can be enhanced by roughening the surface prior to exposure to laser radiation. It seems probable that two-photon absorption is the initiator

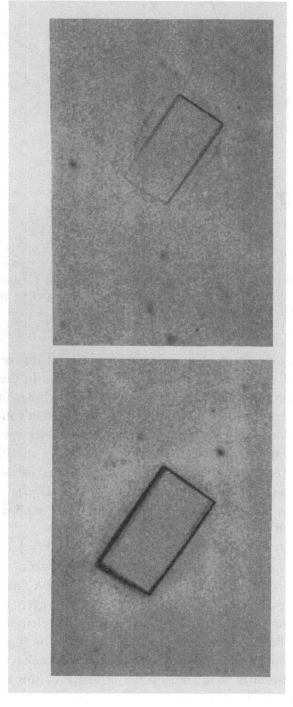

Figure 6.8. A photograph showing smooth clear ablation sites on quartz with 157 nm radiation: left-hand side 62 pulses at $2.3\,\mathrm{J\,cm^{-2}}$; right-hand side six pulses of $5.5\,\mathrm{J\,cm^{-2}}$ (Herman *et al.* 1992).

for this damage, which can appear because trapping of electron–hole pairs occurs at preexisting defects or impurity sites within the sample (Fiori and Devine 1986, Rothschild *et al.* 1989). Such defects are similar to those that would be created by mechanical roughening. Once a high concentration of defects has been created the material will absorb more strongly at 4.0 eV so that single-photon or sequential two-photon absorption will couple further energy into the sample. At the high fluences required for two-photon absorption, subsequent ablation will be efficient. Indeed, Ihlemann *et al.* (1992) reported ablation rates of up to 6 μm per pulse in quartz under these conditions. The ablation rate first increased with fluence and then decreased to about 2 μm per pulse at high fluence (40 J cm^{-2}).

In contrast, irradiation of fused silica at 248 or 193 nm is much less efficient, with ablation rates of about 0.3 μm per pulse for unroughened surfaces and about 1.8 μm per pulse for rough samples. These ablation rates are virtually independent of fluence above threshold. For a polished surface the threshold for ablation at 248 nm is about 11 J cm^{-2} and a clean, well defined ablation structure is generated. Ihlemann *et al.* (1992) found, however, that, once 20–50 pulses have been applied, the ablation rate increased to that of the roughened surface. Under these conditions, the surface undergoes what has been called an 'explosive ablation'.

The ablation threshold for irradiation with a 500 fs, 248 nm pulse is considerably lower (about 1 J cm^{-2}) than that for nanosecond pulses. This is attributed to the enhancement of the two-photon absorption cross-section at the high intensities in these pulses (about 10^{12} W cm^{-2}).

A similar phenomenon has been observed in the ablation of Al$_2$O$_3$ (sapphire) with 20 μs quadrupled Nd:YAG laser pulses at 266 nm (Tam *et al.* 1989, Brand and Tam 1990). Sapphire is highly transparent at this wavelength and so two-photon absorption must dominate the initial interaction. Slow etching (about 0.04 μm per pulse) is observed for about the first 20 pulses. At this point 'explosive' ablation is initiated accompanied roughly by a factor of ten increase in ablation rate per pulse. Material is ejected in the form of a broad angle plume during the slow ablation phase. In the explosive phase, a highly directed forward jet is observed accompanied by a broad angle plume. This plume was found to contain many fine (<20 nm to 3 μm) particles, which appear to have been ejected directly from the surface. This suggests that the explosive ablation phase involves a strong hydrodynamic interaction between vapor and the substrate, which can mechanically remove portions of the substrate.

Since defects are concentrated at the sites where ablation is initiated, it is not unreasonable to find that particulates or large clusters are ejected at high laser fluence. The presence of vacancy aggregates would lead to mechanical weakness, leading to fracture by the ablating vapor.

Webb *et al.* (1992, 1993) and Dickinson *et al.* (1993) have shown that laser-induced damage to cleaved surfaces of MgO is localized at cleavage steps. The damage takes the form of pitting and probably results from explosive ablation localized at these sites. The presence of defects at these points on the surface enhances absorption and results in strong localized interactions. Resonant two-photon absorption across the band-gap is also possible in MgO when the laser photon energy exceeds 4 eV ($\lambda = 308$ nm). Trapping of electrons and holes at cleavage steps would tend to deposit additional energy at these locations.

Dickinson *et al.* (1993) also found that exposure of cleaned MgO to relatively low fluence pulses (about $2\,\mathrm{J\,cm^{-2}}$) results in the formation of 'proto-holes'. These are roughly circular, with diameters of several micrometers. They contain a large number of smaller spheroidal droplets, which appear to have condensed from a melt. Because these droplets are Mg-rich, they are probably the residual material from regions containing a large number of F centers (see Figure 6.3). The aggregation of F^{2+}, F^{+} and F centers produces an area that is rich in Mg. Aggregation would be accompanied by the evolution of oxygen because hole trapping on O^{2-} ions in the MgO lattice reduces their bonding to the surface.

Defects can be created by roughening, by mechanical deformation, or by repetitive exposure to laser radiation. Dickinson *et al.* (1993) found that integrated plume emission as well as strong mass spectrometric signals only appear in cleaved samples after an incubation period, the duration of which depends on the total exposure of the surface to laser radiation. The onset of strong ablative decomposition of MgO also occurs only when the surface has visible damage. They concluded that the repetitive cycling of mechanical deformation in response to multiple laser pulses builds up to the point at which catastrophic damage can occur.

Similar precursor effects have also been observed in silicate glasses (Eschbach *et al.* 1989, Dickinson *et al.* 1990, Braren and Srinivasan 1988). Prior to substantial ablation, pitting is detected at localized surface sites and the area covered by pitting grows with further exposure. This enhances absorption at 248 nm because E centers are formed. Heavily damaged surfaces interact strongly with incident laser radiation and a liquid phase is generated. The morphology of the irradiated area

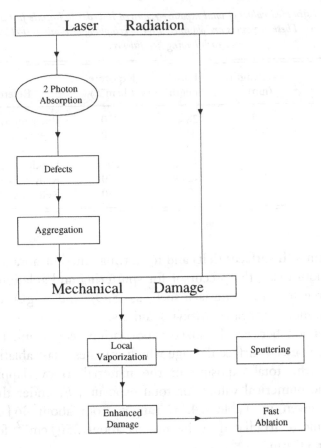

Figure 6.9. The schematic sequence of steps leading to laser sputtering and ablation.

exhibits wavelike structures that are consistent with a hydrodynamic interaction between the plume and this liquid.

Reverse etching, i.e. etching proceeding into the sample from the exit point of the laser beam, has been observed at 193 nm in fused silica (Ihlemann *et al.* 1992). Similar effects can be seen in MgO and have been attributed to the enhancement in laser intensity inside the sample by Fresnel reflection at the back surface. Since the transition probability for two-photon absorption scales as I^2, where I is the total laser intensity, a small increase in intensity due to reflection can give rise to significant changes in absorption, particularly for intensities close to the ablation threshold.

A summary of the major processes involved in laser sputtering and ablation of wide bandgap materials is given in schematic form in Figure 6.9. The two primary channels represent the ablation sequence for

Table 6.4. *Estimated values of total exposure required in order to initiate fast ablation in several oxides. These exposures are obtained after overlapping of many (10–100) pulses, each having low fluence.*

Material	Wavelength (nm)	Pulse length	Exposure ($J\,cm^{-2}$)	Reference
Fused SiO_2	248	28 ns	220	Ihlemann *et al.* (1992)
	248	500 fs	120	Ihlemann *et al.* (1992)
Na trisilicate	248	20 ns	36	Eschbach *et al.* (1989)
MgO	248	20 ns	200	Dickinson *et al.* (1993)
Al_2O_3	248	12 ns	60	Kelly *et al.* (1985)
	532	7 ns	250	Kelly *et al.* (1985)

cleaved or smooth surfaces (left) and for a roughened or abraded surface (right). In both cases, the precursor for sputtering and ablation involves defect aggregates. In materials such as SiO_2 and MgO these are probably vacancy aggregates involving anions.

The transition between localized vaporization accompanied by sputtering and enhanced surface damage which initiates a fast ablation mode depends on the total exposure of the material to overlapping laser pulses. Some numerical values for total exposure, E, under these conditions are given in Table 6.4. E ranges from about $36\,J\,cm^{-2}$ for sodium trisilicate irradiated at 248 nm to about $250\,J\,cm^{-2}$ for Al_2O_3 irradiated at 532 nm.

Since the overall ablation rate and threshold fluence for ablation are strong functions of surface conditions, it is difficult to predict material removal rates with any degree of accuracy for these materials. At short wavelengths ($\lambda = 193\,nm$) most materials can be etched at rates of about 0.01–0.05 µm per pulse at fluences that are 2–3 times the threshold fluence. Threshold fluences are typically $2-5\,J\,cm^{-2}$, which correspond to $(1.9-4.9) \times 10^{18}$ photons cm^{-2} at 193 nm. An ablation rate of 0.05 µm per pulse will remove about ten atomic layers or about 10^{16} atoms cm^{-2} per pulse. For an average binding energy of 10 eV per atom, ablation at this rate requires $1.6 \times 10^{-2}\,J\,cm^{-2}$ or a factor about 200 times smaller than the threshold fluence. This tells us that energy deposition is not the limiting factor in determining the ablation rate. Instead, fluence would seem to play a rôle predominantly in the creation of electron–hole pairs via two-photon absorption. These electron–hole pairs then facilitate ablation by trapping at surface defects.

The density of electron–hole pairs created by two-photon absorption in a single laser pulse is

$$N_{eh} = \alpha^{(2)} F^2 \tau h\nu \tag{4}$$

where $\alpha^{(2)}$, the two-photon absorption coefficient, will depend on laser wavelength and material. A representative value will be taken to be $\alpha^{(2)} \simeq 10\,\text{cm}$ per $10^6\,\text{W}$. Then, with $F = 10^{26}$ photons $\text{cm}^{-2}\,\text{s}^{-1}$, $\tau = 30\,\text{ns}$ and $h\nu = 1.03 \times 10^{-18}\,\text{J}$, one obtains $N_{eh} \simeq 3 \times 10^{21}\,\text{cm}^{-3}$ per pulse as an estimate of the density of electron–hole pairs generated by a single laser pulse at the threshold fluence for ablation. The energy density in the material associated with this excitation is $N_{eh}E_g$, where E_g is the bandgap energy. Then with $E_g = 8\,\text{eV} = 1.28 \times 10^{-18}\,\text{J}$, $N_{eh}E_g = 3.8 \times 10^3\,\text{J}\,\text{cm}^{-3}$ or approximately 10% of the sublimation energy for a typical simple oxide. It is the accumulation of this energy together with the presence of defects that leads to ablation. However, ablation is clearly not a thermal process since the temperature rise, ΔT, induced by this energy deposition is relatively small. With

$$\Delta T = \frac{N_{eh}}{\rho C_p} E_g \tag{5}$$

and taking $\rho = 2\,\text{g}\,\text{cm}^{-3}$, and $C_p = 1.5\,\text{J}\,\text{g}^{-1}\,°\text{C}$ as representative values, $N_{eh}E_g = 3.8 \times 10^3\,\text{J}\,\text{cm}^{-3}$ so that $\Delta T = 1.3 \times 10^3\,°\text{C}$. This suggests that, although surface melting may play a rôle in the ablation of these materials, electronic processes must dominate, with electron–hole trapping at defects being the primary mechanism.

The generation of a high density of electron–hole pairs as well as the ionization of F and F^+ centers by photo-absorption will result in a rapid increase in the density of electrons in the conduction band. Epifanov (1975) has discussed the heating of conduction band electrons by an intense optical field and concludes that electron acceleration can increase electron energy significantly. This excess energy is transferred to the lattice as electrons are scattered by phonons. Damage is initiated when the phonon density exceeds some critical value. The Epifanov process would appear to be likely in highly transparent insulators, although there has been little discussion of this point in relation to the ablation of such materials with intense UV laser radiation (Dickinson et al. 1993). Studies of charged species emitted from $Na_2O\text{–}3SiO_2$ (sodium trisilicate glass) irradiated with 248 nm pulses show that a prompt burst of photoelectron emission coincides with the laser pulse (Langford *et al.* 1990). This would be consistent with the onset of the Epifanov process.

There have been a number of experimental studies of plume spectra, temporal evolution and composition accompanying ablation of insulators with UV laser pulses (Rothenberg and Kelly 1984, Rothenberg and Koren 1984, Walkup *et al.* 1986, Eschbach *et al.* 1989, Tench *et al.* 1991, Salzberg *et al.* 1991, Moiseenko and Lisachenko 1991, Webb *et al.* 1992, Hastie *et al.* 1993, Geohegan 1993, Butt and Wantuck 1993, Roland *et al.* 1993, Rosenfeld *et al.* 1993, Webb *et al.* 1993, Dickinson *et al.* 1993). There have also been extensive studies of plumes generated in the ablation of superconductors for thin film deposition (Chapter 7).

Plasma emission is observed at fluences above threshold although mass spectrometric measurements show that charged particles are present during the pre-ablation phase, particularly with rough or damaged surfaces (Dickinson *et al.* 1993). Below threshold the complexity and amplitude of mass spectra increase gradually with total exposure, i.e. number of overlapping pulses. The intensity of optical emission from the plume also increases gradually with exposure over this range as more and more material is liberated from the surface. In the ablation of Al_2O_3, spectral emission occurs primarily from Al, Al^+ and AlO (Dreyfus *et al.* 1986, Rosenfeld *et al.* 1993). Emission from MgO consists of strong spectral features of Mg and Mg^+, with no emission that could be attributed to MgO. Emission from impurities such as Ca, Si and Al is also commonly observed.

Measurement of ion and molecule velocities shows that the distribution of these is often not Maxwellian. For example, ablation of Al_2O_3 with 248 nm KrF laser radiation was found to yield Al atoms whose kinetic energy increased from about 4 eV at threshold fluence (0.6 J cm^{-2}) to about 20 eV at 3 J cm^{-2} (Dreyfus *et al.* 1986). On the other hand, the kinetic energy of AlO molecules was found to decrease from about 1 eV at threshold to about 0.3 eV at 3 J cm^{-2}. The population of vibrational levels in AlO was, however, found to be in Boltzmann equilibrium at a temperature of 600 K, whereas rotational temperatures ranged from about 600 K at the peak of the flight-time distribution to about 2000 K for molecules having the highest velocities.

Large kinetic energies for atoms and molecules are frequently observed in laser ablation, but the mechanism for the generation of these high energies has not yet been defined. This energy must be

acquired either in the ejection mechanism, or subsequently by acceleration in the gas phase. Acceleration in the gas phase would occur in the
plasma over the target and would probably arise from plasma heating.
However, the fact that large kinetic energies are observed near threshold, when plasma effects are minimal, suggests that atoms may be
ejected from the surface in a highly non-equilibrium state. This would
be compatible with the involvement of electronic energy in the emission
process. Studies of the luminescence emitted from oxides at fluences
near the threshold for ablation show high order excitation processes
that involve more than one electron–hole pair (Dunphy and Duley
1990). The recombination of each electron–hole pair in oxides such as
SiO_2, MgO and Al_2O_3 will liberate about 8 eV of energy. This, together
with the high concentration of electronic excitation in the oxide at and
above threshold, would yield atoms that are ejected from the surface
with an appreciable energy, extending, perhaps, to several times the
electron–hole recombination energy. Thus atoms with kinetic energies
of several tens of electronvolts might be expected. Ion–molecule
reactions in the plume would then generate a variety of other atomic
and molecular species. In this case, since the primary source of excitation of atoms is via electronic excitation in the surface prior to sputtering, there is no requirement that the kinetic energy of ejected species be
limited to the photon energy.

   A low energy component with a velocity distribution that can be
closer to Maxwellian is also usually observed to be emitted from ablating
surfaces (Eschbach *et al.* 1989, Leuchtner *et al.* 1993, Moiseenko and
Lisachenko 1990). Translational energies for particles emitted in this
component are often in the 0.1–0.5 eV ($T = 1160$–5800 K) range, which
are more consistent with the surface temperatures expected during laser
ablation. An even slower component ($T = 350$–550 K) has been identified with emission from surface states (Moiseenko and Lisachenko 1990)

   Figure 6.10 shows a time-of-flight profile for Pb atoms ejected from
$PbZr_xTi_{(1-x)}O_3$ (PZT) in which fast and slow components are both
apparent. The signal from the slow component can be fitted well with
the expression $S(t)$, where

$$S(t) = \frac{A}{t^4} \exp\left[-m\left(\frac{z}{t} - v_s\right)^2 \bigg/ (2kT_s) \right] \qquad (6)$$

which describes an isentropic free jet expansion of vapor away from a
surface. In this equation, A is a normalization constant, t is time, z is the

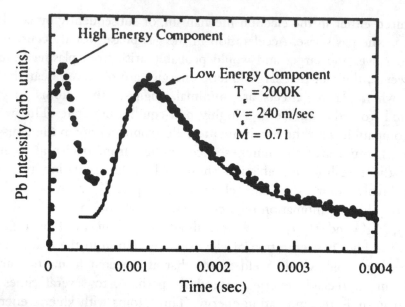

Figure 6.10. A time-of-flight profile for Pb from a fully reacted $PbZr_xTi_{(1-x)}O_3$ target at $1.3\,J\,cm^{-2}$ (248 nm, 30 ns) showing the dual components of the ejected Pb atoms. The solid line represents the fit to the low energy component (longer arrival times) using equation (6.6) (from Leuchtner *et al.* 1993).

flight distance, m is particle mass, v_s is the stream velocity, T_s is the local temperature and k is the Boltzmann constant. T_s and v_s are treated as adjustable parameters in this expression. The Mach number M can be obtained as follows:

$$M = \left(\frac{mv_s^2}{\gamma kT_s}\right)^{1/2} \tag{7}$$

where γ is the heat capacity ratio. In the experiments reported by Leuchtner *et al.* (1993), M was found to be typically 0.5–1.2 for atoms and molecules ejected from fully reacted PZT whereas T_s was in the range $1500 \leq T_s \leq 6500\,K$ depending on fluence (Table 6.5). M, T_s and v_s were all found to *decrease* with increasing fluence. It was suggested that this was due to a higher density of material over the target at high fluence. This material, which might be considered to be a Knudsen layer, would act to redistribute particle velocities.

The observation of a slow emission component characterized by a temperature that is similar to the thermal vaporization temperature of the substrate can be associated with heating of the sample surface underneath

Table 6.5. *Parameters fitting time-of-flight data with equation (7) for 248 nm ablation of fully reacted PZT (from Leuchtner et al. 1993).*

Species	Fluence ($J\,cm^{-2}$)	T_s (K)	v_s ($m\,s^{-1}$)	M
Ti	0.33	3703	1247	1.20
	0.35	3503	1139	1.12
	0.70	2957	1035	1.12
	1.30	2508	833	0.98
Zr	0.33	6471	889	0.90
	0.35	6523	960	0.97
	0.70	6258	1192	1.22
	1.30	4689	806	0.96
Pb	0.35	2659	490	1.23
	0.70	1965	291	0.85
	1.30	1976	253	0.70
PbO	0.35	1874	233	0.72
	0.70	1558	200	0.68
	1.30	1560	147	0.47

the ablated volume (Figure 6.11). This volume is heated by energy deposited from the trailing edge of the incident laser pulse as well as by momentum transfer from the expanding ablation component. This heat source can be approximated by a uniform circular heat source of radius R. The surface temperature $T(t)$ can then be obtained (Duley 1976):

$$T(t) = \frac{2I}{\pi} \frac{(\kappa t)^{1/2}}{K} \left[\frac{1}{\sqrt{\pi}} - \mathrm{ierfc}\left(\frac{R}{2(\kappa t)^{1/2}} \right) \right] \qquad (8)$$

where ierfc is the integral of the error function, I is the incident intensity, which is assumed constant in this approximation, κ is the thermal diffusivity, K is the thermal conductivity and t is time measured from the instant at which the surface is first exposed to I. Since, for $t \gtrsim t_p$, where t_p is the pulse duration (about 10–30 ns), $R \gg (\kappa t)^{1/2}$, $T(t)$ is approximately

$$T(t) = \frac{2I}{K} \left(\frac{\kappa t}{\pi} \right)^{1/2} \qquad (9)$$

Then, with $I = 10^8\,W\,cm^{-2}$, $\kappa = 0.01\,cm^2\,s^{-1}$ and $K = 0.1\,W\,cm^{-1}\,{}^\circ C^{-1}$ as representative values, one obtains

$$T(t) \simeq 1.1 \times 10^8 t^{1/2} \qquad (10)$$

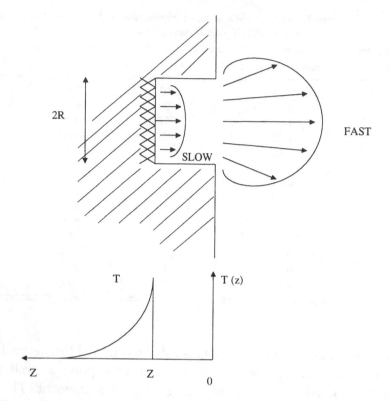

Figure 6.11. The origin of fast and slow components during laser ablation.

so that a surface temperature $T = 5000\,\text{K}$ would be attained in a time $t \simeq 2\,\text{ns}$.

Although this estimate of the substrate temperature is not precise, it does indicate that significant substrate heating can occur from the last part of the laser pulse. It is likely that the slow vaporization component can be associated with this heating. Experimental studies of the surface deformation in Pyrex glass that accompanies ablation with 193 nm laser radiative (von Gutfeld *et al.* 1986) showed that thermal expansion of the substrate occurs during the laser pulse. Such expansion indicates that significant heat transfer to the substrate occurs during the laser pulse.

Although fluence is the dominant parameter in characterizing excitation conditions and particularly in defining the onset of ablation, laser intensity might also be expected to be an important variable. The rôle of peak intensity in laser etching of organic polymers has been discussed in Chapter 5. Peak intensity will be important in determining the two-photon absorption coefficient of inorganic materials that are transparent at the laser wavelength. This would have the effect of making the threshold fluence intensity-dependent. A large decrease in threshold fluence for ablation of clean fused SiO_2 with 248 nm 500 fs pulses has been reported by Ihlemann *et al.* (1992), which illustrates this effect. However, focusing conditions are also important. In experiments on ablation of $LiNbO_3$ with 308 nm radiation Eyett and Bäuerle (1987) found, however, that the threshold fluence is independent of focal spot size but the ablation *rate* increases dramatically at small spot size at constant fluence. In these experiments, the fluence per pulse was kept constant while the focal spot size was varied from 175 to 24 µm. The dependence of etch depth per pulse on spot size was apparent only for spot sizes $\gtrsim 80$ µm. At the smallest spot sizes, the ablation rate per pulse was approximately twice that at 80 µm. This peculiar behavior has been attributed by Eyett and Bäuerle (1987) to an interaction between ejected material and the incident beam. They concluded that a larger spot size permits this material to be ejected over a wider angle so that the incident beam is attenuated by a greater amount than it is when the ejection is constrained to a narrow area. Since the intensity was kept constant at given fluence while the focus was varied, the variation in etching rate with focal spot size cannot be attributed to plasma shielding effects and the mechanism for such attenuation in the ejected plume is uncertain. Other mechanisms such as waveguiding or self-focusing by the plume may also be significant.

6.3.2 AlN, BN and Si_3N_4

These materials can be readily ablated with excimer laser radiation (Table 6.6) although the etching rate is small (10–100 nm per pulse). The etch depth is often directly proportional to the number of applied pulses and is linear with fluence for a limited range above threshold (Figure 6.12). However, at high fluence the ablation rate per pulse can saturate or decrease from its maximum value due to plasma shielding. Etching rate, threshold fluence and the quality of etch are all strong functions of grain size and structure so that the prediction of canonical values for these parameters is difficult. This variability can be seen in the numerical values given in Table 6.6.

The morphology of irradiated nitride ceramics has been studied by Miyamoto and Maruo (1990), Geiger *et al.* (1991), Tönshoff and Gedrat (1990), Doll *et al.* (1991), Schmatjko *et al.* (1988), and Esrom *et al.*

Table 6.6. *Data for etching of nitride ceramics with excimer laser radiation.*

Material	Wavelength (nm)	Threshold fluence ($J\,cm^{-2}$)	Etching rate (nm/pulse/ $J\,cm^{-2}$)	Reference
AlN	193	0.5	8.3	Pedraza *et al.*
	248		43	(1993)
	351		23	
	308	3	11	Schmatjko *et al.* (1988)
Si_3N_4 Hipped	193	1.5	6.2	Gower (1993)
	248	2.0	6.1	
	308	2.5	8.8	
Reaction bonded	193	0.8	6.8	Gower (1993)
	248	1.5	6.8	
	308	2.5	9.5	
Pressureless sintered	193	1.0	6.9	Gower (1993)
	248	1.5	6.8	
	308	2.5	8.3	
Gas pressure sintered	193	1.0	6.9	Gower (1993)
	248	2.0	5.9	
	308	3.0	7.9	
Post-reaction bonded	193	0.8	6.3	Gower (1993)
	248	2.0	6.2	
	308	3.0	9.0	
Hot pressure sintered	193	0.8	6.3	Gower (1993)
	248	2.0	5.3	
	308	3.0	8.2	
Sintered	248	$\simeq 5$	1.5	Miyamoto and Maruo (1990)
Si_3N_4	308	$\simeq 5$	5	Tönshoff and Gedrat (1989)
Si_3N_4	308	5	30	Schmatjko *et al.* (1988)
Si_3N_4	308	5	8	Geiger *et al.* (1991)

(1992). Generally, sharp-edged ablation regions can be produced in all materials with the correct choice of laser wavelength and fluence. Recast material along the vertical sides of deep etched holes as well as on the periphery of holes is a common problem. This material often consists of droplets of elemental metal, which derives from the decomposition of

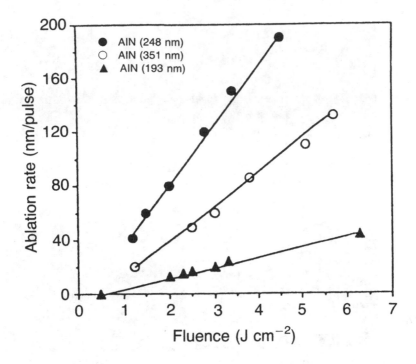

Figure 6.12. Ablation rates of AlN as a function of laser fluence
(from Pedraza *et al.* 1993).

the nitride. For example, droplets of Si are deposited in and around laser etched holes in Si_3N_4 (Miyamoto and Maruo 1990).

In contrast to machining with IR laser radiation, little or no cracking is observed adjacent to holes etched with UV radiation (Schmatjko *et al.* 1988). However, under certain conditions, particularly those near threshold fluence and with a moderate number of overlapping pulses, the ablated area is covered with a multitude of overlapping cones (Figure 6.13). In the case of ablation of Si_3N_4, the cones are largest when ablation occurs with 351 nm laser radiation. The number of cones per cm^2 decreases with fluence although individual cones become larger.

Hontzopoulos and Damigos (1991) have examined the effect of 193 and 248 nm laser radiation on the surface structure of Si_3N_4 at low fluences. They found that, although the untreated Si_3N_4 surface contains a random orientation of 4–5 µm needle-shaped grains, exposure of the surface to $1.5\,J\,cm^{-2}$ has the effect of causing Si crystals to emerge perpendicular to the surface. These crystals grow after exposure at higher laser fluence, but eventually melt to form a smooth surface at

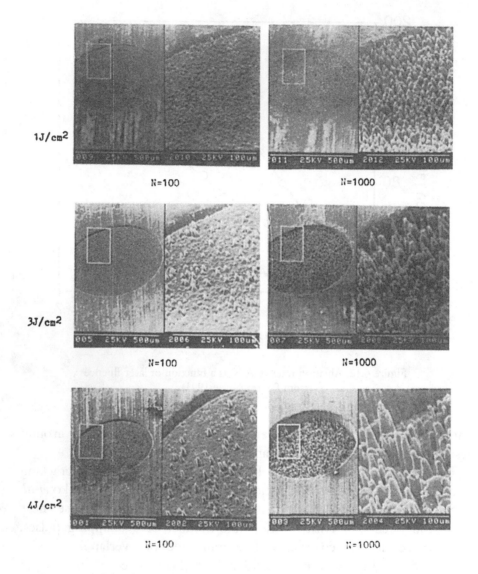

Figure 6.13. The development of conical structures during the ablation of Si_3N_4 ceramic with 248 nm KrF radiation (from Miyamoto and Maruo 1990).

high fluences ($>5\,\mathrm{J\,cm^{-2}}$). These Si crystals are probably the precursors of the Si-rich cones observed at higher levels of exposure (Figure 6.14).

Esrom *et al.* (1992) found that AlN also decomposes on exposure to excimer laser radiation with the deposition of liquid Al. The reaction is

$$AlN + h\nu \rightarrow Al(liquid) + \tfrac{1}{2}N_2 \tag{11}$$

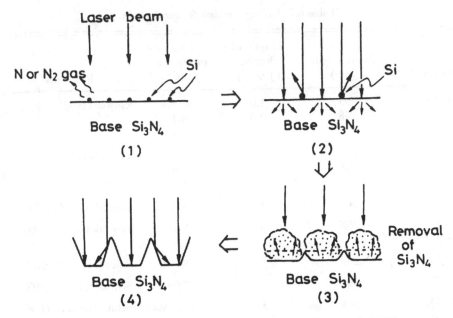

Figure 6.14. A mechanism illustrating uneven processing with cones in Si_3N_4 ceramics at fluences less than $5\,J\,cm^{-2}$ (from Miyamoto and Maruo 1990).

This reaction occurs at the decomposition temperature $T_d = 2573\,K$. It is followed by vaporization of Al:

$$Al(liquid) \rightarrow Al(gas) \tag{12}$$

which occurs at $T_v = 2767\,K$. The overall effect is to produce a thin film of aluminum on the AlN substrate after laser ablation. The thickness of this layer saturates after exposure to several overlapping pulses. Further discussion of this process has been given by Pedraza *et al.* (1993). Roughness in the surface of an AlN ceramic irradiated with overlapping 308 nm pulses has been discussed by Geiger *et al.* (1991).

In BN, UV irradiation results in the liberation of N_2, leaving a B-rich surface in the irradiated area (Doll *et al.* 1991). Spheroidal B-rich particles are also ejected from the irradiated area likely via a hydrodynamic interaction with ejected material.

6.3.3 Simple oxides

UV laser etching of Al_2O_3 ceramics has been the subject of numerous studies. Data is also available for etching of MgO, SiO_2, SiO_x, Y_2O_3

Table 6.7. *Etching rate data for simple oxides.*

Material	Wavelength (nm)	Threshold fluence ($J cm^{-2}$)	Etching rate (nm/pulse/ $J cm^{-2}$)	Reference
Al_2O_3	193	1.0	2.5	Gower (1993)
	248	2.0	4.1	
	308	3.0	6.1	
Al_2O_3 Hipped	193	1.0	2.7	Gower (1993)
	248	2.0	4.0	
	308	3.0	5.5	
Al_2O_3	308	$\simeq 4$	9.2	Lowndes *et al.* (1993)
Al_2O_3 (sapphire)	308	>6	9.2	Lowndes *et al.* (1993)
Al_2O_3	248		8	Sowada *et al.* (1989)
Al_2O_3	308	3	7 (air)	Schmatjko *et al.* (1988)
Al_2O_3	308	3	11.8 (vac)	Schmatjko *et al.* (1988)
SiO_x	248	0.05	6600	Fiori and Devine (1986)
	351	$\simeq 0.5$	500	
SiO_2	157	0.62	23	Herman *et al.* (1992)
Fused SiO_2	193	0.62		Ihlemann *et al.* (1992)
	248	0.62		
	248 (fs)	0.62		
	308			
YSZ (yttria stabilized zirconia)	308	0.5	10 (air) 17 (vac)	Kokai *et al.* (1992)
ZrO_2	193	1.0	4.7	Gower (1993)
	248	2.0	5.0	
	308	4.0	6.4	
ZrO_2	308	3.0	10	Schmatjko *et al.* (1988)

and ZrO_2. A summary of representative etching data for these materials is given in Table 6.7. The volume of ceramic that can be removed in each pulse is strongly fluence-dependent and has a maximum value at fluences that are typically several times the threshold fluence. For Al_2O_3 ceramic, the volumetric removal rate is typically 0.5–1.0 nl per pulse at fluences of 5–12 J cm^{-2} (Gower 1993). Optimized etching of Al_2O_3 is obtained with 308 nm laser radiation at fluences above 10 J cm^{-2}. Aspect

ratios for single holes drilled in Al_2O_3 with 308 nm radiation can be
>5:1 (Schmatjko *et al.* 1988). Typically the walls of these holes have a
small heat-affected zone with a smooth appearance and little cracking.
The melted zone on the periphery of these holes is of higher density
than that of the parent material. Recast material in the form of droplets
is commonly present inside and outside drill holes. Etch depth per pulse
can be increased and the volume of recast material reduced by irradiation in vacuum. An example of a microgear fabricated in alumina using
248 nm laser ablation is shown in Figure 6.15.

At relatively low fluences the surface of Al_2O_3 can be melted, resulting in a glazing or smoothing of the surface layer (Hourdakis *et al.*
1991, Lowndes *et al.* 1993). This decrease in surface roughness is
accompanied by an increase in grain size, a reduction in porosity and in
the density of microcracks. Hourdakis *et al.* (1991) have also reported a
significant decrease in surface resistivity with exposure to overlapping
248 nm pulses at fluences in the $4-8 J cm^{-2}$ range. This decrease is
probably due to an enrichment of Al in the surface layer and an increase
in grain size.

Contrasting behavior has been observed in yttria-stabilized zirconia
(YSZ) ablated with 308 nm pulses (Kokai *et al.* 1992), in which melting
and solidification results in an increase in the number of microcracks
and micropores. This effect was more significant after etching in
vacuum even though the etching rate is enhanced in vacuum. This
was accompanied by an increase in the Zr/O ratio in the irradiated
surface.

Ehrlich *et al.* (1985) have reported on the transformation of Al/O
cermets deposited on SiO_2 and then irradiated with pulses at 193 nm.
These films as deposited were a mixture of Al and Al_2O_3. After
irradiation, the surface composition becomes Al_2O_3-rich while Al is
enriched within the film.

6.3.4 Carbides

SiC ceramic is readily etched with excimer laser radiation. Peak removal
rates are typically 0.8–1.0 nl per pulse at fluences of $5-10 J cm^{-2}$
(Gower 1993). Best results are obtained with 308 nm radiation, but the
cleanest cuts are seen with 193 nm ablation. The threshold fluence
varies in the range $1.5-4 J cm^{-2}$ (Table 6.8). Near threshold the ablated
substrate is covered with conical structures (Tönshoff and Gedrat 1989)

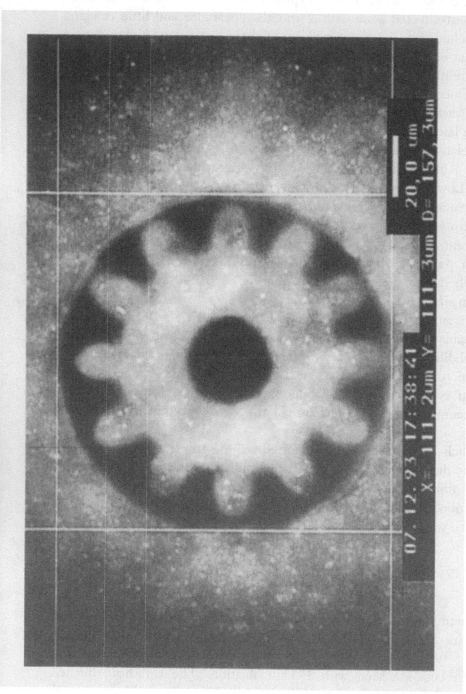

Figure 6.15. A microgear created in alumina by 246 nm laser ablation. The diameter is 120 μm and the part is 600 μm thick (courtesy of Lambda Physik).

Table 6.8. *Ablation data for SiC.*

Wavelength (nm)	Threshold fluence ($J\,cm^{-2}$)	Etching rate (nm/pulse/ $J\,cm^{-2}$)	Reference
193	1.5	5	Gower (1993)
248	2.0	4.7	
308	2.5	7.6	
193	1.5	4.4	Gower (1993)
248	2.3	4.1	
308	4.0	6.4	
308	4	10	Geiger *et al.* (1991)
308		4	Tönshoff and Gedrat (1989)
308	3	12	Schmatjko *et al.* (1988)

but this can be eliminated at higher fluences. However, surface melting and resolidification can result in surface rippling under these conditions. The ablation rate of SiC at 308 nm saturates at about 130 nm per pulse for fluences $\gtrsim 15\,J\,cm^{-2}$, which may be attributable to plasma formation. The mass spectrum of material ablated from SiC (Tench *et al.* 1991) shows evidence for cluster ions under irradiation with a 5 ns, 266 nm Nd:YAG pulse at a fluence of $2\,J\,m^{-2}$. However, at very high fluences no clusters are observed and only C and Si are detected. Condensed material ablated from SiC with 308 nm radiation shows clusters with dimensions of several tens of nanometers (Tench *et al.* 1991).

A comprehensive spectroscopic investigation of the plume vaporizing from ZrC under irradiation at 248 nm (Butt and Wantuck 1993) shows that emission by Zr atoms dominates. The excitation temperature was estimated to be $10^4\,K$.

6.3.5 Other oxides

Numerous other oxide materials have been ablated with UV laser radiation. Primary applications have been in the deposition of thin films and in patterning of oxide solids. Deposition of thin films of high temperature superconducting materials has been extensively studied and the ablation process has been well characterized in some of these materials such as $YBa_2Cu_3O_{7-x}$ (see Chapter 7).

Table 6.9. *Ablation rate data for oxides.*

Material	Wavelength (nm)	Threshold fluence ($J cm^{-2}$)	Etching rate (nm/pulse/ $J cm^{-2}$)	Reference
$BaTiO_3$	308	~0.5		Davis and Gower (1989)
$Bi_4Ti_3O_{12}$	248	0.7		Maffei and Krupanidhi (1992)
$LiNbO_3$	248 (ns)	0.9	40–100	Beuermann *et al.* (1990)
	308 (ns)	0.95	60	
	308 (ps)	0.09	120	
$LiTaO_3$	193	2.0	3.6	Gower (1993)
	248	1.5	4.4	
	308	0.8	7.8	
$PbZrTiO_3$ (PZT)	248	4.0	7.5	Gower (1993)
PZT	308	≃2	12.5	Eyett *et al.* (1987)
PZT	308	3	11.8	Schmatjko *et al.* (1988)
Ni–Zn ferrite	248	≃0.3	10	Tam *et al.* (1991)
Borosilicate glass	193	4	6.0	Gower (1993)
	248	8	39	
	308	10	37	
Float glass	308		200	Buerhop *et al.* (1990)

This section summarizes some of the data obtained on the ablation of other oxide materials. Important materials in this group include piezo-electric, ferroelectric and pyroelectric ceramics. Etching data for some materials in this group are given in Table 6.9.

The etch depth per pulse in $LiNbO_3$ and PZT has been shown to be a strong function of focal spot size (Beuermann *et al.* 1990, Eyett *et al.* 1987, Eyett and Bäuerle 1987). The origin of this dependence has been discussed previously. High aspect ratio tapered holes can be drilled in these materials. Since the etch depth per pulse is 50–100 nm at fluences about $10 J cm^{-2}$ (Table 6.9), etch depths of several hundred micrometers require 10^3–10^4 pulses. There is some evidence for waveguiding at the end of these high aspect ratio holes in that the cross-section at the end of these holes is significantly smaller than that of the laser focus.

Studies of the plume ejected from $LiNbO_3$ and $KNbO_3$ have been reported by Haglund et al. (1991). Spectral lines of K, O and Nb atoms were observed, as expected, from irradiated $KNbO_3$. The concentrations of all species decreased rapidly after the first few pulses, suggesting that surface roughness may be important in initiating ablation. This effect has been seen in other materials.

Beuermann et al. (1990) found that Li and Nb atoms are emitted from the surface of $LiNbO_3$ at fluences significantly lower than the threshold for ablation with nanosecond pulses. This does not occur when ablation is carried out using picosecond pulses. It was estimated that, whereas forty 308 nm photons are required to ablate each $LiNbO_3$ unit with 16 ns pulses, only five photons are required per unit for ablation with picosecond pulses. Below threshold, the velocities of emitted atoms and molecules were found to be given by a Maxwell–Boltzmann distribution with a temperature of 10^4 K. Ablation is accompanied by the deposition of material adjacent to and inside the etch region. The ablation of ferrites (Tam et al. 1991) leads to the formation of a glass 'skin' on the substrate with the presence of many microcracks.

6.4 LASER-ASSISTED CHEMICAL ETCHING

Oxides can be etched by exposure to UV laser light in the presence of gases such as Cl_2 (Beeston et al. 1988), H_2 (Eyett and Bäuerle 1987) and NF_3 (Yokoyama et al. 1985). The etching mechanism is poorly understood, however, and may involve both photoactivated reactions at surface defect sites as well as direct attack by atoms and radicals created in gas phase photochemical processes. Under certain conditions (Hirose et al. 1985), etching rates may be similar when the laser beam directly impacts upon the surface (perpendicular incidence) or is oriented so as to be parallel but separated from the surface (parallel incidence). Incubation or preprocessing may also be required for etching to be initiated (Hirose et al. 1985, Yokoyama et al. 1985). In the 193 nm ArF laser etching of SiO_2 in an NF_3 plus H_2 gas, XPS and infrared spectra show that a variety of new surface oxide and fluoride compounds are created. The overall reaction is then postulated (Hirose et al. 1985) to be

$$SiO_2 + NF_3 + H_2 + h\nu \rightarrow SiF_4 + NO_2 + N_2O + HF$$

with an etching rate of about $1.3\,\mathrm{nm\,s^{-1}}$ at a pulse repetition frequency of 80 Hz (Yokoyama *et al.* 1985).

With CW 257 nm laser radiation Beeston *et al.* (1988) found that $LiNbO_3$ can be etched in Cl_2 gas when the laser intensity, I, is high enough to promote surface melting. They found that etching requires $I \gtrsim 1.1 \times 10^6\,\mathrm{W\,cm^{-2}}$ for a scan speed of $2.7\,\mathrm{\mu m\,s^{-1}}$ and $I \gtrsim 1.2 \times 10^6\,\mathrm{W\,cm^{-2}}$ at a speed of $5\,\mathrm{\mu m\,s^{-1}}$. V-shaped trenches 2–3 μm wide can be created in this way (Figure 6.16). The depth of these trenches can be enhanced by multiple passes of the laser beam over the same region with a limit of 1.7 μm after more than ten passes. This results in the formation of a granular deposit inside the trench, which is depleted in Li. Further deposits are found on both sides of these trenches. $LiNbO_3$ can also be etched by UV laser-induced melting in the presence of a hydrated KF layer (Ashby and Brannon 1986). Laser melting permits the fusion of KF with $LiNbO_3$ to yield a surface composition that is readily soluble in water. The etching rate was found to be about 2×10^{-3} μm per pulse under irradiation with 248 nm KrF laser radiation at an intensity $I \simeq 3.2 \times 10^7\,\mathrm{W\,cm^{-2}}$.

6.5 FORMATION OF BRAGG GRATINGS IN OPTICAL FIBERS

The formation of reflection gratings in optical fibers by doping followed by laser irradiation was first reported by Hill *et al.* (1978) using 488 nm Ar^+ laser radiation in Ge-doped SiO_2 fibers. Irradiation using counter-propagating high intensity laser beams to form a standing wave pattern resulted in the creation of a refractive index modulation with a periodic structure. This structure then resulted in strong back reflection at the laser wavelength. Refractive index variations were found to involve chemical changes induced by two-photon transitions at Ge defect centers (Lam and Garside 1981), suggesting that irradiation at UV wavelengths might also be effective. Meltz *et al.* (1989) then showed that such gratings may be constructed by side-irradiation of Ge doped silica fibers with pulsed UV laser radiation (244 nm) in a configuration in which an interference pattern was projected on the fiber. With an average power of about 20 mW, a refractive index variation of about 3×10^{-5} was observed over a 4.4 mm length of fiber. The resulting grating, acting like a Bragg filter, had a reflectivity of 50–55% and a spectral width at half maximum of 42 GHz.

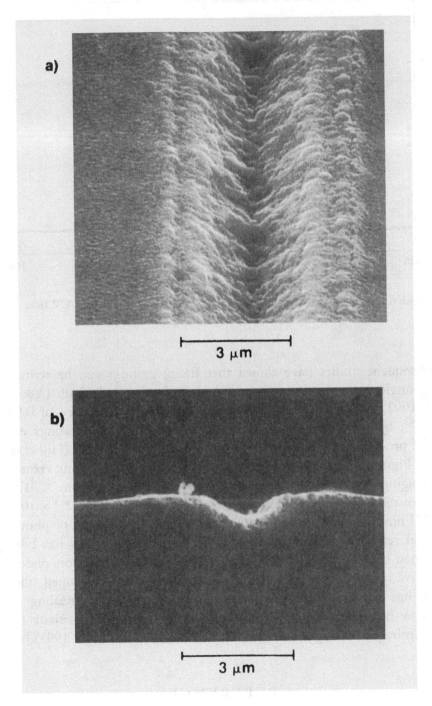

a)

3 μm

b)

3 μm

Figure 6.16. SEM micrographs of a trench in LiNbO₃ etched by eight scans at 1.4 μm s⁻¹ in Cl₂, with laser power of 25 mW. The trench has been washed with methanol and dilute buffered HF: (a) 60° tilted view and (b) edge view (Beeston *et al.* 1988).

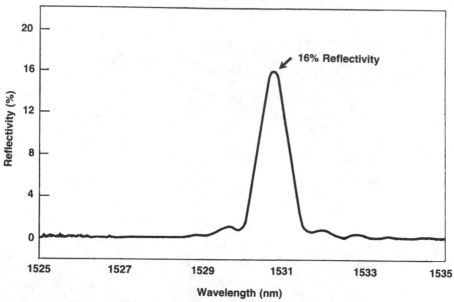

Figure 6.17. The spectral response of a Bragg grating fabricated with a UV laser source and phase mask photolithography (Hill *et al.* 1993).

Subsequent studies have shown that Bragg gratings can be written using single pulses from an excimer laser operating at 248 nm (Askins *et al.* 1992, Archambault *et al.* 1993a,b), typically at a fluence of 0.5–1 J cm^{-2}, by projection of an interference pattern. Bragg gratings can also be produced by irradiation through a phase mask grating (Hill *et al.* 1993). Figure 6.17 shows the spectral response of such a grating created by imaging 6×10^4 KrF pulses at a fluence of 0.1–0.2 J cm^{-2}. The effective refractive index variation produced was found to be 2.2×10^{-4} at 1531 nm. An investigation of the wavelength dependence of photo-induced refractive index variations in Ge-doped silica fibers has been published by Malo *et al.* (1990). The effect of flame heating on photo-refractive index variations created by KrF radiation in Ge-doped silica fibers has been discussed by Bilodeau *et al.* (1993). Annealing of fibers by heating in H$_2$ has been found to enhance the sensitivity for photo-refractive changes (Maxwell *et al.* 1992, Mizrahi *et al.* 1993).

6.6 REFERENCES

Agullo-Lopez F., Lopez F. J. and Jaque F., 1982. *Crystal Lattice Defects Amorphous Mater.* 9, 227.

Archambault J. L., Reekie L. and St J. Russell P., 1993a. *Electron Lett.* **29**, 28.

Archambault J. L., Reekie L. and St J. Russell P., 1993b. *Electron Lett.* **29**, 453.

Ashby C. I. H. and Brannon P. J., 1986. *Appl. Phys. Lett.* **49**, 475.

Askins C. G., Tsai T. E., Williams G. M., Putnam M. A., Bashkansky M. and Friebele E. J., 1992. *Opt. Lett.* **17**, 833.

Arai K., Imai H., Hosono H., Abe Y. and Imagawa H., 1988. *Appl. Phys. Lett.* **53**, 1891.

Ballesteros C., Piqueras J., Llopis J. and Gonzalez R., 1984. *Phys. Stat. Sol.* **83**, 645.

Beeston K. W., Houlding V. H., Beach R. and Osgood R. M. Jr, 1988. *J. Appl. Phys.* **64**, 835.

Beuermann T., Brinkmann H. J., Damm T. and Stuke M., 1990. *Mater. Res. Soc. Symp. Proc.* **191**, 37.

Bilodeau F., Malo B., Albert J., Johnson D. C., Hill K. O., Hibino Y., Abe M. and Kawachi M., 1993. *Opt. Lett.* **18**, 953.

Brand J. L. and Tam A. C., 1990. *Appl. Phys. Lett.* **56**, 883.

Braren B. and Srinivasan R., 1988. *J. Vac. Sci. Technol.* **B6**, 537.

Buerhop C., Blumenthal B., Weissmann R., Lutz N. and Biermann S., 1990. *Appl. Surf. Sci.* **46**, 430.

Butt D. P. and Wantuck P. J., 1993. *Mater. Res. Soc. Symp. Proc.* **285**, 81.

Coluccia S., Deane A. M. and Tench A. J., 1978. *J. Chem. Soc. Faraday Trans.* I, **74**, 2913.

Davis G. M. and Gower M. C., 1989. *Appl. Phys. Lett.* **55**, 112.

Devine R. A. B., Fiori C. and Robertson J., 1986. *Mater. Res. Soc. Symp. Proc.* **61**, 177.

Dickinson J. T., Jensen L. C., Webb R. L., Dawes M. L. and Langford S. C., 1993. *Mater. Res. Soc. Symp. Proc.* **285**, 131.

Dickinson J. T., Langford S. C., Jensen L. C., Eschbach P. A., Pederson L. R. and Baer D. R., 1990. *J. Appl. Phys.* **68**, 1831.

Doll G. L., Perry T. A. and Sell J. A., 1991. *Mater. Res. Soc. Symp. Proc.* **201**, 207.

Dreyfus R. W., Kelly R. and Walkup R. E., 1986. *Appl. Phys. Lett.* **49**, 1478.

Duley W. W., 1976. *CO_2 Lasers: Effects and Applications*, Academic Press, New York.

Duley W. W., 1984. *Opt. Commun.* **51**, 160.

Dunphy K., 1991. Ph.D. Thesis, York University, Toronto.

Dunphy K. and Duley W. W., 1988. *Phys. Stat. Sol.* **B148**, 729.

Dunphy K. and Duley W. W., 1990. *J. Phys. Chem. Solids* **51**, 1070.

Ehrlich D. J., Tsao J. Y. and Bozler C. O., 1985. *J. Vac. Sci. Technol.* **B3**, 1.

Elliott J. R., 1990. *Physics of Amorphous Materials*, Longmans Scientific and Technical, Essex.

Epifanov A. S., 1975. *Soc. Phys. JETP* **40**, 897.

Eschbach P. A., Dickinson J. T., Langford S. C. and Peterson L. R., 1989. *J. Vac. Sci. Technol.* **A7**, 2943.

Esrom H., Zhang J.-Y. and Pedraza A. J., 1992. *Mater. Res. Soc. Symp. Proc.* **236**, 383.

Eyett M. and Bäuerle D., 1987. *Appl. Phys.* **A40**, 235.

Eyett M., Bäuerle D., Wersing W., and Thomann H., 1987. *J. Appl. Phys.* **62**, 1511.

Fiori C. and Devine R. A. B., 1984. *Phys. Rev. Lett.* **52**, 2081.

Fiori C. and Devine R. A. B., 1986. *Phys. Rev.* **B33**, 2972.

Gager W. B., Klein M. J. and Jones W. H., 1964. *Appl. Phys. Lett.* **5**, 131.

Geiger M., Lutz N. and Biermann S., 1991. *Proc. SPIE* **1503**, 238.

Geohegan D. B., 1993. *Mater. Res. Soc. Symp. Proc.* **285**, 27.

Georgiev M. and Singh J., 1992. *Appl. Phys.* **A55**, 175.

Gower M. C., 1993. In *Laser Processing in Manufacturing*, ed. R. C. Crafer and P. J. Oakley, Chapman and Hall, London, p. 191.

Griscom D. L., 1984. *Nucl. Instrum. Methods* **B1**, 481.

Griscom D. L., 1988. *Proc. SPIE* **998**, 30.

Griscom D. J. and Friebele E. J., 1982. *Radiat. Effects* **65**, 63.

Haglund R. F., Arps J. H., Tang K., Niehof A. and Heiland W., 1991. *Mater. Res. Soc. Symp. Proc.* **201**, 507.

Hastie J. W., Bonnell D. W., Paul A. J. and Schenck P. J., 1993. *Mater. Res. Soc. Symp. Proc.* **285**, 39.

Hattori K., Okano A., Nakai Y. and Itoh N., 1992. *Phys. Rev.* B**45**, 8424.

Henderson B. and Wertz J. E., 1977. *Defects in Alkaline Earth Oxides*, Taylor and Francis, London.

Herman P. R., Chen B., Moore D. J. and Canaga-Retnam M., 1992. *Mater. Res. Soc. Symp. Proc.* **236**, 53.

Hill K. O., Fujii Y., Johnson D. C. and Kawasaki B. S., 1978. *Appl. Phys. Lett.* **32**, 647.

Hill K. O., Malo B., Bilodeau F., Johnson D. C. and Albert J., 1993. *Appl. Phys. Lett.* **62**, 1035.

Hirose M., Yokayama S. and Yamakage Y., 1985. *J. Vac. Sci. Technol.* B**3**, 1445.

Hontzopoulos E. and Damigos E., 1991. *Appl. Phys.* A**52**, 421.

Hourdakis G., Hontzopoulos E., Tsetsekou A., Zampetakis Z. and Stournaras C., 1991. *Proc. SPIE* **1503**, 249.

Ichige K., Matsumoto Y. and Nakiki A., 1988. *Nucl. Instrum. Methods* B**33**, 820.

Ihlemann J., Wolff B. and Simon P., 1992. *Appl. Phys.* A**54**, 368.

Itoh N., 1987. *Nucl. Instrum. Methods* B**27**, 155.

Itoh N. and Nakayama T., 1982. *Phys. Lett.* A**92**, 471.

Kelly R., Cuomo J. J., Leary P. A., Rothenberg J. E., Braren R. E. and Aliotta C. F., 1985. *Nucl. Instrum. Methods* B**9**, 329.

Kokai F., Amano K., Ota H. and Umemura F., 1992. *Appl. Phys.* A**54**, 340.

Kuusmann I. L. and Feldbakh E. K., 1981. *Sov. Phys. Solid. State* **23**, 259.

Lam D. K. W. and Garside B. K., 1981. *Appl. Opt.* **20**, 440.

Langford S. C., Jensen L. C., Dickinson J. T. and Pederson L. R., 1990. *J. Appl. Phys.* **68**, 4253.

Leclerc N. *et al.*, 1991a. *Opt. Lett.* **16**, 940.

Leclerc N., Pfleiderer C., Wolfrum J., Greulich K., Leung W. P., Kulkarni M. and Tam A. C., 1991b. *Appl. Phys. Lett.* **59**, 3369.

Lee K. H. and Crawford J. H., 1977. *Phys. Rev.* B**15**, 4065.

Leuchtner R. E., Horowitz J. S. and Chrisey D. B., 1993. *Mater. Res. Soc. Symp. Proc.* **285**, 87.

Liu P., Smith W. L., Loten H., Bechtel J. H. and Bloombergen N., 1978. *Phys. Rev.* B**17**, 4620.

Lowndes D. H., de Silva M., Gudbole M. J., Pedraza A. J. and Geohagen D. B., 1993. *Mater. Res. Soc. Symp. Proc.* **285**, 191.

MacLean S., and Duley W. W., 1984. *J. Phys. Chem. Solids* **45**, 227.

Maffei N. and Krupanidhi S. B., 1992. *Appl. Phys. Lett.* **60**, 781.

Malo B., Vineberg K. A., Bilodeau F., Albert J., Johnson D. C. and Hill K. O., 1990. *Opt. Lett.* **15**, 953.

Maxwell G. D., Kashyap R., Ainslie B. J., Williams D. L. and Armitage J. R., 1992. *Electron. Lett.* **28**, 2106.

McClure D. S., 1959. *Electronic Spectra of Molecules and Ions in Crystals*, Academic Press, New York.

Meltz G., Money W. W. and Glenn W. H., 1989. *Opt. Lett.* **14**, 823.

Miyamoto I. and Maruo H., 1990. *Proc. SPIE* **1279**, 66.

Mizrahi V., Lemaire P. J., Erdogan T., Reed W. A., DiGiovanni D. J. and Atkins R. M., 1993. *Appl. Phys. Lett.* **63**, 1727.

Moiseenko I. F. and Lisachenko A. A., 1990. *Mater. Res. Soc. Symp. Proc.* **191**, 121.

Nagasawa I., Hoshi Y., Ohki Y. and Yahagi K., 1986. *Jap. J. Appl. Phys.* **25**, 464.

Pedraza A. J., Zhang J.-Y. and Esrom H., 1993. *Mater. Res. Soc. Symp. Proc.* **285**, 209.

Roland P. A., La Placa S. and Wynne J. J., 1993. *Mater. Res. Soc. Symp. Proc.* **285**, 117.

Rosenblatt G. H., Rowe M. W., Williams G. P., Williams R. T. and Chen Y., 1989. *Phys. Rev.* **B39**, 10309.

Rosenfeld A., Mitzner R. and König R., 1993. *Mater. Res. Soc. Symp. Proc.* **285**, 123.

Rothenberg J. E. and Kelly R., 1984. *Nucl. Instrum. Methods* **B1**, 291.

Rothenberg J. E. and Koren G., 1984. *Appl. Phys. Lett.* **44**, 664.

Rothschild M., Ehrlich D. J. and Shaver D. C., 1989. *Appl. Phys. Lett.* **55**, 1276.

Salzberg A. P., Santiago D. J., Asmar F., Sandoval D. N. and Weiner B. R., 1991. *Chem. Phys. Lett.* **180**, 161.

Schmatjko K. J., Endres G., Schmidt U. and Banz P. H., 1988. *Proc. SPIE* **957**, 119.

Singh J. and Itoh N., 1990. *Appl. Phys.* **A51**, 427.

Sowada U., Ishizaka S.-I., Kahlert H. J. and Basting D., 1989. *Chemtronics* **4**, 162.

Stathis J. H. and Kastner M. A., 1984a. *Phil. Mag.* **B49**, 357.

Stathis J. H. and Kastner M. A., 1984b. *Phys. Rev.* **B29**, 7079.

Sugioka K., Wada S., Tashiro H., Toyoda K., Sakai T., Takai H., Moriwaki H. and Nakamura A., 1993. *Mater. Res. Soc. Symp. Proc.* **285**, 225.

Tam A. C., Brand J. L., Cheng D. C. and Zapka W., 1989. *Appl. Phys. Lett.* **55**, 2045.

Tam A. C., Leung W. P. and Krajnovich D., 1991. *J. Appl. Phys.* **69**, 2072.

Tench R. J., 1972. *J. Chem. Soc. Faraday I.* **68**, 1181.

Tench R. J., Balooch M., Bernardez L., Allen M. J., Siekhaus W. J., Ollander D. R. and Wang W., 1991. *J. Vac. Sci. Technol.* **B9**, 820.

Tohmon R., Shimogaichi Y., Nunekuni S., Ohki Y., Hama Y. and Nagasawa K., 1989. *Appl. Phys. Lett.* **54**, 1650.

Tönshoff H. K. and Gedrat O., 1989. *Proc. SPIE* **1132**, 104.

Tönshoff H. K. and Gedrat O., 1990. *Proc. SPIE* **1377**, 38.

Triplet B. B., Takahashi T. and Sugamo T., 1987. *Appl. Phys. Lett.* **50**, 1663.

Tsai T. E., Griscom D. L. and Friebele E. J., 1988. *Phys. Rev. Lett.* **61**, 444.

von Gutfeld R. J., McDonald F. A. and Dreyfus R. W., 1986. *Appl. Phys. Lett.* **49**, 1059.

Walkup R. E., Jasinski J. M. and Dreyfus R. W., 1986. *Appl. Phys. Lett.* **48**, 1690.

Webb R. L., Jensen L. C., Langford S. C. and Dickinson J. T., 1992. *Mater. Res. Soc. Symp. Proc.* **236**, 21.

Webb R. L., Jensen L. C., Langford S. C. and Dickinson J. T., 1993. *J. Appl. Phys.* **74**, 2338.

Weeks R. A., 1956. *J. Appl. Phys.* **27**, 1376.

Wertz J. E., Orton J. W. and Auzins P., 1962. *J. Appl. Phys.* (Suppl.) **33**, 322.

Yokoyama S., Yamakage Y. and Hirose M., 1985. *Appl. Phys. Lett.* **47**, 389.

CHAPTER 7

UV laser preparation and etching of superconductors

7.1 INTRODUCTION

Shortly after the announcement of high temperature superconductivity in the La–Ba–Cu–O (Bednorz and Müller 1986) and Y–Ba–Cu–O (Wu et al. 1987a) systems the first reports were published describing in situ preparation of superconducting thin films using laser ablation (Dijkkamp et al. 1987, Wu et al. 1987b, Narayan et al. 1987). The laser ablation method, which is a well known technique for the preparation of thin films of a variety of materials (Duley 1983, Bäuerle 1986, Braren et al. 1993, Chrisey and Hubler 1994), was found to be well suited to the deposition of superconducting films since it permits flexible control over deposition conditions and yields films with good stoichiometry.

Materials such as Y–Ba–Cu–O are, however, complex from both a chemical and a structural point of view (Burns 1992) and therefore vaporization and redeposition of these materials using laser radiation is anticipated to be a complicated process. A full understanding of the physical and chemical mechanisms that accompany laser ablation and in situ deposition has yet to be obtained. Nevertheless, useful progress has been made in the preparation of superconducting films with high zero resistance temperatures (about 90 K) and critical current densities exceeding $10^6 \, \text{A cm}^{-2}$ using the laser ablation method.

7.2 DEPOSITION AND PROPERTIES

The use of excimer laser radiation to prepare thin films of super-conducting material by laser vaporization of the parent compound was first reported in 1987 (Dijkkamp et al. 1987, Wu et al. 1987b, Narayan *et al.* 1987). Laser vaporization was rapidly identified as an excellent technique for the production of thin superconducting films on a variety of substrates (Table 7.1). Nominal preparation conditions are as follows.

Vaporization source	bulk oxide (pellet form)
Ambient atmosphere	vacuum or O_2 gas
Laser fluence	$1-3\,J\,cm^{-2}$
Intensity	about $10^8\,W\,cm^{-2}$
Deposition rate	about 0.1 nm per pulse
Pulse rate	$1-10\,Hz$
Pulse duration	$10-30\,ns$
Substrate temperature	$400-800°C$
Film thickness	$0.1-1\,\mu m$

A schematic diagram of a laser deposition system is shown in Figure 7.1. Most experimental studies have centered on the deposition of $YBa_2Cu_3O_{7-x}$ films, although other materials such as Pb–Sr–Y–Ca–Cu–O (Hughes *et al.* 1991), $Ba_{1-x}K_xBiO_3$ (Moon *et al.* 1991), $Nd_{1.85}Ce_{0.15}CuO_{4-y}$ (Gupta *et al.* 1989), $SmBa_2Cu_3O_7$ (Neifeld *et al.* 1988), $TeSr_2(Ca, Cr)Cu_2O_7$ (Tang *et al.* 1993) Bi–Sr–Ca–Cu–O (Kim *et al.* 1988) and $Tl_2Ba_2Ca_2Cu_3O_{10}$ (Liou *et al.* 1989) have also been prepared. The composition of the resulting films is found to be close to that of the parent material. This occurs because the initial stage of the vaporization process probably involves the ejection of macroscopic clusters from the target material and plasma chemistry.

As-deposited $YBa_2Cu_3O_{7-x}$ films are polycrystalline (Hwang *et al.* 1988) and typically show an onset of superconductivity at 85 K. Zero resistivity is reached at about 30–40 K. However, annealing of deposited films by heating in an oxygen atmosphere followed by slow cooling sharpens the superconducting transition so that typically the width of the superconducting transition in $YBa_2Cu_3O_{7-x}$ is about 2 K and the zero resistance temperature is raised to 85–90 K. A typical resistivity versus T curve for a $YBa_2Cu_3O_{7-x}$ film produced by excimer laser vaporization in vacuum followed by annealing to 900°C in O_2 is shown in Figure 7.2. This film was deposited on an Al_2O_3 substrate using 308 nm laser radiation.

Post-deposition annealing in oxygen has the primary effect of increasing oxygen content. This requirement can be minimized by the preparation of films in an excess of oxygen (Imam *et al.* 1988, Roas *et al.* 1988). Using this technique Roas *et al.* (1988) have reported critical

Table 7.1. *Excimer laser deposition of superconducting films.*

Material	Substrate	T_c (K)	J_c (A cm^{-2})	Note	Reference
YBa$_2$Cu$_3$O$_{7-x}$	SrTiO$_3$	88.6	7×10^5	335 nm	Imam et al. (1988)
	Al$_2$O$_3$	78			
	SrTiO$_3$	90	2.2×10^6	308 nm	Roas et al. (1988)
	SrTiO$_3$	90	4×10^4		Koren et al. (1988)
	Y-ZrO$_2$	85	1×10^4		
	Si	70			
	Si	45	5×10^1	193 nm	Witanachchi et al. (1989)
	MgO/S	70	3×10^3		
	SrTiO$_3$	88	5×10^6	308 nm	Singh et al. (1989)
	Y-ZrO$_2$	89	1×10^6		
	Y-ZrO$_2$	89	1.1×10^5		Norton et al. (1990)
	SrTiO$_3$	89	1×10^6		Chang et al. (1990)
	SrTiO$_3$	85	1.6×10^6	335 and 248 nm	Gupta et al. (1990)
	SrTiO$_3$	75	1×10^6	+0	Gupta and Hussey (1991)
	Y-ZrO$_2$	88	2×10^6		Hwang et al. (1991)
	MgO (100)	82		193 nm, MOCVD	Ushida et al. (1991)
	LaAlO$_3$			308 nm	Holzapfel et al. (1992)
	SrTiO$_3$	88	24×10^6		
	ZrO$_2$/MgO/Al$_2$O$_3$	20		1064 nm	Lynds et al. (1988)
	ZrO$_2$				
	MgO/Al$_2$O$_3$	15		1064 and 532 nm	Kwok et al. (1988)
	SrTiO$_3$				

Substrate		Value	Wavelength	Reference
SrTiO$_3$	86	8×10^2	10.6 μm	Miura et al. (1988)
SrTiO$_3$	88	3×10^5	1064 nm	Koren et al. (1989)
	89	9×10^5	532 nm	
	93	9×10^5	355 nm	
SrTiO$_3$	86.5			Wiener-Avnear et al. (1990)
Y-ZrO$_2$/Si	86	2.2×10^6	308 nm	Fork et al. (1990)
SrTiO$_3$	85			Dijkkamp et al. (1987b)
Al$_2$O$_3$	75			
SrTiO$_3$	85			Wu et al. (1987)
Al$_2$O$_3$	85		308 nm	Narayan et al. (1987)
(100) Si				
(100) MgO				
SrTiO$_3$				
SiO$_2$				
SrTiO$_3$	85	5×10^4		Wu et al. (1988)
Al$_2$O$_3$	73			
SrTiO$_3$	85			Hwang et al. (1988)
SrTiO$_3$	85	1×10^5	193 nm	Witanachchi et al. (1988)
Al$_2$O$_3$	75			
ZrO$_2$	75			
Si	67	5×10^4		Venkatesan et al. (1988a,b)
ZrO$_2$/Si	80			
SiO$_2$/Si				
Al$_2$O$_3$	75			Chang et al. (1988)
SrTiO$_3$	85			
Y-ZrO$_2$/Si	82			Tiwari et al. (1990)
Y-ZrO$_4$/hastelloy	86	7×10^3		Kumar et al. (1990)

Table 7.1. (cont.)

Material	Substrate	T_c (K)	J_c (A cm^{-2})	Note	Reference
	ZrO_2			193 nm	Narumi et al. (1991)
	Y-ZrO_2/Pt-hastelloy				
	Y-ZrO_2/hastelloy	87	3×10^4		Ogale et al. (1991)
	Y-ZrO_2-Ag/55304	88	8×10^4		Reade et al. (1991)
	Y-ZrO_2/Haynes 230	86	3×10^3		Kumar and Narayan (1991)
	$CoSi_2$/Si	83			
	$SiTiO_3$/Si	84.5		193 nm	Sanchez et al. (1992)
	Y-ZrO_2/Haynes 230	92	6×10^5		Reade et al. (1992)
Pb–Sr–Y–Ca–Cu–O	$LaAlO_3$	75		308 nm	Hughes et al. (1991)
$Ba_{1-x}K_xBiO_3$	MgO	22			Moon et al. (1991)
	$SrTiO_3$	23			
	$LaAlO_3$	26			
	Al_2O_3	20			
$Nd_{1.85}Ce_{0.15}CuO_{4-y}$	$SrTiO_3$	20	2×10^5	355 nm	Gupta et al. (1989); Kussmaul et al. (1992)
$SmBa_2Cu_3O_7$				308 nm	Neifeld et al. (1988)
Bi–Sr–Ca–Cu–O	ZrO_2	69		193 nm	Kim et al. (1988)
	$SrTiO_3$	80		308 nm	Fork et al. (1988)
	Y-ZrO_2	82		308 nm	Kumar et al. (1990)
$Tl_2Ba_2Ca_2Cu_3O_{10}$	MgO	110	1×10^4	532 nm	Liou et al. (1989)
$TlSr_2(Ca, Cr)Cu_2O_7$		102	10^6	193 nm	Tang et al. (1993)

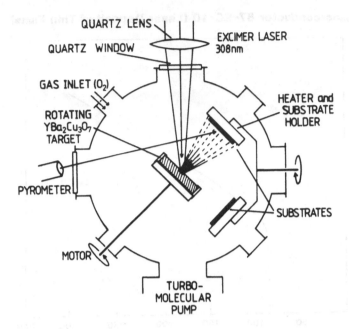

Figure 7.1. A schematic diagram of a laser evaporation system for the
preparation of thin superconducting films (Roas *et al.* 1988).

current densities as high as $2.2 \times 10^6 \, \text{A cm}^{-2}$ at 77 K and a zero resistance temperature of 90 K for $YBa_2Cu_3O_{7-x}$ prepared on $\langle 100 \rangle$
$SrTiO_3$. Singh *et al.* (1989) and Witanachchi *et al.* (1988) find that
substrate temperatures can be reduced even further (to 500–650°C) by
preparing films in an O_2 atmosphere with a DC bias voltage of about
+300–400 V relative to the vaporization source. This procedure would
appear to have the effect of enhancing oxygen content in the deposited
films, probably through an increase in O atom or O_3 concentration in
the gas phase over the deposit. The effect of a DC bias was most apparent at low substrate temperatures.

The effect of oxygen on the quality of deposited films has been evaluated by Gupta *et al.* (1990) and Gupta and Hussey (1991). Gupta and
Hussey (1991) used a supersonic expansive nozzle to release O_2 gas in
synchronization with the arrival of the vaporization front at the substrate surface. They found that superconducting films are formed under
low pressure conditions only when this gas pulse overlapped the arrival
time of the ablated fragments at the deposition surface. A similar effect
was observed when N_2O was used as an oxidant. These results indicate
that a transient increase in oxygen partial pressure over the sample

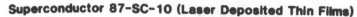

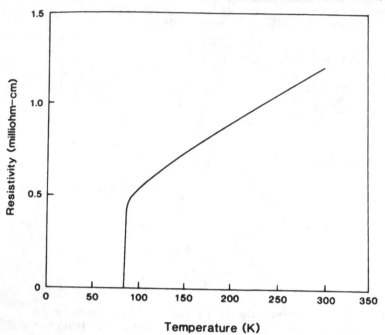

Figure 7.2. Resistivity versus temperature for $YBa_2Cu_3O_{7-x}$ film on sapphire. The film was prepared by vaporization of the bulk superconductor using 308 nm laser radiation (Narayan *et al.* 1987).

while deposition is occurring is sufficient to stabilize the superconducting phase in $YBa_2Cu_3O_{7-x}$. This has been demonstrated in a two-laser experiment (Gupta *et al.* 1990) in which the effect of rarefaction produced by the ablating laser pulse could be probed by a second pulse. It was found that the low density transient behind the first ablation shock front resulted in O_2 depletion. As a consequence, when a second deposition was attempted during this phase, O_2 depletion led to the generation of films with a high oxygen defect concentration and low superconducting transition temperature. A study by Kim and Kwok (1992) has shown that target–substrate distance is also an important variable in these experiments.

$YBa_2Cu_3O_{7-x}$ films may be grown epitaxially on substrates such as [100] $SrTiO_3$ (Roas *et al.* 1988, Singh *et al.* 1989) and [100] yttria-stabilized ZrO_2 (Y-ZrO_2) (Singh *et al.* 1989, Norton *et al.* 1990). Films are highly oriented with their *c* axis aligned along the perpendicular to the substrate surface. This growth occurs as the deposit grows layer by layer. The high surface temperature and relatively slow deposition rate (one pulse per second) probably enhances mobility while allowing atoms to react at the sites required for stoichiometric composition. The interface between the superconducting deposit and the substrate in $SrTiO_3$ can be defined to within 100 nm with little interdiffusion between the substrate and the film (Singh *et al.* 1989).

Deposition on $Y-ZrO_2$ also leads to epitaxial growth but with an interface compound which is semiconducting at 650°C (Hwang *et al.* 1991). This layer provides a pseudo-matched buffer layer between the substrate and the $YBa_2Cu_3O_{7-x}$ deposit.

$YBa_2Cu_3O_{7-x}$ films can also be grown directly on Si by laser deposition (Koren *et al.* 1988) but optimal results are obtained with MgO or $Y-ZrO_2$ buffer layers (Witanachchi *et al.* 1989). Substrate temperature must be kept low (400°C) to minimize reaction and Si diffusion into the deposit.

Buffering of Si with intermediate layers of $Y-ZrO_2$ (Fork *et al.* 1990, Tiwari *et al.* 1990), $SrTiO_3$ (Sanchez *et al.* 1992), and $CoSi_2$ (Kumar and Narayan 1991) has facilitated the *in situ* deposition of high quality $YBa_2Cu_3O_{7-x}$ films. These layers can also be deposited using laser vaporization from the parent compound in the same vacuum system that is used for deposition of the superconductor. Films grow epitaxially on the buffer layer although there is evidence for strain due to the large difference in thermal expansion between Si and $YBa_2Cu_3O_{7-x}$ (Fork *et al.* 1990). Tiwari *et al.* (1990) found that the interface between Si and $Y-ZrO_2$ is smooth whereas that between Si and $YBa_2Cu_3O_{7-x}$ is somewhat rougher with a scale size of about 10 nm. They found that this roughness has no detrimental effect on the growth of $YBa_2Cu_3O_{7-x}$ films.

The requirement for a low resistance contact between $YBa_2Cu_3O_{7-x}$ and Si led Kumar and Narayan (1991) to examine the use of $CoSi_2$ buffer layers. They found that some interdiffusion occurs between $YBa_2Cu_3O_{7-x}$ and $CoSi_2$ at the interface between these materials and that the resulting superconducting film is strained. However, films with a zero resistance temperature of 83 K and a width of the superconducting transition of 8 K were produced. The quality of the buffer layer was found to influence the sharpness of the superconducting transition.

Surprisingly, c axis oriented epitaxial growth of $YBa_2Cu_3O_{7-x}$ films has also been obtained under deposition conditions in which there is no direct path between the target and the deposition surface (Holzapfel *et al.* 1992). Off-axis deposition (Figure 7.3) occurs via diffusion of parent species out of the plume, resulting in droplet-free homogeneous

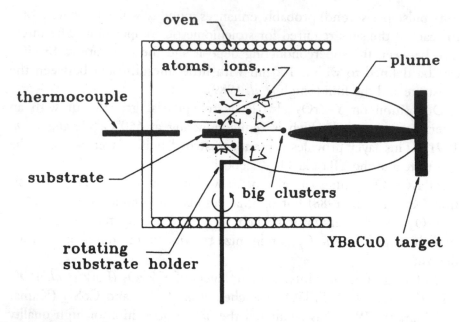

Figure 7.3. A schematic diagram of the off-axis laser deposition method
developed by Holzapfel *et al.* (1992).

films. Deposition rates of $0.3-1\,\mathrm{nm\,s^{-1}}$ were reported with the possibility of simultaneous deposition on two sides of the same substrate. A 115 nm thick $YBa_2Cu_3O_{7-x}$ film deposited on $SrTiO_3$ at 760°C under these conditions was found to have a critical current density of $2.4 \times 10^6\,\mathrm{A\,cm^{-2}}$ at 77 K and $T_c = 88$ K (Holzapfel *et al.* 1992).

It has been demonstrated (Ushida *et al.* 1991) that superconducting $YBa_2Cu_3O_{7-x}$ films can also be deposited on a substrate ([100] MgO) by laser-induced metal–organic chemical vapor deposition (LIMOCVD). Organometallic compounds of Y, Ba and Cu were irradiated with 193 nm excimer laser pulses during deposition on a MgO substrate maintained at 650–700°C. A control area was not exposed to laser radiation. Irradiated material was found to have a higher zero resistance temperature (72 K) than unirradiated material (45 K). In addition, irradiated surfaces were very smooth whereas unirradiated material was rough. Irradiated regions were also found to be strongly *a* axis oriented and of higher density than unirradiated material. A heating of the substrate/film by laser radiation was ruled out as the cause of these differences. Indeed, at a laser wavelength of 193 nm, a rich photochemistry is expected.

Superconducting films have also been prepared *in situ* using a variety of other laser sources (Lynds *et al.* 1988, Miura *et al.* 1988, Kwok *et al.* 1988, Koren *et al.* 1989). $YBa_2Cu_3O_{7-x}$ films prepared using pulses from a 10.6 μm TEA CO_2 laser at intensities of $1.5 \times 10^8\,W\,cm^{-2}$ were found to have zero resistance temperatures in the 83–86 K range (Miura *et al.* 1988). The substrate temperature was 400°C and preparation occurred in O_2. A notable difference between films prepared with 10.6 μm radiation and those deposited using excimer laser vaporization was that the former were found to be deficient in Y. Films of Sm–Ba–Cu–O and Er–Ba–Cu–O deposited using 10.6 μm laser pulses were also found to be deficient in Sm and Er, respectively. This difference may be due, in part, to the increased penetration depth for 10.6 μm versus 248 nm laser radiation in the parent materials. As a result, a larger volume of the target is vaporized per pulse. The excitation temperature might then be reduced relative to that obtained under irradiation at shorter wavelengths.

A comprehensive comparison of the effect of laser wavelengths in the 1064–193 nm range on deposition of $YBa_2Cu_3O_{7-x}$ films on (100) $SrTiO_3$ substrates has been reported by Koren *et al.* (1989). Rougher surfaces with an increase in particulates were observed when irradiation occurred at 1064 nm. In addition films deposited at this wavelength were thicker under standard conditions than those deposited using shorter wavelength laser radiation. The critical current density, J_c, at a given temperature was found to be a strong function of laser wavelength, with those films prepared at 355 nm having much higher values of J_c than those prepared at 1064 and 532 nm. In fact, a film prepared at 355 nm was found to have a value of J_c that exceeded that of one prepared by irradiation at 193 nm. All films were found to contain a significant concentration of particulates.

For many applications of high temperature superconductors it is desirable to have the superconductor deposited as a thin film on a metallic substrate. Preparation of such films is difficult because of interdiffusion between the substrate and the superconducting film. Nevertheless, high quality $YBa_2Cu_3O_{7-x}$ films have been deposited by laser vaporization onto a variety of metal substrates using buffer layers to minimize this problem (Witanachchi *et al.* 1990, Kumar *et al.* 1990, Narumi *et al.* 1991, Reade *et al.* 1991, 1992, Ogale *et al.* 1991). The use of a buffer layer yields improved *c* axis alignment, higher zero resistance temperature and higher critical current density than is possible with direct deposition in the absence of a buffer layer. On Hastelloy, textured films can be deposited with good c axis orientation perpendicular to the substrate (Kumar *et al.* 1990). These films had a zero resistance temperature of 86 K and a critical current density at 77 K of about $7 \times 10^3 \, A \, cm^{-2}$.

Narumi *et al.* (1991) found that a coating of Pt applied to the Hastelloy surface prior to deposition of the Y-ZrO$_2$ buffer layer further improved these properties and particularly the critical current density at high magnetic field strength. In a similar study, Ogale *et al.* (1991) found that deposition on a Y-ZrO$_2$–Ag composite film on 304 stainless steel yielded YBa$_2$Cu$_3$O$_{7-x}$ films with a transition temperature of 88 K and a critical current density of about 105 A cm^{-2} at 20 K. Such composite buffer layers were found to be more effective in promoting the formation of high quality superconducting films than a bilayer buffer of 240 nm of Ag followed by 240 nm of Y-ZrO$_2$. The properties of the superconducting film could be maintained even after the substrate had been subjected to bending through a radius of 2 cm. The pinhole density in films deposited on the bilayer buffer was about 10^6 cm^{-2} whereas few pinholes were observed for deposition on the composite layer.

7.3 VAPORIZATION PRODUCTS

Atomic, ionic and molecular species ejected from superconductor surfaces during laser ablation can be studied using optical spectroscopic (Auciello *et al.* 1988a, Weimer 1988, Girault *et al.* 1989a, Sakeek *et al.* 1994), direct imaging (Gupta *et al.* 1991, Geohagen 1992) or mass spectrometric (Chen *et al.* 1990, Fukushima *et al.* 1993, Izumi *et al.* 1991a,b, Becker and Pallix 1988) techniques. These observations showed that a variety of species exist in the plasma that accompanies ablation, and that the composition of the ejected material may change as material moves from the ablation source to the surface on which deposition occurs.

Optical spectra obtained from light emitted by the plume created by ablation of YBa$_2$Cu$_3$O$_{7-x}$ surfaces in vacuum show emission lines of Y, Y$^+$, Ba and Ba$^+$ as well as of Cu and Cu$^+$ and several molecules (Table 7.2). Emission from atomic oxygen is also seen near 777 nm (Girault *et al.* 1989a) and band emission from the simple diatomics, BaO, CuO and YO is observed. Similar spectra are seen from other target materials (Deshmukh *et al.* 1988, Girault *et al.* 1990).

The overall emission spectrum from YBa$_2$Cu$_3$O$_{7-x}$ is complex and depends on excitation conditions (Geyer and Weimer 1989) as well as

Table 7.2. *Wavelengths of strong emission features from the plasma accompanying ablation of* $YBa_2Cu_3O_{7-x}$ *in vacuum.*

Emitter	Wavelength (nm)
Ba	553.55
Ba$^+$	455.40
	649.69
Cu	578.21
Cu$^+$	490.97[a]
	493.16[a]
O	777.2
	777.41
	844.65
BaO	604
CuO	643
	649
YO	597
	599
	600
	616.5

[a] Uncertain identification (Ying *et al.* 1988).

on location in the plume (Girault *et al.* 1989a, Dyer *et al.* 1990). This complicates line identification and the assignment of individual spectral features to specific emitting species. In particular, the identification of Cu$^+$ from plasma spectra is uncertain because of contamination by emission from Y and Y$^+$ (Ying *et al.* 1988). However, other identifications are secure and have been verified by studies of emission spectra from plasmas created by ablation of individual CuO, BaCO$_3$ and Y$_2$O$_3$ targets (Geyer and Weimer 1989).

Girault *et al.* (1989a) have studied the temporal evolution of Cu and YO emission at different distances from a $YBa_2Cu_3O_{7-x}$ target irradiated at 248 nm and find that Cu emission decreases rapidly at distances greater than 1 mm. This behavior was also reported by Dyer *et al.* (1988). The emission from YO was found to consist of two temporal components when emission was recorded from points close to the target ($\lesssim 1$ mm). A fast component was identified with YO molecules ejected directly from the target, whereas the longer component which persisted at large distances from the target was associated with reactions in the plume (Girault *et al.* 1989a). In this case, emission could be associated with the formation of YO in the plume followed by chemiluminescence or by collisional excitation.

When several spectral lines are observed from the same species, an excitation temperature can often be obtained from quantitative measurements of line intensity together with the assumption of local thermodynamic equilibrium (LTE). Ying *et al.* (1988) have applied this method to obtain Boltzmann electronic excitation temperatures, T^*, for Ba, Cu and Y atoms in the plasma over YBCO. For Cu atoms, T^* ranged from about 8800 K just above the ablating surface to about 6800 K at the deposition surface 0.1 m away. T^* for Y atoms was found to be about 10^3 K lower under the same conditions. Excitation temperatures in this range have also been reported by Dyer *et al.* (1988).

A comprehensive study of translational rotational and vibrational energy distributions in YO after ablation from $YBa_2Cu_3O_{7-x}$ has been reported by Otis and Goodwin (1993). They find that excitation temperatures inferred from vibrational and rotational sequences in YO are in the 1000–1300 K range (about 0.1 eV). On the other hand, translational energies of YO under these conditions are found to be 2.4–6.5 eV confirming that a strong disequilibrium exists between translation and vibration/rotation excitation channels. The translational energy of YO is compatible with a pulsed supersonic adiabatic expansion.

The expansion velocity of individual ablated species can be obtained from time of flight (TOF) mass spectra (Izumi *et al.* 1991a,b, Fried *et al.* 1994) or by recording time-resolved emission at a specific distance from the irradiated area (Dyer *et al.* 1988). The expansion of the plasma as a whole can be followed with a streak camera (Dyer *et al.* 1990, Eryu *et al.* 1989) or by using a synchronized pulsed laser to image the ablation front (Gupta *et al.* 1991).

Expansion velocities for Cu, Y, Y^+ and Ba rise rapidly with laser fluence above threshold to $(1-1.2) \times 10^4 \, m\,s^{-1}$ at fluences of $2\,J\,cm^{-2}$ (Dyer *et al.* 1988). However, Izumi *et al.* (1991b) showed that the composition of the plasma is a strong function of irradiation history, with high molecular weight species being emitted after repetitive irradiation of the same area on the YBCO target. Retarding potential measurements indicate that atomic ions may have kinetic energies exceeding 200 eV, whereas molecular ions such as BaO^+ and YO^+ have energies of several tens of electronvolts (Izumi *et al.* 1991a). These energies would appear to arise from acceleration of ions by a local field, which develops as electrons with low mass rapidly expand away from the interaction volume (Izumi *et al.* 1991a). Venkatesan *et al.* (1988) have observed that the angular dependence of ejected material consists of a $\cos^n \theta$ component, where $n = 3-12$, compatible with thermal vaporization with a superimposed highly forward directed component that is consistent with a secondary emission mechanism.

Eryu et al. (1989) have used a streak camera with an image intensifier to obtain spatially resolved temporal measurements of plume intensity. These show that a highly forward directed fast component was dominant at low fluence ($0.14 J cm^{-2}$). A slower spatially broader component appeared at higher fluence and became dominant. Expansion velocities were found to be about $5 \times 10^4 m s^{-1}$ for the fast component and about $0.6 \times 10^4 m s^{-1}$ for the slow component. Mass spectra under these conditions showed that no atomic species were present at fluences $<0.4 J cm^{-2}$. However, clusters with significant mass were observed under these conditions. Such clusters undoubtedly interact further with the trailing edge of the excimer laser pulse during ejection (Singh and Viatella 1994). Eryu et al. (1989) found that atomic peaks together with large clusters appear in the mass spectrum at high fluence. They suggested that cluster decomposition is responsible for the generation of highly excited atomic, ionic and molecular fragments at high incident laser fluence.

Becker and Pallix (1988), in a study of $YBa_2Cu_3O_{7-x}$ irradiated at 1064 and 532 nm, also found that large clusters are ejected from the surface up to fluences that result in the initiation of a luminous plume. When the plume first becomes luminous, clusters with masses of up to 10^6 amu were observed. These then disappear at higher laser fluence as visible emission appears from the plume. This result again suggests that clusters may act as the precursor for atomic and molecular plume species, and that the primary dissociative interaction at high laser fluence occurs when incident laser radiation heats these clusters. In a related study Koren et al. (1990) reported that heating of clusters ejected from $YBa_2Cu_3O_{7-x}$ with a second laser pulse acts to reduce cluster size and improve the quality of deposited films.

Direct measurement of the stress pulse induced in the $YBa_2Cu_3O_{7-x}$ target during laser irradiation (Dyer et al. 1992) confirms this finding. Peak pressures as high as $2.5 \times 10^6 Pa$ lasting for about 20 ns (FWHM) were observed at the fluence of $0.25 J cm^{-2}$. The relation between surface pressure P_s, surface ablation velocity $\dot{x}$ and surface temperature T_s can be written (Dyer et al. 1992)

$$\dot{x} = \frac{\beta P_s}{\rho} \left(\frac{\bar{m}}{2\pi k T_s} \right)^{1/2} \tag{1}$$

where ρ is sample density, $\bar{m}$ is the average mass of a vaporizing species and k is the Boltzmann constant. β is a factor arising from redeposition of vaporized species, which was taken to be 1.37 by Dyer *et al.* (1992). The average mass of vaporizing species can be estimated from this expression when $\dot{x}$, P_s and T_s are known.

With $\dot{x} = 4 \times 10^{-2}$ µm per pulse, $\rho = 4.65 \times 10^3 \, \text{kg m}^{-3}$, $2700 \leq T_s \leq 5300 \, \text{K}$ and $P_s = 2.5 \times 10^6 \, \text{Pa}$ one obtains $\bar{m} = (1-2) \times 10^3$ amu. This mass is significantly larger than would be expected from a thermal vaporization source at $T_s = 2700$–$5300 \, \text{K}$. This result then again suggests that the initial vaporization mechanism involved the ejection of large clusters or droplets of material (Yang *et al.* 1991). Venkatesan *et al.* (1989) have used an additional laser to ionize clusters in the plume and have studied the mass spectrum of the products of this interaction. A quadrupled Nd:YAG laser (266 nm, 5 ns pulse) was used to yield intensities of $(5.5-12) \times 10^{13} \, \text{W m}^{-2}$ at the plume. Under these conditions, a rich spectrum of products was observed with evidence for many diatomic and suboxide species. Larger molecules such as YBa_2O_3, $BaCuO$, $YBaO_2$, $YBa_2O_6H_2$ and Ba_2O were found to be abundant under conditions for which any high mass neutral cluster would be dissociated.

A careful study of the relationship between ablation volume and fluence has been reported by Neifeld *et al.* (1991). With irradiation at 248 nm, they found that the ablation volume per pulse first rises rapidly with fluence above threshold, peaking at about $7.5 \times 10^{-13} \, \text{m}^3$ at $4 \, \text{J cm}^{-2}$. At higher fluences the ablated volume per pulse is found to decrease, reaching $4 \times 10^{-13} \, \text{m}^3$ per pulse at a fluence of $19 \, \text{J cm}^{-2}$. This effect is due to a rapid decrease in the ablated area per pulse at high fluence. The ablation depth per pulse continues to increase almost linearly with fluence up to $19 \, \text{J cm}^{-2}$.

Surprisingly, the fraction of material appearing at the detector in ionized form was found to decrease rapidly between fluences of 1–$2 \, \text{J cm}^{-2}$. Above $2 \, \text{J cm}^{-2}$ the ionization fraction was found to be approximately constant at about 0.2. It was suggested that this decrease in fractional ionization might arise from enhanced electron–ion recombination in the plasma formed over the surface at high fluences. However, geometrical effects might also be important since the angular dependence of emitted species becomes more diffuse at higher fluences (Eryu *et al.* 1989). Neifeld *et al.* (1991) also found that the average kinetic energy per particle tends to saturate (i.e. remain constant) for fluences exceeding $6 \, \text{J cm}^{-2}$.

The effect of ablation on the target itself has been investigated by Foltyn *et al.* (1991) and Cohen *et al.* (1991). Both experiments were conducted using 308 nm XeCl laser pulses at fluences of several $J\,cm^{-2}$. The morphology of the target surface was found to be greatly altered after repetitive pulsing under conditions that led to the deposition of proto-stoichiometric $YBa_2Cu_3O_{7-x}$ films. A dominant effect was found to be the creation of columnar structures in the irradiated target surface. The tops of these columns were observed to be rich in Y and were therefore resistant to ablation. Such a concentration of Y results in a strong decrease in vaporization rate with cumulative exposure of the target. However, this enrichment in the surface was not reflected in the composition either of the plasma or of the deposited YBCO layer. In their analysis of these effects, Cohen *et al.* (1991) concluded that the etched surface of the target never reaches a temperature at which thermal decomposition occurs. Since decomposition occurs above about 1050°C, surface temperatures probably never become this great. This observation suggests that Y enrichment effects are then the result of a true ablative photo-decomposition process and are not due to thermal heating. Redeposition of material from a Knudsen layer is also important under these conditions.

The development of the shock front associated with the expansion of ejected material into vacuum or into a gas has been studied by Scott *et al.* (1990), Dyer *et al.* (1990), Gupta *et al.* (1991), Singh *et al.* (1990) and by Geohagen (1992). At low ambient pressure or in vacuum two components are observed (Figure 7.4(a)–(f)). The first is attached to the target and persists for about 2 µs. Subsequently, it expands in a forward directed pattern with a constant leading edge velocity of about $10^4\,m\,s^{-1}$.

This morphology changes significantly in the presence of 100 mTorr of O_2 (Figures 7.4(g)–(l)). Under these conditions, a new component in the form of a contact front appears at the leading edge of the expanding plasma after about 1 µs. This becomes the dominant source of optical emission for times greater than 2 µs. As this shock front propagates into the ambient gas, it slows to about $0.06 \times 10^4\,m\,s^{-1}$ after 5 µs. Geohagen (1992) found that the expansion velocity of this shock front is well described by both of the following functions

$$d = at^{0.4} \tag{2}$$

$$d = x_f[1 - \exp(-\delta t)] \tag{3}$$

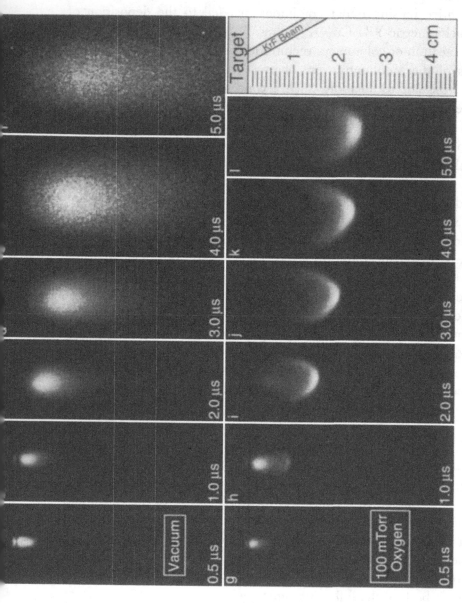

Figure 7.4. ICCD photographs of the visible plasma emission (exposure times 20 ns) following 1.0 J cm⁻² KrF/YBCO ablation into (a)–(f) 1 × 10⁻⁶ Torr and (g)–(l) 100 mTorr oxygen at the indicated delay times from the arrival of the laser pulse. The 0.2 cm × 0.2 cm 248 nm laser pulse irradiated the YBCO target at an angle of 30°, as shown (Geohagen 1992).

where d is the distance from the target, t is time and x_f, a and δ are empirically determined parameters. Under the conditions of the experiments performed by Geohagen (1992), $a = 1.26 \times 10^4\,(\mathrm{m\,s^{-1}})^{-0.4}$, $x_f = 0.03\,\mathrm{m}$ and $\delta = 0.36\,\mu\mathrm{s}^{-1}$. Similar results have been reported by Scott $et\ al.$ (1990).

Scott $et\ al.$ (1990) also found that the shock front can become unstable with the expansion velocity becoming angle-dependent and the front developing undulations, which are probably due to hydrodynamic instabilities along the interface with the denser surrounding medium. The pressure dependence of the shock wave expansion has been discussed by Gupta $et\ al.$ (1991) using the standard theory for blast wave formation (Zeldovich and Raiser 1986). This predicts that

$$R(t) \propto \left(\frac{E}{\rho_0}\right)^{1/5} t^{0.4} \tag{4}$$

for a spherical wave and

$$R(t) \propto \left(\frac{E}{\rho_0}\right)^{1/3} t^{0.67} \tag{5}$$

for a plane wave, where $R(t)$ is the time-dependent radius of the shock front far away from the point of explosion. E is the amount of energy released in the explosion, which in this case is the amount of energy released from ablation. ρ_0 is the ambient gas density. Such shocks are initiated when the thickness of the compressed gas layer exceeds the mean free path of molecules in the ambient gas. Under standard conditions the mean free path for collisions is $l \simeq 3.7 \times 10^{-6}/P$, where P is the gas pressure in atmospheres. Because l increases as P^{-1}, formation of a shock front is inhibited at low gas pressure. This effect can be seen in the plume morphology shown in Figure 7.4. Gupta $et\ al.$ (1991) found that $R \propto P^{-0.4}$ for expansion of the ablation products of $YBa_2Cu_3O_7$ into O_2 at higher pressure ($P \gtrsim 0.25\,\mathrm{atm}$). This is consistent with expansion of a quasi-planar shock front and is compatible with the strongly forward directed nature of the ablative interaction. In addition, the relatively larger focal area of the excimer laser beam on the target would tend to favor, at least initially, the creation of a planar expansion.

Dyer $et\ al.$ (1990) have obtained streak camera images of the ablation front in the presence of O_2. Their data (Figure 7.5) show that $R \propto t^{0.4}$

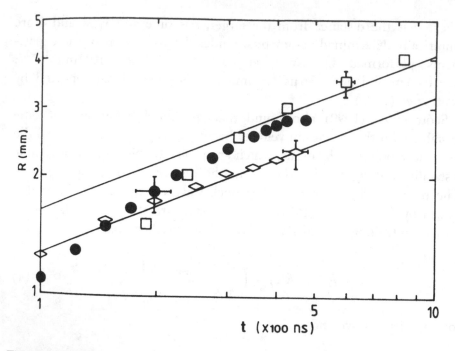

Figure 7.5. Distance (R)–time (t) plots of the luminous front for KrF laser ablation in oxygen derived from: ($\bullet$), streak photography: $P_0 = 6.5$ mbar and 5.5 J cm^{-2}; ($\square$), time-resolved recording of Y^+ emission: $P_0 = 6.5$ mbar and 6 J cm^{-2}; and ($\diamond$), streak photography: $P_0 = 20$ mbar and 5 J cm^{-2}. The solid lines have the form $R \propto t^{0.4}$ (Dyer *et al.* 1990).

at long times (about $0.5\,\mu$s). The larger slope observed at shorter times arises since the mass of ablated material is much larger than the mass of gas swept up in the shock. Under these conditions, the ambient gas is relatively unimportant and expansion proceeds as in vacuum. Wu *et al.* (1989) and Girault *et al.* (1989b) have studied the optical emission from the plasma formed by ablation of $YBa_2Cu_3O_7$ in oxygen. Wu *et al.* (1989) found that the pressure of O_2 has the effect of increasing the intensity of almost all emission lines, with emission from the slow velocity components exhibiting the strongest effect. Overall the intensities of atomic emission lines were found to scale as $P - kP^2$, where k is a constant, with O_2 pressure, whereas the intensity of molecular emission lines scaled as P. Wu *et al.* (1988) suggested that oxides may form in the plume via reactions between electronically excited O atoms and elemental metal species. They found that this enhanced oxidation is carried over into the deposited material, resulting in a higher transition temperature in samples not subjected to post-deposition annealing.

In general, the vaporization of superconducting material such as $YBa_2Cu_3O_7$ in response to intense excimer laser radiation can be described qualitatively by the following sequence (Figure 7.6).

1. Defect formation at fluences below the threshold for ablation (about $0.12\,J\,cm^{-2}$ (Dyer *et al.* 1992)). These defects may be created by electron–hole recombination and could result in the formation of vacancy aggregates, perhaps associated with trapped oxygen molecules. These precursor effects would be the incubator sites for further more substantial damage at higher fluence.

2. Target dissolution and cluster formation. This process would occur for fluences exceeding the threshold for ablation and results in the evolution of large clusters or shattered components of the target. Initiation of this process occurs at sites containing a high concentration of defects together with trapped gas. Expansion of this gas would act as the means by which clusters and fragments are ejected. Their motion is sharply directed away from the laser focal spot.

3. Heating and dissociation of clusters. At higher fluences ($\gtrsim 1\,J\,cm^{-2}$) clusters created by the leading edge of the excimer laser pulse are further heated by incident laser photons and dissociate. The products of their dissociation consist of a wide range of atomic, ionic and molecular species and form a Knudsen layer. They expand away from the point of interaction in all directions. Plasma chemistry leading to large molecules may become important under these conditions. In addition, charge separation leads to high electric fields and the acceleration of ions to high kinetic energies (about $200\,eV$).

4. Formation of a propagating shock front. This depends on laser fluence and ambient gas pressure. The shock front detaches from the initial plasma over the irradiated surface and propagates at supersonic speed ($\gtrsim 10\,km\,s^{-1}$) into the surrounding gas. As the shock front accumulates a larger mass of gas it slows with the displacement of the shock front from its point of origin given by an expression of the form $R(t) \simeq at^{0.4}$.

The overall effect is to transfer the chemical composition of the target to the deposition surface. Stoichiometry is maintained in this process since the primary mechanism for removal of material from the target appears to be removal in the form of large clusters, which

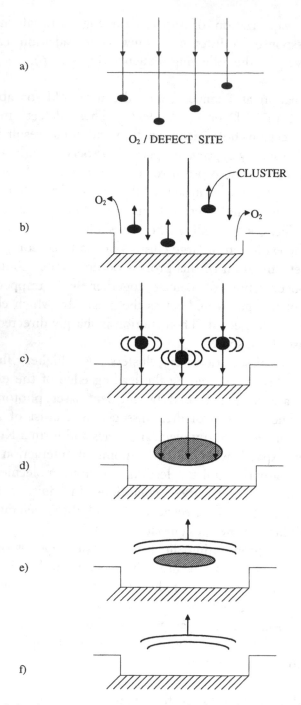

Figure 7.6. A schematic representation of the ablation/vaporization sequence in YBCO. (a) Incubation and defect formation. (b) Ejection of clusters. (c) Heating and dissociation of clusters, to form a Knudsen layer. (d) Plasma formation and heating. (e) Formation of shock front. (f) Detachment and propagation of shock front.

presumably maintain the composition of the target (Yang *et al.* 1991). Redeposition from a Knudsen layer may also contribute to this effect. Metal-rich regions in the target surface remaining after prolonged ablation then presumably are the locations where vacancy formation has led to oxygen depletion. Once formed, such regions would be resistant to subsequent ablation since they would lack the oxygen required to initiate the explosive removal of clusters from the surface.

7.4 PATTERNING AND ETCHING

The fabrication of thin film superconducting devices and the use of superconducting material as interconnects in microelectronic circuits requires that a technique be developed for patterning and etching of deposited films. Laser ablation is an attractive candidate for this application. A number of studies have demonstrated that dry etching with excimer or other laser radiation is feasible and is capable of providing highly resolved structures on the surface of $YBa_2Cu_3O_{7-x}$ films (Imam *et al.* 1987, Scheurmann *et al.* 1987, Auciello *et al.* 1988a,b, Mannhart *et al.* 1988, Savva *et al.* 1989, Singh *et al.* 1990, Heitz *et al.* 1990, Schwab *et al.* 1991, Shen *et al.* 1991, von der Burg *et al.* 1992).

Patterning can be accomplished during deposition through the use of a contact mask. For example, Singh *et al.* (1989) used a steel mask for deposition of $YBa_2Cu_3O_{7-x}$ films on $SrTiO_3$ substrates. They found that the superconducting properties of the films deposited depend on the size of the mask. However, for transverse dimensions greater than 400 μm, no differences between microstructural and superconducting properties were observed.

Savva *et al.* (1989) reported on the deposition of 85 μm × 85 μm squares of $YBa_2Cu_3O_{7-x}$ on MgO using a Cu contact mask. They found that the superconducting phase transition occurs at about 80 K whereas the zero resistance temperature is near 40 K. These deposits could be prepared with micrometer resolution.

An alternative approach is to use a mask for dry etching using laser ablation. Savva *et al.* (1989) showed that a chrome-on-quartz mask can be utilized to etch patterns with micrometer-sized resolution over an area as large as $1 \, cm^2$. The limit to resolution was provided by the diffraction of light by the edges of the mask.

Important parameters for dry etching include the etch depth per pulse, the fluence dependence of the etching rate, and the linearity of

Table 7.3. *Etch depth per pulse at 248 nm for* $YBa_2Cu_3O_{7-x}$ *films.*

Etch depth (μm/pulse)	Fluence (J cm^{-2})	Substrate	Reference
0.1	0.2	Al$_2$O$_3$	Imam *et al.* (1987)
5.6	0.3		
8.2	0.7		
15.1	3.8		
0.18	2.2	(100) MgO	Heitz *et al.* (1990)
0.25	4.2		
0.09	1.5	Silica	Savva *et al.* (1989)

the total etch depth with number of overlapping laser pulses. All of these parameters are expected to be wavelength-dependent.

Heitz *et al.* (1990) found that the ablation of thin films of $YBa_2Cu_3O_{7-x}$ on (100) MgO is characterized by a threshold fluence at 248 nm of $0.04 \pm 0.01 \, \text{J cm}^{-2}$. For fluences in the range 0.27–0.75 J cm^{-2} non-stoichiometric ablation is observed, with the unevaporated material being rich in Y. A similar effect has also been reported by Foltyn *et al.* (1991). When the fluence exceeded 0.75 J cm^{-2} stoichiometric ablation was observed. In the stoichiometric (high fluence) range, the etch depth was found to increase linearly with the number of laser pulses. Table 7.3 summarizes some empirical values of etch depth per pulse for ablation of $YBa_2Cu_3O_{7-x}$ films using 248 nm laser radiation.

Figure 7.7 shows the fluence dependence of the etching rate at 248 nm for $YBa_2Cu_3O_{7-x}$ films on fused silica. These data suggest that the rate of ablation may depend significantly on laser wavelengths, with the largest etch depth per pulse occurring for ablation with 193 nm radiation. However, the substrate may also play a rôle in determining etch depth if laser radiation can penetrate through the superconductor film. For example, Table 7.3 shows that the etch depth per pulse at 248 nm is significantly less for films deposited on silica than for those deposited on Al$_2$O$_3$ or on MgO.

In general, the etch depth per pulse has been found to depend on $\ln E$, where E is the laser fluence (Imam *et al.* 1987). This suggests a Beer–Lambert law for attenuation of laser radiation in the film. Then

$$x(E) = \frac{1}{\alpha} \ln \left(\frac{E}{E_{TH}} \right) \tag{6}$$

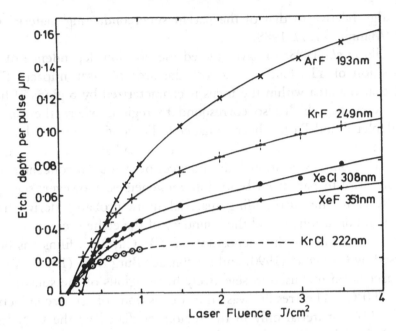

Figure 7.7. Etch depth per pulse or fluence for $YBa_2Cu_3O_{7-x}$ films on fused silica with ablation at several excimer laser wavelengths (after Savva *et al.* 1989).

where $x(E)$ is the etch depth per pulse at fluence E and E_{TH} is the threshold fluence. α is the absorption coefficient at the laser wavelength. From an analysis of their ablation data at 248 nm, Imam *et al.* (1987) concluded that $\alpha(248\,\text{nm}) = 2.3 \times 10^5\,\text{cm}^{-1}$. Savva *et al.* (1987) estimated α at several laser wavelengths and found that α at a given wavelength is a strong function of post-deposition annealing conditions. For example, at 308 nm they found $\alpha(308\,\text{nm}) = 1.35 \times 10^4\,\text{cm}^{-1}$ before annealing and $\alpha(308\,\text{nm}) = 1.75 \times 10^4\,\text{cm}^{-1}$ after annealing. They also concluded that the etch depth per pulse does not follow the wavelength dependence of the absorption coefficient. This suggests that photochemistry may also play a part in the excitation and ablation process.

During ablation, material ejected from the laser focus is redeposited to some extent over adjacent areas of the target. In addition, material at the boundary of the ablated area can be melted. Mannhart *et al.* (1988) find that this melted region can extend about 0.5 µm into the adjacent unirradiated material. Nevertheless, a bridge of about 1 µm width could be created in superconducting $YBa_2Cu_3O_{7-x}$ using laser ablation. No difference in the resistance or temperature properties of the material in the bridge relative to that of the parent superconducting material was noted (Mannhart *et al.* 1988).

Auciello *et al.* (1988a,b) have studied the position dependence of the composition of $YBa_2Cu_3O_{7-x}$ across the area of laser impact. They found that the area within the focus is characterized by regions of high oxygen content, which also correspond to regions where the Cu has been depleted and Ba has been enhanced. The region surrounding the primary laser impact area was found to have an enhanced Cu concentration. These complex positional dependences of elemental concentrations probably arise in part from beam inhomogeneities. However, segregation effects in the surrounding material more probably derive from melting and diffusion around the boundary of the etched region.

In situ patterning during deposition of $YBa_2Cu_3O_{7-x}$ films has been discussed by Chu *et al.* (1990) and by von der Burg *et al.* (1992). A CW CO_2 laser beam was used to selectively heat regions of the substrate to about 700°C. The result was the deposition of superconducting $YBa_2Cu_3O_{7-x}$ material only in the region irradiated by the CO_2 laser beam. This high quality superconducting material was surrounded by an intermediate region of lower quality superconductor $(T(R=0) < 90\,K)$. Outside this region only semiconducting material was deposited. Spatial resolution was obtained by imaging a mask in the CO_2 laser beam. However, diffusion of heat from the irradiated area over the time required for deposition (about 8 min) resulted in a further degradation in edge quality. It is likely that this resolution could be improved by careful design of this CO_2 laser beam profile as well as mask geometry.

Because the conductive properties of $YBa_2Cu_3O_{7-x}$ depend on the oxygen content of deposited films, modification of this content offers the possibility of creating semiconducting lines or elements in otherwise superconducting material. Dye *et al.* (1990) and Shen *et al.* (1991) have shown how the oxygen content of $YBa_2Cu_3O_{7-x}$ films can be modulated by post-deposition irradiation with 488/515.5 nm Ar^+ laser light. Irradiation in vacuum or in an N_2 atmosphere was shown to yield semiconducting material because oxygen is depleted through laser heating. The resolution was 4 μm. The transition from superconductor to semiconductor could be reversed by reirradiation in the presence of an atmosphere of oxygen gas. Optimal intensities for writing and rewriting were found to be in the range $(0.35-1.0) \times 10^6\,W\,cm^{-2}$ and scan speeds were typically $5\,\mu m\,s^{-1}$. This process may be useful in the production of semiconducting weak links for superconducting quantum interference devices.

Other experiments (Singh *et al.* 1991, Harkness and Singh 1994) have utilized radiation from a second excimer laser ($\lambda = 308$ nm) for post deposition processing of $YBa_2Cu_3O_{7-x}$ films formed *in situ* by vaporization at 248 nm. This treatment performed at low fluence ($\lesssim 100$ mJ cm^{-2}), was found to improve the crystallinity of the film with a resultant increase (by about 15%) in the critical current density at 77 K. Under optimum conditions, the critical current density was increased from 5×10^6 to 5.6×10^6 A cm^{-2}. However, when the fluence of the laser beam at 308 nm exceeded the threshold for melting, the critical current density dropped dramatically.

7.5 DEVICE FORMATION

Although there has been much emphasis on the deposition of superconducting films for large scale applications such as coatings on wires and on microwave components, many commercial applications of this technology may derive instead from the growth of infrared detectors or superconducting quantum interference devices (SQUIDS). A key element in the development of these applications in the formation of the Josephson junction. These junctions exhibit a current at zero voltage, or alternatively, when biased, act as emitters of electromagnetic radiation. A common construction for such devices consists of a 'weak link' connecting two superconducting regions. The weak link can be formed by providing a normal conducting path (metal or semiconductor) between two superconducting elements. Such paths exist naturally as grain boundaries in materials such as $YBa_2Cu_3O_{7-x}$. However, emphasis has been placed on the use of laser techniques to optimize such boundary conditions (Char *et al.* 1991) or to form junctions directly using laser deposition (Koren *et al.* 1991) or ablation (Mannhart *et al.* 1988) techniques. Post-deposition processing using focused Ar$^+$ laser radiation to create non-superconducting tracks has also been used to create such interconnecting paths (Dye *et al.* 1990, Shen *et al.* 1991).

Control over grain boundary conditions so as to produce the required grain boundaries along photolithographically defined lines has been achieved by Char *et al.* (1991) using different epitaxial growth on laser-deposited seed and buffer layers. The geometry is as shown in Figure 7.8(a).

Epitaxial relations for the growth plane in this geometry are $SrTiO_3[110]//Al_2O_3[11\bar{2}0]$ to the right-hand side of the interface and

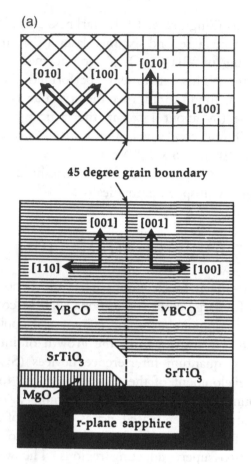

(a)

45 degree grain boundary

Figure 7.8. The geometry of the junction region in $YBa_2Cu_3O_{7-x}$ DC SQUID devices as developed by (a) Char *et al.* (1991) and (b) Koren *et al.* (1991).

$SrTiO_3[100]//MgO[100]//Al_2O_3[11\bar{2}0]$ to the left-hand side. The growth of $YBa_2Cu_3O_{7-x}$ on this layered structure then occurs epitaxially with the appearance of a 45° grain boundary at the interface. Fabrication of a dc SQUID device with this material shows modulation at temperatures as high as 88 K. Typical current densities achievable with these junctions were reported to be 10^2–10^3 A cm^{-2} at 77 K.

Figure 7.8(b) shows the geometry adopted by Koren *et al.* (1991). The first $YBa_2Cu_3O_{7-x}$ layer was created on $SrTiO_3(100)$ by laser ablation with a substrate temperature of 730°C in 0.2 Torr of O_2. BaF_2 was then deposited as shown, using laser ablation. This was followed by the creation of a barrier layer via a plasma discharge in CF_4 gas. The $YBa_2Cu_3O_{7-x}$ overlayer was then deposited on top of this spectrum. A narrow junction was then created by post-processing using laser

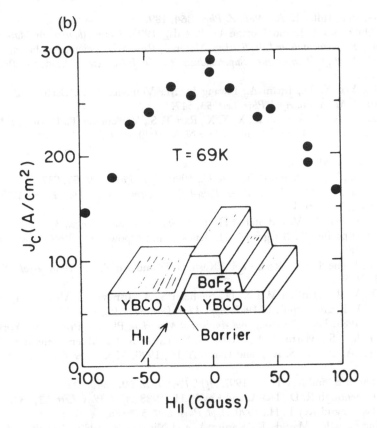

(b)

ablation. This yielded a 10 μm × 60 μm microbridge with the junction at its center. The junctions formed exhibited a critical current density of 300 A cm^{-2} at 70 K. The current–voltage curve of these junctions at 67 K when irradiated at 10.7 GHz showed voltage steps (Shapiro steps) in integral multiples of $h\nu/(2l)$, where $\nu = 10.7$ GHz. Suppression of the critical current by a magnetic field was observed in the region of low critical current. More complex SQUID devices incorporating flux transformers have been created in laser deposited devices using ion milling and photolithography (Oh *et al.* 1991).

7.6 REFERENCES

Auciello O., Athavale S., Hankins O. E., Sito M., Schreiner A. F and Biunno N., 1988a. *Appl. Phys. Lett.* **53**, 72.
Auciello O., Krauss A. R., Santiago-Aviles J., Schreiner A. F. and Gruen D. M., 1988b. *Appl. Phys. Lett.* **52**, 239.
Bäuerle D., 1986. *Chemical Processing with Lasers*, Springer-Verlag, New York.
Becker C. H. and Pallix J. B., 1988. *J. Appl. Phys.* **64**, 5152.

Bednorz J. G. and Müller K. A., 1986. *Z. Phys.* **B64**, 189.

Braren B., Dubowski J. J. and Norton D. P. (eds), 1993. *Laser Ablation in Materials Processing: Fundamentals and Applications*, Materials Research Society, Pittsburgh.

Burns G., 1992. *High Temperature Superconductivity. An Introduction*, Academic Press, New York.

Chang C. C., Wu X. D., Imam A., Hwang D. M., Venkatesan T., Barboux P. and Tarascom J. M., 1988. *Appl. Phys. Lett.* **53**, 517.

Chang C. C., Wu X. D., Ramesh R., Xi X. X., Rari T. S., Venkatesan T., Hwang D. M., Muenchausen R. E., Folteyn S. and Nogar N. S., 1990. *Appl. Phys. Lett.* **57**, 1814.

Char K., Colclough M. S., Garrison S. M., Newman N. and Zaharchuk G., 1991. *Appl. Phys. Lett.* **59**, 733.

Chen C. H., Murphy T. M. and Phillips R. C., 1990. *Appl. Phys. Lett.* **57**, 937.

Chrisey D. B. and Hubler G. K., 1994. *Pulsed Laser Deposition of Thin Films*, Wiley-Interscience, New York.

Chu W., Przychowski M. V. and Stafast H., 1990. *Supercond. Sci. Technol.* **3**, 497.

Cohen A., Allenspacher P., Brieger M. M., Jeuck I. and Opower H., 1991. *Appl. Phys. Lett.* **59**, 2186.

Deshmukh S., Rothe E. W., Reck G. P., Kushida T. and Xu Z. G., 1988. *Appl. Phys. Lett.* **53**, 2698.

Dijkkamp D., Venkatesan T., Wu X. D., Shaheen S. A., Jisrawi N., Min-Lee Y. H., McLean W. L. and Croft M., 1987. *Appl. Phys. Lett.* **51**, 619.

Duley W. W., 1983. *Laser Processing and Analysis of Materials*, Plenum Press, New York.

Dye R. C., Foltyn S., Martin J. A., Nogar N. S., Estler R. C., Muenschausen R. E., Mukherjee A., Brueck S. R. J. and Carim A. H., 1990. *Mater. Res. Soc. Symp. Proc.* **191**, 193.

Dyer P. E., Farrar S. and Key P. H., 1992. *Appl. Phys. Lett.* **60**, 1890.

Dyer P. E., Greenough R. D., Issa A. and Key P. H., 1988. *Appl. Phys. Lett.* **53**, 534.

Dyer P. E., Issa A. and Key P. H., 1990. *Appl. Phys. Lett.* **57**, 186.

Eryu O., Murakami K., Masuda K., Kasuya A. and Nishina Y., 1989. *Appl. Phys. Lett.* **54**, 2716.

Foltyn S. R., Dye R. C., Ott K. C., Peterson E., Hubbard K. M., Hutchinson W., Muenshausen R. E., Estler R. C. and Wu X. D., 1991. *Appl. Phys. Lett.* **59**, 594.

Fork D. K., Boyce J. B., Ponce F. A., Johnson R. I., Anderson G. B., Connell G. A. N., Eom C. B. and Geballe T. H., 1988. *Appl. Phys. Lett.* **53**, 337.

Fork D. K., Fenner D. B., Barton R. W., Phillips J. M., Connell G. A. N., Boyu J. B. and Gebralle T. H., 1990. *Appl. Phys. Lett.* **57**, 1163.

Fried D., Jodeh S., Beck G. P., Rothe E. W. and Kushida T., 1994. *J. Appl. Phys.* **75**, 522.

Fukushima K., Kanke Y. and Morishita T., 1993. *J. Appl. Phys.* **74**, 6948.

Geohagen D. B., 1992. *Appl. Phys. Lett.* **60**, 2732.

Geyer T. J. and Weimer W. A., 1989. *Appl. Phys. Lett.* **54**, 469.

Girault G., Damiani D., Aubreton J. and Catherinot A., 1989a. *Appl. Phys. Lett.* **55**, 182.

Girault G., Damiani D., Aubreton J. and Catherinot A., 1989b. *Appl. Phys. Lett.* **54**, 2035.

Girault G., Damiani D., Champeaux C., Marchet P., Mercurio J. P., Aubreton J. and Catherinot A., 1990. *Appl. Phys. Lett.* **56**, 1472.

Gupta A., Braren B., Casey K. G., Hussey B. W. and Kelly R., 1991. *Appl. Phys. Lett.* **59**, 1302.

Gupta A. and Hussey B. W., 1991. *Appl. Phys. Lett.* **58**, 1211.

Gupta A., Hussey B. W., Kussmaul A. and Segmüller A., 1990. *Appl. Phys. Lett.* **57**, 2365.

Gupta A., Koren G., Tsuei C. C., Segmüller A. and McGuire T. R., 1989. *Appl. Phys. Lett.* **55**, 1795.

Harkness S. D. and Singh R. K., 1994. *J. Appl. Phys.* **75**, 669.

Heitz J., Wang X. Z., Schwab P., Bäuerle D. and Schultz L., 1990. *J. Appl. Phys.* **68**, 2512.

Holzapfel B., Roas B., Schultz L., Bauer P. and Seamann-Ischerko G., 1992. *Appl. Phys. Lett.* **61**, 3178.

Hughes R. A., Lu Y., Timusk T. and Preston J. S., 1991. *Appl. Phys. Lett.* **58**, 762.

Hwang D. M., Nazar L., Venkatesan T. and Wu X. D., 1988. *Appl. Phys. Lett.* **52**, 1834.

Hwang D. M., Ying Q. Y. and Kwok H. S., 1991. *Appl. Phys. Lett.* **58**, 2429.

Imam A., Hegde M. S., Wu X. D., Venkatesan T., England P., Miceli P. F., Chase E. W., Chang C. C., Tarascon J. M. and Wachtman J. B., 1988. *Appl. Phys. Lett.* **53**, 908.

Imam A., Wu X. D., Venkatesan T., Ogale S. B., Chang C. C. and Dijkkamp D., 1987. *Appl. Phys. Lett.* **51**, 1112.

Izumi H., Ohata K., Sawada T., Morishita T. and Tanaka S., 1991a. *Appl. Phys. Lett.* **59**, 597.

Izumi H., Ohata K., Sawada T., Morishita T. and Tanaka S., 1991b. *Appl. Phys. Lett.* **59**, 2950.

Kim B. F., Bohandy J., Phillips T. E., Green W. J., Agnostinelli E., Adrian F. J. and Moorjani K., 1988. *Appl. Phys. Lett.* **53**, 321.

Kim H. S. and Kwok H. S., 1992. *Appl. Phys. Lett.* **61**, 2234.

Koren G., Aharoni E., Polturak E. and Cohen D., 1991. *Appl. Phys. Lett.* **58**, 634.

Koren G., Baseman R. J., Gupta A., Lutwyche M. I. and Laibowitz R. B., 1990. *Appl. Phys. Lett.* **56**, 2144.

Koren G., Gupta A., Baseman R. J., Lutwyche M. I. and Laibowitz R. B., 1989. *Appl. Phys. Lett.* **55**, 2450.

Koren G., Polturak E., Fisher B., Cohen D. and Kimel G., 1988. *Appl. Phys. Lett.* **53**, 2330.

Kumar A., Ganapathi L., Kanetkar S. M. and Narayan J., 1990. *Appl. Phys. Lett.* **57**, 2594.

Kumar A., Ganapathi L. and Narayan J., 1990. *Appl. Phys. Lett.* **56**, 2034.

Kumar A. and Narayan J., 1991. *Appl. Phys. Lett.* **59**, 1785.

Kussmaul A., Moodera J. S., Tedrow P. M. and Gupta A., 1992. *Appl. Phys. Lett.* **61**, 2715.

Kwok H. S., Mattocks P., Shi L., Wang X. W., Witanachchi S., Ying Q. Y., Zheng J. P. and Shaw D. T., 1988. *Appl. Phys. Lett.* **52**, 1825.

Liou S. H., Aylesworth K. D., Ianno N. J., Johs B., Thompson D., Meyer D., Woollam J. A. and Barry C., 1989. *Appl. Phys. Lett.* **54**, 760.

Lynds L., Weinberger B. R., Peterson G. G. and Krasinski H. A., 1988. *Appl. Phys. Lett.* **52**, 320.

Mannhart J., Scheuermann M., Tsuei C. C., Oprysho M. M., Chi C. C., Umbach C. P., Koch R. H. and Miller C., 1988. *Appl. Phys. Lett.* **52**, 127.

Miura S., Yoshitake T., Satoh T., Miyasaka Y. and Shohata N., 1988. *Appl. Phys. Lett.* **52**, 1008.

Moon B. M., Platt C. E., Schweinfurth R. A. and van Harlingen D. J., 1991. *Appl. Phys. Lett.* **59**, 1905.

Narayan J., Biunno N., Singh R., Holland O. W. and Auciello O., 1987. *Appl. Phys. Lett.* **51**, 1845.

Narumi E., Song L. W., Yang F., Patel S., Kao Y. H. and Shaw D. T., 1991. *Appl. Phys. Lett.* **58**, 1202.

Neifeld R. A., Gunapala S., Liang C., Shaheen S. A., Croft M., Priu J., Simons D. and Hill W. T., 1988. *Appl. Phys. Lett.* **53**, 703.

Neifeld R. A., Potenziani E., Sinclair W. R., Hill W. T., Turner B. and Pinkas A., 1991. *J. Appl. Phys.* **69**, 1107.

Norton D. P., Lowndes D. H., Budai J. D., Christen D. K., Jones E. C., Lay K. W. and Tkaczyk J. E., 1990. *Appl. Phys. Lett.* **57**, 1164.

Ogale S. B., Koinkar V. N., Visanathau R., Roy S. D. and Kanetkar S. M., 1991. *Appl. Phys. Lett.* **59**, 1908.

Oh B., Koch R. H., Gallagher W. J., Robertazzi R. P. and Eidelloth W., 1991. *Appl. Phys. Lett.* **59**, 123.

Otis C. E. and Goodwin P. M., 1993. *J. Appl. Phys.* **73**, 1957.

Reade R. P., Berdahl P., Russo R. E. and Garrison S. M., 1992. *Appl. Phys. Lett.* **61**, 2231.

Reade R. P., Mao X. L. and Russo R. E., 1991. *Appl. Phys. Lett.* **59**, 739.

Roas B., Schultz L. and Endres G., 1988. *Appl. Phys. Lett.* **53**, 1557.

Sakeek H. F., Morrow T., Graham W. G. and Walmsley D. G. 1994. *J. Appl. Phys.* **75**, 1138.

Sanchez F., Varela M., Queralt X., Aguiar R. and Morenza J. L., 1992. *Appl. Phys. Lett.* **61**, 2228.

Savva N., Williams K. F., Davis G. M. and Gower M. C., 1989. *J. Quant. Electron.* **25**, 2399.

Scheuermann M., Chi C. C., Tsuei C. C., Yee D. S., Cuomo J. J., Laibowitz R. B., Koch R. H., Braren B., Srinivasan R. and Pulechaty M. M., 1987. *Appl. Phys. Lett.* **51**, 1951.

Schwab P., Heitz J., Proyer S. and Bäuerle D., 1991. *Appl. Phys.* A**53**, 282.

Scott K., Huntley J. M., Phillips W. A., Clarke J. and Field J. E., 1990. *Appl. Phys. Lett.* **57**, 922.

Shen Y. Q., Freltoft T. and Vase P., 1991. *Appl. Phys. Lett.* **59**, 1365.

Singh R. K., Bhattacharya D., Harkness S., Narayan J., Diwari P., Jahnacke C., Sparks R. and Paesler M., 1991. *Appl. Phys. Lett.* **59**, 1380.

Singh R. K., Holland O. W. and Narayan J., 1990. *J. Appl. Phys.* **68**, 233.

Singh R. K., Narayan J., Singh A. K. and Krishnaswanny J., 1989. *Appl. Phys. Lett.* **54**, 2271.

Singh R. K. and Viatella J., 1994. *J. Appl. Phys.* **75**, 1204.

Tang Y. Q., Chen K. Y., Chan I. N., Chen Z. Y., Shi Y. J., Salamo G. J., Chan F. T. and Sheng Z. Z., 1993. *J. Appl. Phys.* **74**, 4259.

Tiwari P., Kanetkar S. M., Sharan S. and Narayan J., 1990. *Appl. Phys. Lett.* **57**, 1578.

Ushida T., Higa H., Higashiyama K., Hirabayashi I. and Tanaka S., 1991. *Appl. Phys. Lett.* **59**, 860.

Venkatesan T., Chase E. W., Wu X. D., Imam A., Chang C. C. and Shokoohi F. K., 1988a. *Appl. Phys. Lett.* **53**, 243.

Venkatesan T., Wu X., Imam A., Chang C. C., Hegde M. S. and Datta B., 1989. *IEEE J. Quant. Electron.* **25**, 2388.

Venkatesan T., Wu X. D., Imam A. and Wachtman J. B., 1988b. *Appl. Phys. Lett.* **52**, 1193.

von der Burg E., Diegel M., Stafast H. and Grill W., 1992. *Appl. Phys.* A**54**, 373.

Weimer W. A., 1988. *Appl. Phys. Lett.* **52**, 2171.

Wiener-Avnear E., Kerber G. L., McFall J. E., Spargo J. W., and Toth A. G., 1990. *Appl. Phys. Lett.* **56**, 1802.

Witanachchi S., Kwok H. S., Wang X. W. and Shaw D. T., 1988. *Appl. Phys. Lett.* **53**, 234.

Witanachchi S., Patel S., Kwok H. S. and Shaw D. T., 1989. *Appl. Phys. Lett.* **54**, 578.

Witanachchi S., Patel S., Zhu Y. Z., Kwok H. S. and Shaw D. T., 1990. *J. Mater. Res.* **5**, 717.

Wu M. K., Ashburn J. R., Torng C. J., Hor P. H., Meng R. L., Gao L., Huang Z. J., Wang Y. Q. and Chu C. W., 1987a. *Phys. Rev. Lett.* **58**, 908.

Wu X. D., Dijkkamp D., Ogale S. B., Imam A., Chase E. W., Miceli P. F., Chang C. C., Tarascon J. M. and Venkatesan T., 1987b. *Appl. Phys. Lett.* **51**, 861.

Wu X. D., Dutta B., Hegde M. S., Imam A., Venkatesan T., Chase E. W., Chang C. C. and Howard R., 1989. *Appl. Phys. Lett.* **54**, 179.

Wu X. D., Imam A., Venkatesan T., Chang C. C., Chase E. W., Barboux P., Tarascon J. M. and Wilkens B., 1988. *Appl. Phys. Lett.* **52**, 754.

Yang Y. A., Xia P., Junkin A. L. and Bloomfield L. A., 1991. *Phys. Rev. Lett.* **66**, 1205.

Ying Q. Y., Shaw D. T. and Kwok H. S., 1988. *Appl. Phys. Lett.* **53**, 1762.

Zeldovich Y. B. and Raiser Y. P., 1986. *Physics of Shock Waves and High Hydrodynamics Phenomena*, Academic Press, New York.

CHAPTER 8

Interactions and effects in semiconductors

8.1 LASER SPUTTERING

Direct removal of semiconductor material by laser light without a reactive intermediary has been reported for a number of solids (see reviews by Bäuerle 1986 and Ashby 1991). Direct ablation of small quantities of semiconductor material using UV laser radiation has specific applications in link breaking for circuit restructuring (Smith *et al.* 1981, Raffel *et al.* 1985). However, the majority of ablation studies have been carried out in a reactive atmosphere or medium.

A summary of work carried out on the direct ablation or photo-sublimation of semiconductor materials is given in Table 8.1. In general ablative effects are limited to excitation with high intensity pulses whereas photosublimation occurs by exposure of a semiconductor to CW, visible laser radiation at somewhat lower intensity. Figure 8.1 shows the intensity threshold for ablation in several semiconductors.

Early experiments on the ablation of Si at 193 and 248 nm (Shinn *et al.* 1986) showed that neither the ablation rate nor the ablation threshold fluence were strongly dependent on laser wavelength, for excitation in this wavelength range. Above the threshold fluence the etching rate was found to be 0.2–0.5 µm per pulse at intensities of 10^7–10^8 W cm^{-2}. The threshold fluence was about 1.3 J cm^{-2} both at 193 and at 248 nm and was found to be independent of ambient gas pressure for pressures between 0–1000 Torr. A strong plasma emission showing a wide variety of Si, Si$^+$ and Si^{2+} spectral lines was observed at the highest intensities used (about 10^8 W cm^{-2}). These spectral features decayed over a time scale of about 420 ns and were pressure-broadened, suggesting that strong plasma heating was present. The Si surface temperature at the threshold for ablation was estimated to be 2700–3000°C. A time-of-flight

Table 8.1. *A summary of experimental studies of direct etching of semiconductors.*

Material	Wavelength (nm)	Reference
CdS	488	Rothschild *et al.* (1987)
	337	Namiki *et al.* (1986)
	248	Brewer *et al.* (1991)
CdTe	458–515	Uzan *et al.* (1984)
		Arnone *et al.* (1986)
	488	Rothschild *et al.* (1987)
	248	Brewer *et al.* (1990, 1991)
GaAs	694.3	Pospieszczyk *et al.* (1983)
	540	Nakayama *et al.* (1984)
	193–351	Davis *et al.* (1988)
	337	Namiki *et al.* (1991)
	193	Herman *et al.* (1992)
GaInAsP	193	Herman *et al.* (1992)
GaN	337	Namiki *et al.* (1991)
GaP	580–610	Okano *et al.* (1991)
	540	Nakayama *et al.* (1984)
	540–920	Hattori *et al.* (1991, 1992)
	600	Georgiev and Singh (1992)
	337	Namiki *et al.* (1991)
Ge	248	Bialkowski *et al.* (1990)
HgCdTe	488	Rothschild *et al.* (1987)
InP	540	Nakayama *et al.* (1984)
Si	248	Hayasaka *et al.* (1986)
	193, 248	Shinn *et al.* (1986)
	649.3	Pospieszczyk *et al.* (1983)
	248	Bialkowski *et al.* (1990)
	530	Hanabusa (1993)
ZnO	337	Nakayama *et al.* (1982)
		Nakayama (1983)
ZnS	308	Arlinghaus *et al.* (1989)

mass spectrometric study of Si emission following heating with a 20 ns ruby laser pulse ($\lambda = 694.3$ nm) at fluences in the $1.2\,\mathrm{J\,cm^{-2}}$ range was found to yield kinetic temperatures in the range 1700–2400°C, which would support a quasi-equilibrium vaporization process (Stritzker *et al.* 1981) in which Si atoms are emitted from the Si surface at temperatures close to the normal boiling point (2477°C). However, a study of the time dependence of charge emission from Si irradiated with 248 nm pulses

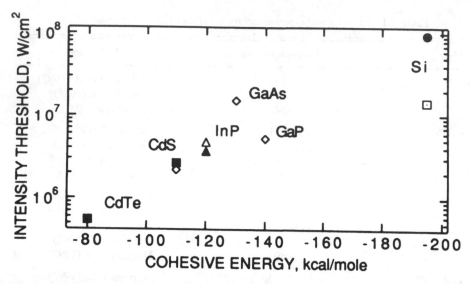

Figure 8.1. Laser intensity thresholds for desorption of semiconductor surface layers plotted as a function of cohesive energy (i.e. heat of atomization). Semiconductors were irradiated under the following conditions: CdTe:KrF laser (248 nm, 20 ns); CdS:KrF laser (248 nm, 20 ns), nitrogen laser (337 nm, 5 ns); InP:ArF laser (193 nm, 20 ns); nitrogen laser/dye laser (337 nm, 5 ns); GaAs, GaP:nitrogen laser (337 nm, 5 ns); Si:ruby laser (694 nm, 20 ns); and dye laser (450 nm, 3 ns) (Brewer *et al.* 1991).

from a KrF laser (Bialkowski *et al.* 1990) showed that charge separation occurs with electron emission following the laser pulse (about 60 ns) whereas positive ion emission persists for several microseconds at fluences close to the threshold for melting. This behavior would be consistent with the presence of a Knudsen layer over the surface (Kelly and Dreyfus 1988, Kelly 1990). At relatively low vaporization rates, the effect of this Knudsen layer is to enhance the collision rate for particles emitted from the surface, which would reduce the drift velocity in an external field and lengthen the delay time for the passage of Si and Si^+ through this region. This retarding effect would be expected to dominate only at low laser intensities, since strong heating of the Knudsen layer will occur when fluences exceed the ablation threshold (Kelly and Dreyfus 1988). Little heating of the Knudsen layer would be expected at the fluences used by Bialkowski *et al.* (1990), which were close to the melting threshold (about $0.7\,\mathrm{J\,cm^{-2}}$). However, heating is undoubtedly important under the conditions of the experiments of Shinn *et al.* (1986).

The creation of a Knudsen layer depends on the ablation rate, which in a thermal process near threshold, will be highly dependent on vapor pressure. In Ge the vapor pressure at the melting point is about 10^{-3}

that of Si at its melting point so that vaporization rates will be reduced. Bialkowski *et al.* (1990) found that positive ion emission from Ge persists only for the duration of the laser pulse near the melting threshold, which would be compatible with the absence of a Knudsen layer under these conditions.

Hanabusa (1993) has reported additional time-of-flight measurements for Si, Si^+ and Si^{2+} emitted from Si irradiated with 20 ns pulses from a frequency doubled Nd:YAG laser ($\lambda = 530$ nm). The fluence was about $1\,J\,cm^{-2}$. Under these conditions, particle speeds of $(1.3 \pm 0.2) \times 10^6$, $(2.7 \pm 0.3) \times 10^6$ and $(4.4 \pm 0.5) \times 10^6\,cm\,s^{-1}$ were found for Si, Si^+ and Si^{2+}, respectively, at a location of 5 mm from the point of irradiation. These speeds were found to be independent of fluence and are large enough to imply significant non-equilibrium effects in the emission process.

The origin of this non-equilibrium may occur in the emission of clusters by the surface followed by heating of these clusters in the Knudsen layer over the surface. Such processes have been observed in the UV laser ablation of superconducting oxides (Chapter 7) and are characterized by fast and slow velocity atomic/ionic components and a complex spatial and time dependence in the plume. A study of plume morphology during ablation of Si with 248 nm laser pulses (Kasuya and Nishina 1990) supports this interpretation; however, direct measurements of the composition of the deposit from Si showed only clusters that probably appeared by aggregation of smaller species on condensation (Tench *et al.* 1991). Tench *et al.* (1991) found that the vapor over Si ablated with 5 ns 266 nm quadrupled Nd:YAG pulses contained only monomer and dimer ions. This contrasted with samples collected from the plume over irradiated SiC, which showed quasi-spherical clusters with about 0.5 nm diameters as well as irregularly shaped clusters having dimensions up to 2.2–6.2 nm.

Laser desorption from compound semiconductors has been extensively studied at photon energies that exceed the band gap energy (Nakayama *et al.* 1984, Namiki *et al.* 1987, Ichige *et al.* 1988, Herman *et al.* 1992) and at sub-bandgap energies (Hattori *et al.* 1990, Nakai *et al.* 1991, Okano *et al.* 1991). For a compound MX, where M is the metal and X is a non-metallic atom, excitation leads to the sputtering reaction

$$(MX)_S + h\nu \rightarrow M + X + X_n$$

where $X_n \equiv X_2$, X_3, etc. are non-metallic clusters. Overall, etching rates are strongly fluence-dependent, as is typical for ablative material removal (Figure 8.2). A summary of threshold fluences for ablation of some compound semiconductors is given in Table 8.2.

Surprisingly, complex sputtering effects are observed for photon energies well below the bandgap energy in several materials (Hattori *et al.* 1990, Okano *et al.* 1991). The mechanism for emission of particles from the irradiated surface under these conditions is still uncertain (Hattori *et al.* 1992). However, particle emission would appear to be defect-initiated, with several types of surface defect center indicated (Hattori *et al.* 1991). These defects are excited directly by photons of sub-band-gap energies. Higher energy photons with energies greater than the valence band to unoccupied surface band energy difference but less than the indirect bandgap energy create electron–hole pairs that are localized near the surface and can be trapped at surface defects. Sputtering results from the instability of surface defects excited in this way. At higher photon energies, i.e. those greater than the energy of the indirect band-gap, electron–hole pairs are formed in the bulk of the material and are then trapped at surface defects. A comprehensive study of the ablation of GaP(110) at various photon energies in the above range has been reported by Hattori *et al.* (1992). The yield of Ga was measured as a function of fluence, number of pulses and annealing conditions at several excitation wavelengths. A complex parametric dependence was observed when excitation occurred with photons having energies less than that of the indirect bandgap. In particular, for fluences above the ablation threshold the yield per pulse was found to depend on the number of pulses superimposed on the surface. The ablation threshold fluence for sub-bandgap photons was also found to have a wide scatter, whereas it was well defined for higher energy photons. These results are compatible with a model (Ichige *et al.* 1988) in which dangling bonds are created by localization of electron–hole pairs at atoms near surface defects. Lowering of the ablation threshold in damaged surfaces is consistent with the involvement of surface defects in the emission process.

Following Nakai *et al.* (1991) and Hattori *et al.* (1992), Figure 8.3 shows a schematic representation of some surface defects, which may be involved in particle emission. If dN_E is the number of sputtered atoms per laser pulse and n is the number of pulses applied

$$\frac{dN_E}{dn} = gN_D(n) \tag{1}$$

where g is the probability of emission of a defect and $N_D(n)$ is the number of the requisite defect atoms after $n - 1$ pulses. g will contain

(a)

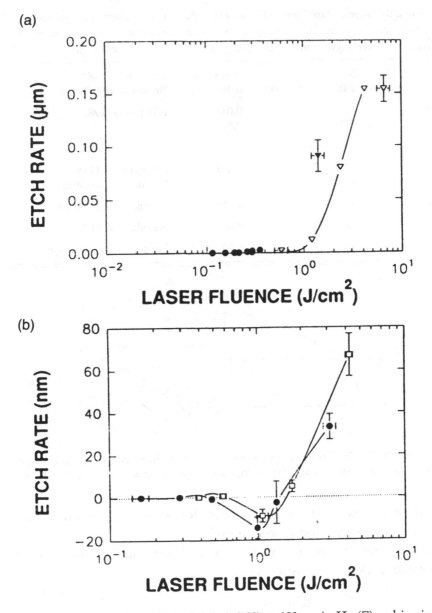

Figure 8.2. (a) Photoablation rates for GaAs(100) at 193 nm in He (∇) and in air (▼). (●), Non-zero ablation rates in air. (b) Photoablation rates at 157 nm for InP (100) (□) and GaAs (100) (●) are similar. Redeposition and thermal damage led to negative ablation rates at near $1 \, \mathrm{J \, cm^{-2}}$ fluence (Herman *et al.* 1992).

Table 8.2. *Representative threshold fluences for ablation in compound semiconductors.*

Material	Wavelength (nm)	Fluence ($J\,cm^{-2}$)	Reference
CdS	337	0.015	Nakayama (1983)
	248	0.06	Brewer *et al.* (1991)
CdSe	510	0.05	Nakayama (1983)
	410	0.014	
CdTe	248	0.015	Brewer *et al.* (1990)
GaAs	337	0.075	Namiki *et al.* (1987)
	193	$\simeq 1$	Herman *et al.* (1992)
GaN	337	0.015	Namiki *et al.* (1987)
GaP	337	0.055	Namiki *et al.* (1987)
InP	540	0.20	Moison and Bensoussan (1982)

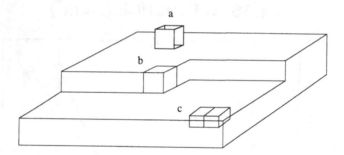

Figure 8.3. A simplified representation of surface defects involved in particle emission during laser sputtering: (a) an adatom, (b) a kink-type defect and (c) a vacancy defect.

both the cross-section for energy localization at the appropriate site and the efficiency for ejection following excitation.

For emission of an adatom (Figure 8.3) the number of defects is reduced and $dN_E = -dN_D(n)$. Then

$$\frac{dN_E}{dn} = gN_D(0)\exp(-gn) \qquad (2)$$

where $N_D(0)$ is the number of defects prior to irradiation. In this case, the number of sputtered atoms per pulse decreases with irradiation as the number of adatoms is reduced.

For emission of an atom from a kink-type defect, the number of defect sites remains constant (to zeroth order). Thus $dN_D(n) \simeq 0$ and $dN_E(n)/dn = $ constant.

Emission of an atom from a vacancy site (Figure 8.3) does not destroy the defect and can result in an increase in the surface concentration of vacancy sites (Okano *et al.* 1993). Thus $dN_E(n) \simeq dN_D(n)$ so that

$$\frac{dN_E(n)}{dn} = gN_D(0) \exp(gn) \qquad (3)$$

and the number of sputtered atoms per pulse should increase rapidly with the number of overlapping pulses. This would be consistent with the increase in sputtering yield observed after multiple pulsing at fluences near the threshold for ablation.

One expects that emission of an adatom will be more closely related to direct deposition of photon energy within surface states than to that by transfer from the substrate. Electrons in adatoms are associated with surface states and therefore can be excited directly by sub-bandgap photons. The rapid decay of the sputtering yield for Ga^0 from GaP(110) excited with multiple 1.43 eV pulses has been associated with direct surface excitation and subsequent removal of a limited number of adatoms (Hattori *et al.* 1992). Multiple excitation or trapping of at least two electron–hole pairs at a defect site is required for desorption (Singh and Itoh 1990, Georgiev and Singh 1992).

Surface composition and morphology following laser sputtering of compound semiconductors have been studied by Moison and Bensoussan (1982), Brewer *et al.* (1990, 1991) and by Herman *et al.* (1992). Figure 8.4 shows the fluence dependence of surface morphology and structure in (100) InP under irradiation with 2.3 eV laser photons in a 3 ns pulse (Moison and Bensoussan 1982). This early work showed that structure, morphology and composition are all related to sputtering and attempt to establish new equilibrium/metastable states as sputtering proceeds. Extensively disordered surfaces are typically found at high laser fluence. However, surface reconstruction may occur at lower fluences (Brewer *et al.* 1990, 1991). A thorough study of the composition of CdTe and CdS irradiated at 248 nm has shown that particle emission occurs at fluences lower than those that result in surface melting (Brewer *et al.* 1991). Cd and Te_2 are emitted from CdTe under these conditions whereas Cd^+, Te^+ and Te_2^+ are emitted at higher fluences. A stoichiometric CdTe surface irradiated at fluences ≥ 35 mJ cm^{-2} was found to preferentially desorb Cd so that the surface became enriched in Te. This Te-rich layer was found to be $\lesssim 2$ nm thick. The Cd/Te ratio could be returned to unity by subsequent exposure to a large

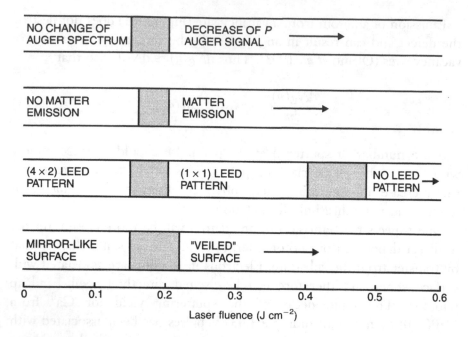

Figure 8.4. A schematic summary of effects induced by 3 ns, 2.3 eV ruby laser pulses incident on a (100) InP surface (after Moison and Benssoussan 1982).

number of lower fluence pulses. This was accompanied by a partial restoration of equilibrium surface structure. It appears that reversibility occurs when subsequent exposure to many low fluence pulses removes the original Te-rich (about 2 nm thick) layer exposing the underlying CdTe surface, which has the normal 1:1 Cd/Te ratio. This can occur as a thermal mechanism if it is assumed that the rate constants for Cd desorption and Te_2 formation are largest at highest surface temperature (high fluence), and exceed the rate constant for Te_2 desorption. At low surface temperature (low fluence), the rate constant for Te_2 desorption exceeds those for Cd desorption and Te_2 formation so that the enriched Te layer is preferentially sputtered.

Similar behavior has also been observed in CdS irradiated at 248 nm (Brewer *et al.* 1991) with the difference that sulfur is preferentially desorbed, leading to Cd enrichment in the surface. The Cd/S ratio can be returned to stoichiometry by subsequent exposure to a low fluence pulse. Mass spectra show that S_1, S_2 and Cd are emitted (Namiki *et al.* 1986). However, the velocity distribution can be described in terms of a single Maxwellian function only at low laser fluence. At higher fluence, both fast and slow components are observed, with the slow components

Table 8.3. *Characteristics of waveguides generated in CdTe using CW Ar$^+$ laser light. α_{exp} is the experimental loss coefficient and α_{calc} is the loss coefficient calculated for propagation in a hollow Te waveguide (after Arnone et al. 1986).*

Wavelength (nm)	Width (μm)	α_{exp} (μm^{-1})	α_{calc} (μm^{-1})
458	0.4	0.30 ± 0.10	0.48
515	0.4	0.34 ± 0.11	0.65
488	1.0	0.050 ± 0.017	0.033

emitted over a wider scattering angle (Namiki *et al.* 1986). This behavior is similar to that observed in the laser sputtering of other materials (e.g. superconductors) and suggest that a Knudsen layer may be present over the surface under these conditions (Kelly 1990). However, other experiments on sputtering of GaN, GaP and GaAs with 337 nm laser radiation (Namiki *et al.* 1991) showed that the velocity distribution of sputtered Ga and X_2 is inconsistent with the Kelly (1990) Knudsen layer model. Instead, sputtered species may be ejected with a range of non-Maxwellian velocities, which reflect their formation on the surface and excitation via collision at this stage. A similar effect may occur in the ejection of Zn atoms from ZnS below the threshold fluence for ablation (Arlinghaus *et al.* 1989), for which high kinetic temperatures have also been observed.

CW Ar$^+$ ion laser radiation with wavelengths in the 458–515 nm range has also been shown to initiate rapid sputtering of CdTe at temperatures well below the melting point (Uzan *et al.* 1984, Arnone *et al.* 1986). The activation energy for laser-induced decomposition was found to be 0.49 eV, which is significantly lower than the activation energy that is associated with thermal sublimation (1.93 eV) (Uzan *et al.* 1984). The difference between these two activation energies (1.44 eV) is similar to the bandgap energy in CdTe and indicates that electron–hole pair generation by photoabsorption is important in the sputtering process.

An analysis of the vapor evolving from CdTe under irradiation with CW Ar$^+$ laser light (Arnone *et al.* 1986) showed that the Cd/Te ratio was greatly enhanced in the sputtered material. This implies that Te is enriched in the CdTe surface; a similar effect to that observed when sputtering occurs with UV radiation (Brewer *et al.* 1991). Under these conditions, self-focusing also occurs, resulting in the propagation of a narrow, waveguide-like structure with a high aspect ratio. Arnone *et al.* (1986) concluded that propagation occurs in a hollow Te waveguide (Table 8.3).

8.2 LASER-INDUCED CHEMICAL ETCHING

Semiconductors can be etched by means of reactive intermediaries generated either directly or indirectly by laser radiation. The overall etching process often involves aspects of both photochemical and thermally driven reactions and etching can be carried out in a gaseous environment (dry etching) or in the presence of a liquid. In the latter case, both photoelectrochemical and carrier-driven photochemical etching can be used.

Laser-induced dry chemical etching has been widely investigated and is attractive because it is potentially free of the limitations inherent in wet etching and the substrate damage caused by high energy processes such as reactive ion etching (RIE). It also offers the advantage of low contamination and is not intrinsically a process that results in the creation of surface or bulk defects. Both of these properties are compatible with the requirement for a reduction in minimum feature size in VLSI devices.

Although laser-induced dry chemical etching is a promising technique, several aspects of the etching process are poorly understood and pattern resolution is limited by such factors as reactant diffusion and carrier diffusion as well as thermal heating effects. Laser-induced chemical etching of Si by Cl_2, which should be one of the simplest systems, illustrates this point. Laser etching in the Si/Cl_2 system has been extensively studied both at UV and at visible wavelengths and with pulsed as well as CW sources (Bäuerle 1986, Ashby 1991). Figure 8.5 shows the dependence of etching rate for Si in Cl_2 on laser fluence when irradiation occurs at 308 nm (Kullmer and Bäuerle 1987). A comparison of etching rates at 308 and 248 nm is also shown in Figure 8.6 (Sesselmann 1989).

The data of Kullmer and Bäuerle (1987) clearly show that laser-induced chemical etching in this system occurs in several distinct stages, although this is not as apparent in the data of Sesselmann (1989). These stages are:

(i)   a linear increase in etching rate with fluence at low fluence, up to a limiting fluence that is dependent on the Cl_2 pressure;

(ii)   a subsequent superlinear dependence of etching rate on fluence over the range $0.15–0.4 \, J \, cm^{-2}$; and

(iii)   saturation at high fluence at an etching rate that is relatively independent of Cl_2 pressure.

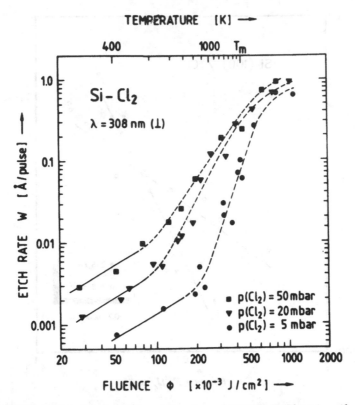

Figure 8.5. The etching rate of Si as a function of 308 nm XeCl excimer laser fluence for various Cl_2 pressures. The temperatures indicated on the upper scale were calculated (from Kullmer and Bäuerle 1987).

The wavelength dependence of etching rate at constant fluence and Cl_2 pressure (Kullmer and Bäuerle 1987) shows that the overall etching rate is largest for laser wavelengths near 308 nm, at least at low fluence. For higher fluences the etching rate at 248 nm greatly exceeds that at 308 nm.

Observation of an etching effect that peaks near 308 nm is a strong indication that direct photodissociation of Cl_2 can be of primary importance. Photodissociation of Cl_2 in this spectral region occurs following absorption in the $^1E_g^+ \rightarrow {}^1\Pi_u$ band (Heaven and Clyne 1982). The cross-section for absorption into this band decreases rapidly at wavelengths below 300 nm and is very small at 248 nm so that few gas phase Cl atoms are generated. On the other hand, since the bandgap energy of Si is 1.12 eV, all laser wavelengths in this range can result in electron–hole pair generation (Kullmer and Bäuerle 1988), which leads in turn to substrate heating and to the possibility of electron transfer to

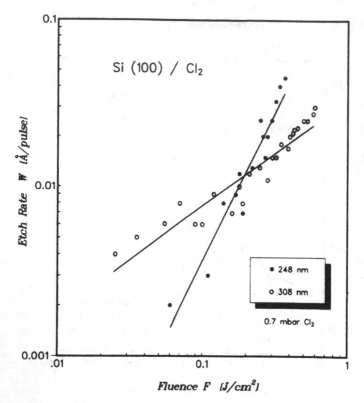

Figure 8.6. The etching rate W as a function of laser energy fluence F for a $\langle 100 \rangle$ Si surface in a 0.7 mbar Cl_2 ambient with excimer laser radiation at 308 and 248 nm and a 20 pulse s^{-1} repetition rate. The straight lines are least squares fits (from Sesselmann 1989).

adsorbed Cl atoms. In addition, photons with energies of 4 eV (308 nm) and 5 eV (248 nm) can lead to direct bond excitation and photodissociation of adsorbed species (Baller *et al.* 1986, Sesselmann 1989).

Reksten *et al.* (1986), in a study of the laser-induced enhancement of etching of Si by CF_4/O_2 or NF_3 in a conventional plasma reactor, found that CW laser radiation at 350, 514 and 647 nm has a strong effect on the etching rate for 1 Ω cm p- and n-type Si. No significant difference in etching rate was observed at different laser wavelengths. In 10^{-3} Ω cm p- and n-type Si, however, the greatest enhancement in etching rate was observed at 350 nm. These results suggest that the etching rate must be proportional to the flux of photogenerated carriers at the Si surface and the ratio of the absorption length for incident laser radiation to the depth of the space charge region at the surface. At UV wavelengths, the decreased absorption length ensures that photogenerated carriers are created primarily within this space charge region, which enhances the probability of interaction with adsorbed species.

Sesselmann (1989) concluded that the dramatic difference between etching rates at 308 and 248 nm (Figure 8.6) can be attributed to the manner in which photons of these energies interact with Cl_2 and Si. The primary interaction at 308 nm is to dissociate gaseous Cl_2, which results in an enhancement of the adsorption of Cl atoms at the Si surface. The binding energy for Cl on Si is 1–2 eV (Seel and Bagus 1983) and each Cl atom is bonded to a single Si atom. The surface then consists of a chemisorbed layer of SiCl, which is formed, in part, by direct reaction with photolytically generated Cl and, in part, by the dissociative adsorption of Cl_2 molecules. The latter reaction is strongly temperature-dependent.

The dissociation of gaseous Cl_2 at 248 nm is inefficient, so that surface Cl can be formed only by dissociative adsorption of Cl_2. As a result, the 248 nm etching rate at low fluences relies on the spontaneous reaction between Cl_2 and Si and is therefore less efficient than at 308 nm. Kullmer and Bäuerle (1987) suggest that etching under these low fluence conditions occurs when Cl^- ions formed from adsorbed Cl by electron transfer from photogenerated electron–hole pairs diffuse and eventually recombine with Si atoms to form $SiCl_4$ or some other volatile $SiCl_x$ compound. Baller et al. (1986) detected Si, SiCl, $SiCl_2$ and $SiCl_3$ during 308 and 248 nm laser etching of Si/Cl_2 under these conditions.

At higher laser fluences, the etching rate at 248 nm can exceed that at 308 nm. Since the thermal heating effect in Si at 308 and 248 nm should be similar, enhanced etching at 248 nm must be attributable to the appearance of a new channel for photodesorption. This may involve direct interaction with adsorbed SiCl molecules (Baller et al. 1986) or may be due to bond breaking by trapping of Frenkel excitons (Susselmann 1989). It should be noted that these effects occur at fluences that are well below those for melting (about $0.7 \, J \, cm^{-2}$) or ablation (about $1.3 \, J \, cm^{-2}$) of Si with excimer laser pulses.

At laser fluences that approach or exceed that for surface melting, the etching rate is observed to saturate (Figure 8.5) and is only slightly dependent on Cl_2 pressure. Since the etching rate is virtually independent of Cl_2 pressure, the rate limiting step in the overall etching reaction cannot be the adsorption of gaseous atoms or molecules. Thus gaseous phase photochemical processes cannot be important, and the reaction probably proceeds by activation of an adsorbed surface layer.

Kullmer and Bäuerle (1987) also found that the etching rate under these high fluence conditions is also not a strong function of laser wavelength. These observations suggest that the etching reaction is then a thermal process in which adsorbed Cl and Cl_2 moieties react with the Si substrate in response to the heating effect of the incident laser radiation.

A comprehensive study by Horike *et al.* (1987) of UV laser-induced etching in Si/Cl_2 systems provided further evidence for an etching mechanism in which adsorbed Cl atoms accept electrons from photo-excited carriers, which are then pulled into the substrate. When electrons are available, such as for example in n^+ type Si, spontaneous electron transfer occurs without the involvement of photogenerated carriers. F atoms, generated by photodissociation of NF_3 using 193 nm ArF radiation, were found to spontaneously trap electrons both from n- and from p-type Si.

Etching in the Si/Cl_2 system with CW laser radiation has been shown to be useful in the production of vias (Treyz *et al.* 1989). Smooth, cylindrical vias with diameters in the 50–100 μm range could be etched in n-type 4–6 Ω cm Si wafers with a thickness of 250 μm in 1000 s using 80 mW of UV Ar^+ laser radiation ($\lambda = 350$–360 nm). At 0.8 W the etching time was reduced to 300 s. Under these conditions, there is little heating of the Si and etching proceeds by the photochemical generation of Cl atoms which diffuse to the Si surface. The area of the entrance hole of the via, $A = \pi R^2$, was found to be linearly dependent on etching time and on laser power. This suggests a simple physical model in which the focused laser beam is assumed to create a line source of strength S_0 extending into the via (Figure 8.7). This line source represents a source of Cl atoms that diffuse to the walls of the via, where they react to yield volatile $SiCl_x$ species. With

$$S_0 = 2\frac{\alpha P_s}{h\nu} \tag{4}$$

where α is the absorption coefficient of Cl_2 gas at the laser wavelength, P_s is the laser power at the entrance surface of the via and $h\nu$ is the photon energy, the Cl flux at radius R is

$$\mathcal{J}_{surface} = \frac{S_0}{2\pi R} \tag{5}$$

$$= \mathcal{J}_{etch} + \mathcal{J}_{rec} \tag{6}$$

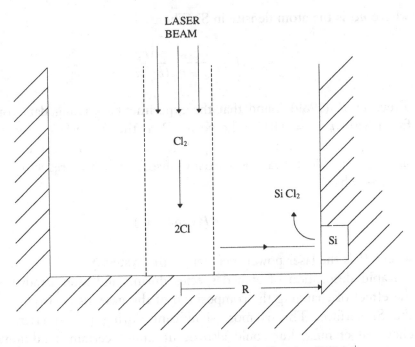

Figure 8.7. A model for etching of vias in Si with Cl_2 using CW Ar^+
laser radiation (after Treyz *et al.* 1989).

where $\mathcal{J}_{etch}$ and $\mathcal{J}_{rec}$ are etching and recombination fluxes, respectively.
Then

$$\mathcal{J}_{etch} = \frac{\eta_e n_R}{4} \bar{v} \tag{7}$$

$$\mathcal{J}_{rec} = \frac{\eta_r n_R}{4} \bar{v} \tag{8}$$

where η_e and η_r are the reaction efficiencies for etching and recombination, respectively, $\bar{v}$ is the Cl atom thermal velocity and n_R is the Cl atom density at the via surface. The etching reaction is taken to be

$$Si + 2Cl \rightarrow SiCl_2\uparrow \tag{9}$$

so that the rate of increase of the via radius is

$$\frac{dR}{dt} = \frac{1}{2}\frac{\mathcal{J}_{etch}}{n_{Si}} \tag{10}$$

where n_{Si} is the atom density in Si. Then

$$A = \frac{\eta_e}{\eta_e + \eta_r} \frac{\alpha P_s}{(h\nu)n_{Si}} \qquad (11)$$

Treyz *et al.* (1989) found that the experimental etching data could be fitted with $\eta_e/\eta_t = 0.011$. The power P_s at the via surface is dependent on the gas pressure and optical path length, l, through the etching gas since the incident beam is partially absorbed before reaching the via surface. Then

$$P_s = P_i \exp(-\alpha l) \qquad (12)$$

where P_i is the laser power incident on the system.

Rapid desorption of etched species during and after irradiation has the effect of changing the composition of the gaseous environment over the Si surface. This is most significant during pulsed laser assisted chemical etching, but could also occur under certain conditions with CW laser etching, for example, during the creation of deep holes. This situation is similar in many ways to plume formation in laser ablative etching, in which a Knudsen layer can be created over the vaporizing surface. This layer causes the velocity distribution of emitted species to be changed and can result in the redeposition of ablated species back onto the original surface. The rôle of post-desorption collisions in the Si/Cl_2 system when etching occurs with excimer laser pulses has been discussed by Boulmer *et al.* (1992).

Etching of extended structures in a preferred direction can be inhibited by sidewall reactions involving desorbed products or by the diffusion of photogenerated Cl atoms. Hayasaka *et al.* (1986) reported that, when laser-assisted etching is carried out in a Cl_2 plus $Si(CH_3)_4$ gas, reaction of $Si(CH_3)_4$ with Cl forms a non-volatile methylated polymer that coats the sidewalls of the ablated area and protects these sidewalls from further etching. This polymer is ablated in the region of direct laser impact so that chemical etching can occur at that point. Hayasaka *et al.* (1986) also report on the use of methyl-methacrylate gas to form sidewall protective layers during Cl atom ablation of poly-Si.

Laser-assisted chemical etching of compound semiconductors has been studied since the reports of Ehrlich *et al.* (1980a,b) on the CW etching of GaAs and InP with halogen atoms liberated from

Table 8.4. *Laser-induced etching rates for InP and GaAs using a CW 257.2 nm laser with a spot size of 19 µm (FWHM) (from Ehrlich et al. 1980a,b). The laser intensity was 100 W cm^{-2}.*

Substrate	Etchant	Pressure (atm)	Rate (nm s^{-1})
n-GaAs(100)	CH$_3$Br	0.99	0.52
n-InP(100)	CH$_3$Br	0.99	0.94
GaAs (amorphous)	CH$_3$Br	0.99	0.97
GaAs (amorphous)	CF$_3$I	0.033	9.9×10^{-3}
n-InP(100)	CH$_3$Cl	1.32	

methyl-halides. Table 8.4 lists etching rate data obtained by Ehrlich *et al.* (1980a,b) in these early experiments at a laser wavelength of 257.2 nm. The incident laser intensity on the semiconductor surface was in the 15–1000 W cm^{-2} range. Under these conditions, and with a reactant pressure of about 1 atm, the etching rate was found to scale linearly with laser intensity. At constant laser intensity, the etching rate for CH$_3$Br etching of GaAs was observed to rise linearly with gas pressure, up to $p \simeq 0.15$ atm. It then increased more slowly at higher gas pressure. This behavior suggests that the etching rate is limited by the availability of reactants at low pressure and by photoexcitation rates or an optical depth effect at high pressure. Ehrlich *et al.* (1980a,b) proposed that the etching reaction is

$$CH_3Br + h\nu \rightarrow CH_3 + Br \tag{13}$$

$$n(Br) \rightarrow n(Br):(GaAs)_{ads} \tag{14}$$

$$\{(Ga, As)Br_n\}_{ads} \rightarrow \{(Ga, As)Br_n\}_{vapor} \tag{15}$$

where the nature and composition of the volatile products was not assigned. In view of the removal mechanisms identified for UV laser etching in the Si/Cl$_2$ system, it is likely that electron transfer from photogenerated electron–hole pairs to surface halogen atoms is also important in the etching of GaAs and InP. This possibility is supported by reaction rate studies (Ehrlich *et al.* 1980a,b, Ashby 1984), which show that laser radiation does not simply act to increase the rate of the thermally activated reaction between the halogen and the semiconductor, but involves the direct generation of electron–hole pairs by excitation across the bandgap.

Subsequent studies of laser-assisted chemical etching of compound semiconductors have centered on the use of excimer laser radiation (Ashby 1991 and references therein). Reactants include the methyl-halogens (Brewer *et al.* 1984), Cl_2 (Koren and Hurst 1988, Qin *et al.* 1988, Maki and Ehrlich 1989, Donnelly and Hayes 1990), HBr (Brewer *et al.* 1984, 1985, 1986) and ozone (Koren and Hurst 1988). The advantage is that all these gaseous precursors can be photodissociated at one or more excimer laser wavelengths to liberate atomic halogens. Furthermore, the large area of the excimer laser beam permits simultaneous etching over an extended area on the semiconductor together with imaging or patterning.

With precursors such as CH_3Br or CF_3Br, etching occurs when either Br or the CH_3 or CF_3 radical interacts with the surface of the semiconductors. Methyl-halide compounds both with Ga and with As are volatile and can desorb under room temperature conditions. Some less volatile compounds such as $GaBr_3$ and $AsBr_3$ can be desorbed by modest heating of the substrate (Brewer *et al.* 1984). Brewer *et al.* (1984) found that the etching rate of GaAs with CH_3Br at 193 nm increases exponentially with equilibrium substrate temperature and attributed this effect to the thermal desorption of more tightly bonded surface species created in the laser-assisted reaction. They found that the etching rate can be as high as $0.5 \, \mu m \, min^{-1}$ at a laser pulse repetition rate of 50 Hz.

The generation of etchants using photochemical decomposition of precursor gases that contain carbon can lead to contamination of the surface by the carbonaceous byproducts of this decomposition. These products can include simple reactive species such as C atoms or C_2 molecules as well as particulates formed by the recombination of gaseous products. Such contamination is obviously unacceptable at the high level of cleanliness required for VLSI circuits. As a result, precursors such as Cl_2 or HBr are preferred. Both these molecules can be photodissociated with 193 nm radiation.

Brewer *et al.* (1985) examined laser-assisted dry etching of GaAs with HBr both in parallel and in perpendicular irradiation geometries. They found that overall etching rates are greatly enhanced in the perpendicular incidence configuration. This effect probably arises in part due to the enhanced reactivity at the laser-irradiated surfaces and in part from the laser-assisted desorption of the reaction products. The etching rate was found to be a linear function of time at constant laser intensity and substrate temperature. For constant laser intensity, repetition rate and HBr pressure, the etching rate of GaAs(100) was found to peak at temperatures in the range 50–70°C and then decrease at higher temperatures. This complex behavior was ascribed to a temperature-dependent variation in surface structure and composition. At low temperature ($\lesssim 50°C$) the GaAs surface was found to be covered with a liquid bromide salt of Ga and As. This coating was resistant to etching.

A phase change at about 50°C resulted in the disappearance of the liquid and enhanced etching rates. At high temperatures a non-volatile Br- and O-rich layer was created by reaction at the surface, leading to a reduction in etching efficiency. Under optimum conditions, at a substrate temperature of 60°C, the etching rate was found to be as large as 2.2 nm per pulse.

In a subsequent study, Brewer *et al.* (1986) reported that the addition of reactive or non-reactive buffer gases to HBr during laser etching of GaAs can have a significant effect on the overall resolution of etch lines. This effect is illustrated in Figure 8.8. Non-reactive buffer gases seem to improve etch resolution by decreasing the mean free path for reactants over the surface. This is accompanied by an overall decrease in etching rate because the enhanced collision rate leads to the thermalization of reactants and to radical recombination. Gaseous chemical quenchers such as C_2H_4 also improve the resolution of etched features. However, this is probably due to the redeposition of a passivating layer back onto the substrate. A resolution of about 0.4 μm was observed in these experiments.

Laser-assisted dry chemical etching of GaAs in Cl_2 has been found to proceed in the sequence (i) photochemical dissociation of Cl_2, (ii) diffusion of Cl atoms to the surface, (iii) adsorption, (iv) formation of $GaCl_x$ species and (v) laser desorption of volatile compounds. A summary of etching rate data for the GaAs/Cl_2 and InP/Cl_2 systems is given in Table 8.5. In all systems the etching rate is strongly dependent on substrate temperature once a threshold temperature has been reached. This effect can be seen clearly in Figure 8.9, in which the etching rate is plotted versus temperature for n-GaAs(100) (Berman 1991). These data show that the etching rate is virtually independent of substrate temperature up to 75°C. At higher temperatures, the etching rate exhibits an Arrhénius dependence with an activation energy of 49.3 kJ mol^{-1} and is independent of Cl_2 pressure.

A detailed study of the effect of the pulse repetition rate on etching in the GaAs/Cl_2 system at low fluence by Maki and Ehrlich (1989) for

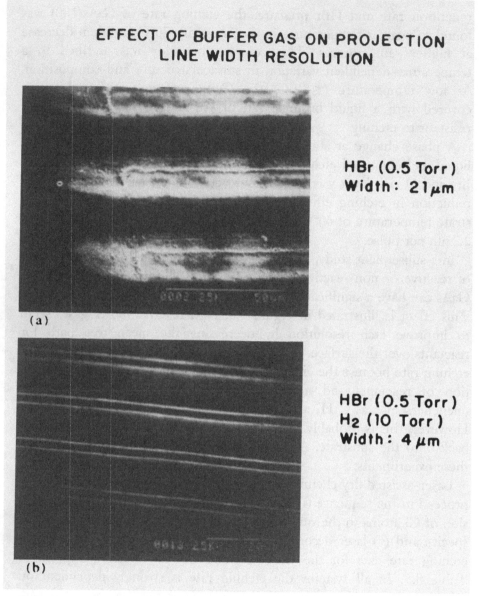

EFFECT OF BUFFER GAS ON PROJECTION LINE WIDTH RESOLUTION

HBr (0.5 Torr)
Width: 21 μm

HBr (0.5 Torr)
H₂ (10 Torr)
Width: 4 μm

Figure 8.8. A comparison of etched lines in GaAs using 193 nm laser-assisted etching in HBr gas (a) without buffer gas and (b) with H_2 buffer gas (from Brewer *et al.* 1986).

epitaxial layers of GaAs grown by molecular beam epitaxy on GaAs substrates has shown that there are three distinct etching regimes. These are defined by the pulse repetition frequency f (Figure 8.10). For $10 \leq f \leq 150$ Hz and a fluence of $8\,\text{mJ}\,\text{cm}^{-2}$ etching removes about 0.1 monolayer per pulse. This rate is independent of f and is termed a

Table 8.5. *Etching rates for excimer laser-assisted etching of GaAs and InP in Cl_2.*

Substrate	Pressure (atm)	Wavelength (nm)	Rate (nm/pulse)	Fluence ($J\,cm^{-2}$)	Reference
n-GaAs(100)	6.5×10^{-4}	248 (10 Hz)	3 (44°C)	1	Koren and Hurst (1988)
n-GaAs(100)	$1.3 \times 10^{-3}+$ 1.3×10^{-2} Ar	193 (10 Hz)	0.2 (150°C)	0.05	Tejedor and Briones (1991)
	$1.3 \times 10^{-6}+$ 1.3×10^{-3} Ar	193 (10 Hz)	0.01 (120°C)	0.04	
n-GaAs(100)	2.6×10^{-3}	308 (10 Hz)	1 (50°C) 9 (155°C)	0.0175 0.0175	Berman (1991)
n-InP(100)	2.5×10^{-6}	193 (10 Hz)	0.2 (139°C)	0.10	Donnelly and Hayes (1990)

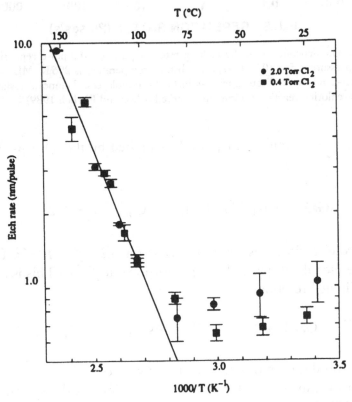

Figure 8.9. The etching rate of samples of (100)-oriented GaAs etched by 0.4 and 2.0 Torr of Cl_2 as a function of substrate temperature. The solid line is a least squares fit to all of the data at temperatures above 100°C. This fit corresponds to an activation energy of 11.8 kcal mol^{-1} (Berman 1991).

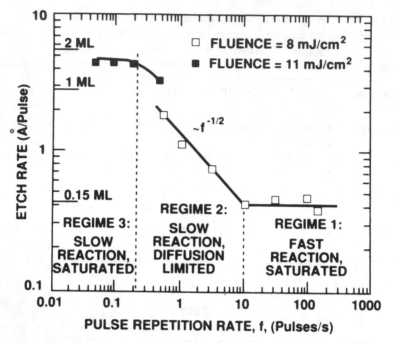

Figure 8.10. The dependence of the etching rate per pulse on the pulse repetition rate f. The Cl_2 pressure is 5×10^{-3} Torr; the substrate temperature is 296 K. ML notations along the left-hand axis indicate thickness in GaAs monolayers. Temporal assignments of the GaAs–chlorine reaction regimes are marked (Maki and Ehrlich 1989).

fast reaction saturation regime, which is limited by the adsorption rate of Cl, i.e.

$$GaAs + Cl_2 \downarrow \xrightarrow{\text{fast}} GaCl_n + AsCl_n \qquad n = 1, 2 \qquad (16)$$

The second regime was found to occur for $0.2 \le f \le 10\,\text{Hz}$. In this regime the etch depth per pulse is proportional to $f^{-1/2}$. This is a slow, diffusion-limited reaction

$$GaCl_n + AsCl_n + Cl_2 \rightarrow AsCl_3\uparrow + GaCl_3 \qquad (17)$$

which involves diffusion of Cl_2 or Cl to active sites.

The third regime occurs for $f \le 0.2\,\text{Hz}$ and follows the slow desorption of $GaCl_3$:

$$GaCl_3 \rightarrow GaCl_3\uparrow + \text{free surface site} \qquad (18)$$

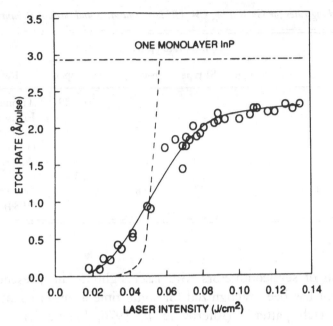

Figure 8.11. The InP etching rate as a function of laser fluence. The laser repetition rate is 10 Hz, the Cl_2 pressure is 1.9 mTorr, and the wafer temperature is 139°C. The dashed line represents a computation of the etching rate near threshold if etching were limited by sublimation of $InCl_3$ (Donnelly and Hayes 1990).

Although the etch depth per pulse is large in this range, the overall etching rate is reduced due to the low pulse repetition frequency.

In the InP/Cl_2 system (Donnelly and Hayes 1990), the etching rate using 193 nm ArF laser radiation was found to be directly proportional to f for $f < 10\,Hz$ at a constant fluence of $0.1\,J\,cm^{-2}$. From 10–40 Hz the etching rate increased superlinearly. Under these conditions, which correspond to an average laser intensity of up to $4\,W\,cm^{-2}$, the equilibrium temperature at the center of the focal spot was estimated to rise to 212°C, facilitating spontaneous etching of InP by Cl_2 and accounting for the superlinear dependence of etching rate on f. The etching rate per pulse at 10 Hz and a fluence of $0.1\,J\,cm^{-2}$ was found to saturate at high Cl_2 pressure. This suggests that the rate limiting step in etching involves the desorption of reaction products. If this were the case, then the etching rate should saturate at about one monolayer per pulse at high laser fluence. A comparison between theory and experiment based on this model is given in Figure 8.11 (Donnelly and Hayes 1990). A simple model, based on the laser-induced sublimation of $InCl_3$ is seen to satisfactorily predict the observed fluence threshold for etching, together with the saturation in etching rate at high fluence.

Table 8.6. *Etching rates for GaAs using CW UV laser radiation and various etchant solutions.*

	Etching rate ($\mu m\,min^{-1}$)				
Solutions	n-type	SI type	p-type	Cr-doped	Reference
$HNO_3:H_2O$ (1:10)				60–120	Tisone and Johnson (1983)
$KOH:H_2O$ (1:20)	8	6	0.5		Podlesnik *et al.* (1984)
$HNO_3:H_2O$ (1:20)	12	10	1.0		Podlesnik *et al.* (1984)
$H_2SO_4:H_2O_2:H_2O$ (1:1:100)	18	13	0.8		Podlesnik *et al.* (1984)

Irradiation of semiconductors with laser light in the presence of a liquid etchant has been recognized for some time as an efficient means of creating etch patterns (Alferov *et al.* 1976, Haynes *et al.* 1980, Ostermayer and Kohl 1981, Osgood *et al.* 1982). Some early work by Alferov *et al.* (1976) used radiation from an He–Cd laser (440 nm) to create gratings with a spacing of about 4 µm and a depth of about 0.7 µm in n-GaAs by projecting a holographic interference pattern onto the substrate submerged in a photoetch solution. Laser-enhanced etching of a variety of compound semiconductors in $H_2SO_4/H_2O_2/H_2O$ and other solutions was subsequently reported by Osgood *et al.* (1982) and Podlesnik *et al.* (1983). Vias, gratings and other structures can be produced using this technique. Etching rates were found to be dependent on crystallographic orientation and the level of doping. Maximum etching rates were found to be about 2 µm min^{-1}, under conditions in which little laser-induced substrate heating occurred. In Si GaAs the etching rate depended on crystal orientation, with different orientations giving different rates, (111A) > (110) $\simeq$ (100). A 30–60 µm via could be etched in 125 µm thick GaAs with 100 mW of 488 nm Ar$^+$ laser radiation at the rate of about 30 µm min^{-1}. The diameter of these vias tapered from about 30 µm at the entrance to about 60 µm at the exit hole. A summary of etching rates is given in Table 8.6.

A remarkable effect, involving the etching of vertical high aspect ratio holes and slits in GaAs by 257 nm CW laser irradiation in etching solutions has been reported by Podlesnik *et al.* (1984, 1986a,b). They reported that a 1 µm diameter hole with perfectly vertical walls could be etched through a 100 µm thick GaAs wafer with a CW laser intensity of

$100\,\mathrm{mW\,cm^{-2}}$. At somewhat higher intensity a $4\,\mu\mathrm{m}$ diameter hole was drilled through a $250\,\mu\mathrm{m}$ thick sample.

Liquid phase etching under these conditions appears to occur in three stages. In the first stage, a hole whose profile mimics that of the Gaussian beam is created on the GaAs surface. With increasing irradiation time this hole evolves into a tubular structure and then into the extended cylindrical hole that propagates with vertical sides into the material. This hole acts as a hollow cylindrical waveguide filled with the etching solution. Podlesnik *et al.* (1986a) found that this structure supports a hybrid (EH_{11}) electromagnetic mode with an attenuation coefficient for 257 nm laser radiation in the range <1–$3.5\,\mathrm{mm^{-1}}$. The largest value of the attenuation coefficient was found for hole diameters approaching $1.6\,\mu\mathrm{m}$. The ultimate limit to the etching rate in these high aspect ratio structures is probably due to mass transport effects.

Laser etching with 257 nm radiation in $\mathrm{HF:HNO_3:H_2O}$ (4:1:50) solutions has been used to create rib waveguides on GaAs/AlGaAs heterostructures (Podlesnik *et al.* 1986b, Willner *et al.* 1989). The simplest structure is created by etching two trenches about $20\,\mu\mathrm{m}$ apart and about $1\,\mu\mathrm{m}$ thick in the $2\,\mu\mathrm{m}$ thick top GaAs layer. The transmission efficiency of such a structure increases dramatically for trench depths exceeding about $0.5\,\mu\mathrm{m}$. When the etch depth is shallow, a significant portion of a propagating beam is contained in side regions. This effect has been used to fabricate a compact waveguide coupler by reducing the depth of a common trench over a limited length in two parallel waveguides (Figure 8.12).

Photoelectrochemical etching of gratings in InP using 442 nm He–Cd laser radiation has been reported by Bäcklin (1987). A holographic technique was used to etch gratings with the 200–250 nm pitch required for first order operation of $1.55\,\mu\mathrm{m}$ DFB/DBR lasers. These gratings were created over areas of up to about $2\,\mathrm{cm^2}$ on the ($\bar{1}10$) plane of InP.

8.3 PHOTOOXIDATION

Photooxidation of Si to form SiO_2 has been studied at a variety of laser wavelengths extending from $10.6\,\mu\mathrm{m}$ (Boyd and Wilson 1982) to 193 nm (Slaoui *et al.* 1988) and both with CW and with pulsed laser sources (Bäuerle 1986). Although all laser wavelengths are effective in enhancing the thermal oxidation rate of Si in an O_2 atmosphere, much

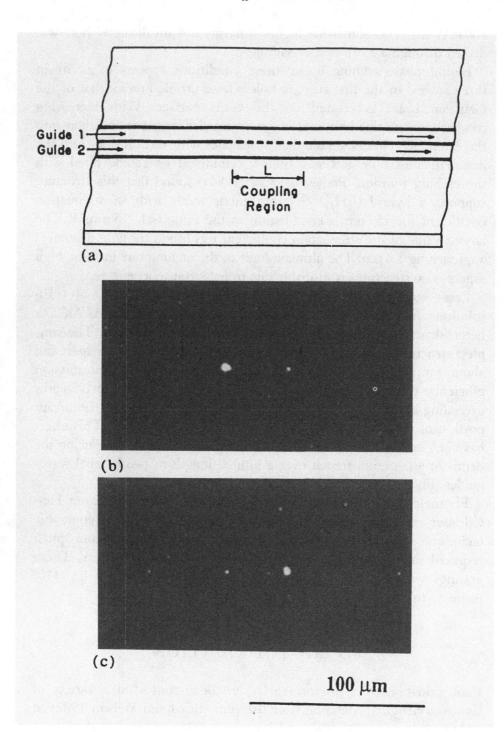

(a)

Guide 1

Guide 2

L

Coupling
Region

(b)

(c)

100 µm

of this enhancement at visible and IR wavelengths is probably attributable primarily to laser-induced heating of the Si substrate. Growth rates for SiO_2 on Si irradiated with CW visible or IR laser radiation are only slightly enhanced over thermal equilibrium rates. A careful study of the wavelength dependence of growth rate (Schafer and Lyon 1982) shows, however, that UV laser radiation exhibits a significantly larger enhancement in growth rate under the same irradiation conditions than does visible laser radiation. The origin of this enhancement at UV wavelengths is unclear, but photoelectron excitation to form surface O_2^- as well as defect generation would seem to be important with photon energies that exceed 3.5 eV.

Rapid growth of SiO_2 on Si at 400°C irradiated with 308 nm XeCl pulses in one atmosphere of O_2 has been reported by Orlowski and Richter (1984) and Richter et al. (1984). For an oxide thickness of 30–180 nm, the growth rate was linear at about $10\,nm\,s^{-1}$ or 0.1 nm per pulse at a repetition rate of 100 Hz. At higher thicknesses, the growth kinetics were found to be quadratic with a parabolic rate constant of $8.5\,\mu m^2\,h^{-1}$, which is 30 times the thermal equilibrium value at an equilibrium temperature of 1000°C in 1 atm of O_2. They suggested that this enormous enhancement might have arisen in part due to the photodissociation of O_2 (in a two-photon process) to form O atoms, which more easily diffuse through the oxide layer. Field-enhanced tunneling of anions has also been thought to be important in the laser-enhanced chemical etching of Si with Cl_2 (Kullmer and Bäuerle 1987). SiO_2 layers of up to about 1 μm thickness were prepared in this way. The bandgap energy for these SiO_2 layers was estimated to be 9.1 eV, which is similar to that measured for thermally deposited films. The electrical properties of these laser-deposited films could be improved by a short thermal anneal at 900°C in 1 atm of O_2. The breakdown voltage was found to be $>5 \times 10^7\,V\,m^{-1}$. A similar enhancement in the oxidation of Si by O_2 under UV laser irradiation has been reported by Boyd (1989) using 248 nm KrF laser pulses and by Liu et al. (1981) using 266 nm radiation from a quadrupled Q-switched Nd:YAG laser. IR spectra of SiO_2 films produced in this way have been obtained by Liu et al. (1981)

Figure 8.12. (a) A top view schematic diagram of a directional coupler. The dashed line indicates a shallow trench (coupling region) and bold lines indicate deep trenches (non-coupling region). Optical outputs for a directional coupler of coupling length 1 mm are shown both for the right- and for the left-hand waveguides: (b) the wave launched into the right-hand guide, and (c) the wave launched into the left-hand guide (Willner et al. 1989).

and by Slaoui *et al.* (1988). In the latter case ion-implanted Si films were oxidized by laser irradiation. A narrow band at $1075\,\mathrm{cm}^{-1}$ characteristic of the Si–O stretch in SiO_2 was observed. However, this band was accompanied by others at $950\,\mathrm{cm}^{-1}$ and about $1175\,\mathrm{cm}^{-1}$, which could be attributed to Si–OH and defect Si–O stretching bands, respectively. The positions of these satellite features was found to be strongly fluence-dependent, particularly for fluences high enough to produce surface melting of Si. These observations suggest that solid–liquid interfacial phenomena as well as O_2 incorporation into the liquid Si melt may be significant and may lead to a variety of Si–O complexes in addition to SiO_2. The melting threshold of Si exposed to 193 nm ArF laser radiation has been studied by Foulon *et al.* (1988). They found the threshold melting fluence for crystalline Si to be about $0.4\,\mathrm{J\,cm}^{-2}$. For Si amorphized by implantation of Si^+ this threshold was 0.1–$0.2\,\mathrm{J\,cm}^{-2}$, dependent on the thickness of the amorphized layer. Fogarassy *et al.* (1988) observed no oxidation in crystalline Si irradiated with 193 nm ArF pulses at fluences $\lesssim 0.4\,\mathrm{J\,cm}^{-2}$ whereas rapid oxidation of Si-implanted material occurred at 0.1–$0.12\,\mathrm{J\,cm}^{-2}$ under similar irradiation conditions.

Figure 8.13 shows the intensity of the Si–O IR absorption peak as a function of laser fluence at 193 nm for Si crystals implanted with various atoms. Oxygen incorporation to form the oxide is seen to be most efficient in As-implanted material. In all cases the largest incorporation of oxygen occurs for laser fluences near $0.3\,\mathrm{J\,cm}^{-2}$ and no oxygen incorporation occurs when samples are irradiated at high fluence (about $0.5\,\mathrm{J\,cm}^{-2}$). This somewhat contradictory behavior can be understood in terms of a simple thermal annealing model.

The threshold fluence for melting in these amorphized Si layers is 0.1–$0.2\,\mathrm{J\,cm}^{-2}$ (Foulon *et al.* 1988) and no oxidation is observed at fluences smaller than these values. This suggests that melting and recrystallization are required for the formation of oxide. This effect increases with laser fluence up to fluences at which the undamaged Si substrate begins to melt. When this occurs, epitaxial recrystallization proceeds from the substrate into the amorphous layer. Oxidation seems to require that a large-grained polycrystalline layer be formed during the first few laser pulses with oxide formation occurring at the grain boundaries of this material. A summary of various thermal oxidation models for Si has been given by Boyd (1989). Oxidation of Si using CW Hg lamp radiation has been discussed by Kazor and Boyd (1994).

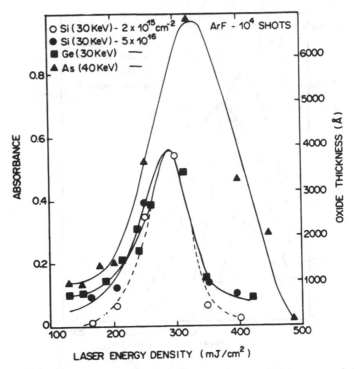

Figure 8.13. The intensity of the IR absorption peak (recorded in terms of absorbance and in terms of oxide thickness) for 30 keV Si- and Ge- and 40 keV As-implanted Si, irradiated in air with a pulsed ArF laser (10^4 shots) as a function of the laser energy density (Fogarassy *et al.* 1988).

The oxidation of SiO in O_2 using 193 nm ArF pulses has been discussed by Blum *et al.* (1983). The threshold for the formation of SiO_2 as reflected in the appearance of the IR Si–O absorption at $1075\,cm^{-1}$ was found to be about $0.02\,J\,cm^{-2}$. This is significantly lower than that for amorphized Si and may arise from the amorphous nature of the SiO precursor. Stoichiometric SiO_2 layers were created after exposure to 10^3 pulses at a fluence of $0.11\,J\,cm^{-2}$.

The re-emission of Si-implanted SiO_2 under exposure to 308 nm XeCl laser pulses has been reported by Shimuzu *et al.* (1988). The Si distribution in As-implanted samples is found to shift toward the SiO_2 surface with increasing laser fluence. This is accompanied by a narrowing of the distribution function and finally by a significant decrease in the total amount of Si retained in the SiO_2 (Figure 8.14). This behavior is opposite to that observed in Si/SiO_2 samples thermally annealed at 1100°C, in which Si is found to move away from the SiO_2

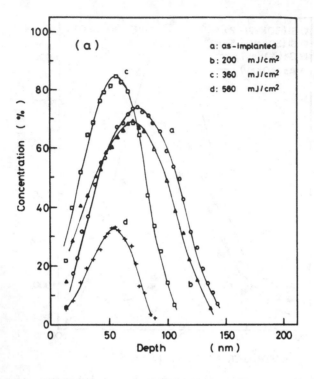

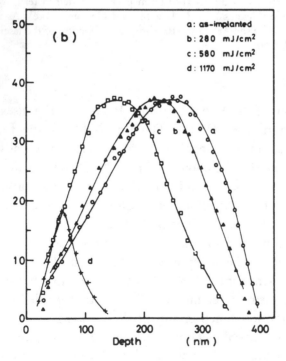

surface. This difference can be attributed to the strong non-uniform thermal gradient induced by laser heating. A study of different oxide phases formed on Si during 698 nm ruby laser irradiation has shown that surface composition changes from SiO_x to SiO_2 as oxidation increases (Cros *et al.* 1982). Other studies of oxygen incorporation by laser irradiation of Si at 1.06 µm have been reported by Hoh *et al.* (1980) and by Chiang *et al.* (1981). Studies of the oxidation of crystalline Ge with 193 nm radiation have been reported by Vega *et al.* (1993).

The rôle of oxygen incorporation during excimer laser irradiation of GaAs oxides has been studied using isotopic substitution of ^{16}O for ^{18}O (Siejka *et al.* 1985). GaAs oxide films were prepared on GaAs by anodization in ^{18}O-rich solution. On exposure to 1000 pulses of 193 nm radiation in O_2 at a fluence of $0.045 \, J \, cm^{-2}$ per pulse, efficient exchange of ^{16}O for ^{18}O was observed to occur without any reduction in the total oxygen content of the oxide. No surface ablation was observed at this fluence and the thermal heating effect was minimal. This suggests that O atom exchange proceeds due to hole trapping at surface defects and involves the mechanism suggested by Itoh and Nakayama (1982). Further studies of ^{16}O incorporation in GaAs oxides at longer laser wavelengths have been given by Cohen *et al.* (1984).

8.4 LASER-ASSISTED DOPING

The rapid heating and cooling rates possible with laser sources has provided a new method for the creation of thin doped surface layers on semiconductors. Incorporation of the dopant occurs when laser radiation is used to heat the semiconductor surface to a temperature at which diffusion of dopant species can occur. The dopant can be obtained from a preplaced or deposited layer on the surface of the semiconductor or from adsorbed gaseous, liquid or solid precursors. In some cases the

Figure 8.14. Concentration profiles, obtained from RBS measurements, of Si atoms implanted into SiO_2 to a dose of 5×10^{17} ions/cm² at energies of (a) 50 keV and (b) 180 keV and subsequently irradiated with a laser pulse. In each figure curves marked 'a' are the As-implanted depth profile and other curves are after subsequent excimer laser irradiation to doses described in the figures. 0% means that there are no additional Si atoms implanted into SiO_2, and 100% means that the density of implanted Si atoms after implantation is equal to that of amorphous Si (Shimuzu *et al.* 1988).

dopant may be implanted prior to laser treatment. Laser heating may proceed to the point at which surface melting occurs, enhancing dopant diffusion.

Ultraviolet laser heating can be accompanied by electronic and photo-chemical as well as thermal effects. In particular, the deposition of dopant species by photodecomposition of gaseous precursors offers a wide range of doping possibilities whereas control over surface tempera-ture can be used to limit dopant diffusion and the depth of doped layers. Excimer laser radiation also permits doping to be carried out over an extended area on the semiconductor surface whereas scanned CW UV laser radiation can be used to create localized areas of doping with dimensions extending to the sub-micrometer range. In some instances, the width of doped regions created with CW laser radiation can be smaller than the focal area when the laser intensity and exposure time are tailored to limit the extent of the heated area on the semi-conductor surface.

Some advantages of laser-assisted doping have been summarized by Bentini *et al.* (1988). These include:

1)   a dislocation-free doped layer,
2)   no surface roughness under optimized irradiation conditions,
3)   high dopant incorporation rates,
4)   shallow heavily doped layers are obtainable,
5)   excellent lateral resolution when masks or imaging used, and
6)   precise control over p–n junction properties.

Early studies of laser-assisted doping by Fairfield and Schwuttke (1968) and by Harper and Cohen (1970) explored the relation between pulse energy, pulse duration and beam size and the properties of the doped region. Later work by Narayan *et al.* (1978a,b) showed that p–n junctions could be formed in Si by laser-assisted B doping. B was deposited as a thin layer on Si prior to irradiation with pulses from a ruby laser (694 nm). Doping of Si and InP using UV laser radiation was subsequently demonstrated by Deutsch *et al.* (1980, 1981a,b) and by Ehrlich *et al.* (1980). Ehrlich *et al.* (1980) used superimposed 514.5 and 257.2 nm CW laser beams to introduce Cd atoms into InP. Cd atoms were obtained by UV photodecomposition of $Cd(CH_3)_2$. The laser beam at 514.5 nm was then used to heat the surface to tem-peratures in the 600–1000°C range to permit Cd atoms to diffuse into the InP substrate. This diffusion was found to occur much more rapidly than the rate of deposition, so the rate limiting step was the photodecomposition of $Cd(CH_3)_2$. Dopant concentrations of up to $10^9\,cm^{-3}$ could be introduced. The level of dopant concentration was found to be linearly dependent on UV laser intensity and $Cd(CH_3)_2$ pressure under conditions for which no surface melting occurred ($T < 1060°C$).

Ohmic contacts can be introduced into InP by laser-assisted doping with Cd or Zn (Deutsch *et al.* 1980). Gaseous precursors such as $M(CH_3)$, where $M = Cd$ or Zn, were found to be good sources of the dopant when irradiated with 193 nm ArF laser radiation. Irradiation at 351 nm (XeF laser) was found to be ineffective, confirming that photochemical decomposition rather than pyrolytic decomposition is the dominant process in the formation of the metal dopant. The resistance of the contacts produced in this way was found to decrease dramatically after incubation with about ten overlapping 193 nm pulses. This was attributed to a surface cleaning effect and to the need to seed the surface with the appropriate nucleation sites for subsequent growth of the doped layer. It is also likely that melting occurs under these conditions.

Subsequent studies have focused on doping of Si with various atoms including Al (Deutsch *et al.* 1981b), B (Deutsch *et al.* 1981a,b, Ibbs and Lloyd 1983, Carey *et al.* 1985, Kato *et al.* 1987a, Sameshima *et al.* 1987, Slaoui *et al.* 1990, Inui *et al.* 1991), P (Deutsch *et al.* 1981a, Kato *et al.* 1987b, Bentini *et al.* 1988) and Sb (Fogarassy *et al.* 1985). Gaseous precursors for B include $B(CH_3)_3$, B_2H_6, BCl_3, BF_3 and $B(C_2H_5)_3$ whereas those for P are PCl_3 and $POCl_3$. Doping occurs when atoms created by photochemical decomposition of the precursor collide with the Si surface and are incorporated into the melt (Figure 8.15) or when adsorbed molecules are dissociated photolytically or pyrolytically. The overall process has been termed gas immersion laser doping (GILD) (Carey *et al.* 1985). When deposition occurs from a gaseous precursor or laser induced melting of predeposited impurity doping (LIMPID) (Sameshima *et al.* 1987) when doping is produced by laser irradiation of a predeposited layer.

Some general characteristics of GILD technique include:

1) Increasing dopant concentration with number of overlapping laser pulses (Figure 8.16).

2) Increasing thickness of the doped layer with number of overlapping pulses.

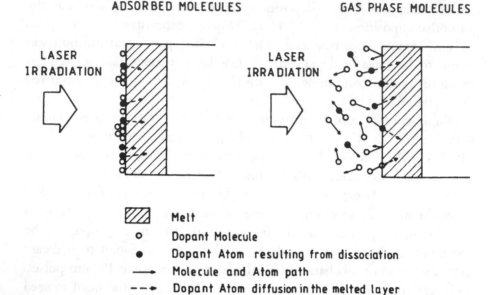

Figure 8.15. A schematic diagram of the two dopant sources considered in the GILD process. In the adsorption case, the dopant is incorporated from an adsorbed (chemisorbed or physisorbed) layer of molecules on the silicon surface. In the impingement case, the dopant is incorporated via gas impingement on the molten silicon (Slaoui *et al.* 1990).

3)   The total incorporated dopant dosage increases linearly with total melting time, i.e. number of overlapping laser pulses.

4)   Sheet resistance decreases with number of overlapping pulses (Figure 8.17).

5)   Sheet resistance decreases rapidly with laser fluence for fluences above the melting threshold.

6)   Sheet resistance rapidly decreases with increasing precursor gas pressure and then remains relatively constant for pressures near 1 atm.

7)   Doping levels achievable with ambient gaseous precursors are higher than those obtained when only adsorbed layers are present.

Several of these characteristics are also observed in the LIMPID technique, specifically the tendency of the dopant concentration and depth distribution to increase with laser pulse fluence and with the number of overlapping pulses applied. The incorporation of dopants in Si in this case has been shown (Fogarassy *et al.* 1985) to depend on the velocity of the solid–liquid interface during resolidification. For Sb

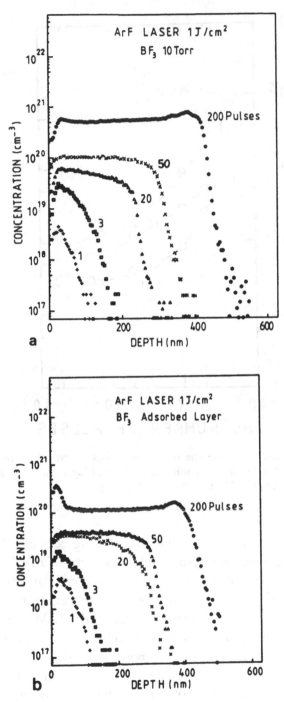

Figure 8.16. Boron concentration profiles by SIMS measurements: (a) doping in 10 Torr BF_3 and (b) doping using the BF_3 adsorbed layer (Slaoui *et al.* 1990).

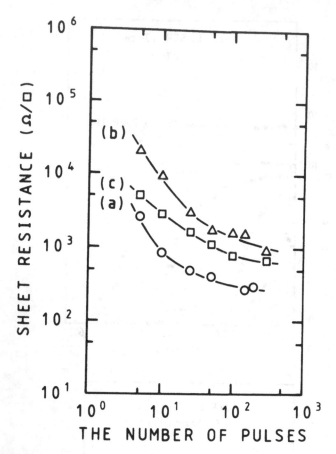

Figure 8.17. Sheet resistance versus the number of pulses. Curves show the doping (a) in POCl$_3$ ambients at 1.2 J cm^{-2}, (b) using the adsorbed layers at 0.8 J cm^{-2}, and (c) using the adsorbed layers at 1.2 J cm^{-2} (Kato *et al.* 1987b).

doping of Si from an evaporated layer using 248 nm KrF laser radiation, this velocity was estimated at about 5.5 m s^{-1}. Interface instabilities during regrowth lead to well defined cellular structures. B-rich layers with a thickness $\lesssim 0.2 \, \mu$m and a B concentration as high as 2×10^{21} cm^{-3} have been fabricated in Si using the LIMPID technique (Sameshima and Usui 1988). The GILD technique can also be used to form heavily doped shallow surface regions; Yoshioka *et al.* (1989) reported on the formation of a shallow junction (about 0.1 μm) with a B concentration of about 5×10^{20} cm^{-3} using ArF irradiation in the presence of B$_2$H$_6$ gas.

Fabrication of p$^+$–n junctions using the GILD technique with depths of only 0.03 μm has been described by Weiner *et al.* (1992). The ideality

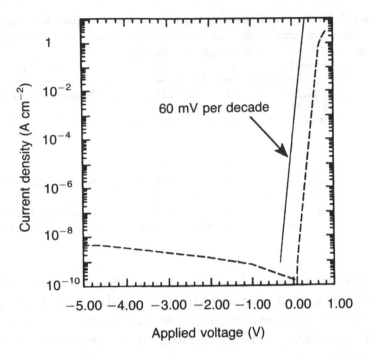

Figure 8.18. Current density–voltage (I–V) characteristics for a 300 Å p$^+$–n diode doped using the laser process (Weiner et al. 1992).

factors for these diodes were found to be 1.01–1.05 over seven decades of current. Figure 8.18 shows an I–V characteristic for a diode created in this way. Fabrication of CMOS (Carey et al. 1985, Tomita et al. 1989), PMOS (Carey et al. 1988), MOSFET (Carey et al. 1986) and narrow base n–p–n transistors (Weiner and Sigmon 1989) using GILD has also been described.

8.5 LASER ANNEALING AND AMORPHIZATION

Historically, laser annealing has been viewed as a means of eliminating ion implantation damage associated with doping in semiconductors (Antonenko et al. 1976, Shtyrkov et al. 1976, Narayan et al. 1978a,b). Both CW radiation (Gat and Gibbons 1978) and pulsed (Narayan et al. 1978a,b) laser radiation have been shown to be effective in annealing out this damage with little or no degradation of the substrate. This early work was primarily carried out with CW Ar or Kr, Nd:YAG and ruby lasers. Annealing with excimer laser radiation was first reported by Yaron and Hess (1980).

Table 8.7. *The fluence threshold for melting of Si surfaces with excimer laser pulses.*

Wavelength (nm)	Material	Pulse length $(n\,s^{-1})$	Fluence $(J\,cm^{-2})$	Reference
308	a–Si:Si	35	0.15	Narayan *et al.* (1984)
	c–Si:Si		0.85	
	B:Si	25	0.5	
		50	0.62	
		70	1.0	
	As:Si	70	0.5	
248	a–Si:Si	24	0.16 ± 0.02	Narayan *et al.* (1984)
	c–Si:Si		0.75 ± 0.05	
193	Si	14	0.13	Gorodetsky *et al.* (1985)

Problems associated with the laser annealing technique involve dopant diffusion, the formation of surface irregularities such as ripples and waves and stress generation. These have limited the large scale industrial application of laser annealing in the processing of semi-conductors. Recent advances, however, have shown that laser annealing with excimer laser radiation may be useful in the fabrication of high performance thin film transistors (TFTs) (Sera *et al.* 1989, Zhang *et al.* 1992). This has renewed interest in laser annealing particularly for post-formation processing of TFTs in integrated circuits, image sensors and active matrix liquid crystal displays.

A number of studies have shown that the absorption of an excimer laser pulse by Si leads to strong localized surface heating (Narayan *et al.* 1984, 1985, Gorodetsky *et al.* 1985, Jellison and Lowndes 1985, Ong *et al.* 1986, Lukes *et al.* 1992). The fluence threshold for melting depends on the laser pulse duration and to some extent on wavelength (Table 8.7). In general, the threshold for melting of amorphous layers is signi-ficantly lower than that for melting of crystalline Si (Figure 8.19). Recrystallization of amorphous layers has been found to yield a mixture of large and fine polycrystals (LP and FP, respectively, in Figure 8.19) (Narayan *et al.* 1985). The FP phase was found to be characterized by crystals having a grain size of about 10 nm. This phase would appear to be formed by explosive recrystallization (Thompson *et al.* 1984), whereas the LP phase is consistent with resolidification of the laser-induced melt back to the free Si surface.

The distribution of implanted dopants is, as expected, dependent on laser fluence. Typically, some broadening of the distribution function

Table 8.8. *Comparison of equilibrium solubility limits and maximum substitutional dopant concentrations after laser ($\lambda = 694\,nm$, ruby) annealing (White et al. 1980).*

Dopant	Equilibrium solubility (cm^{-3})	Maximum substitution concentration (cm^{-3})	Ratio
As	1.5×10^{21}	6×10^{21}	4
Sb	7×10^{19}	1.3×10^{21}	18
Bi	8×10^{17}	4×10^{21}	500
Ga	4.5×10^{19}	4.5×10^{20}	10
In	8×10^{17}	1.5×10^{20}	188

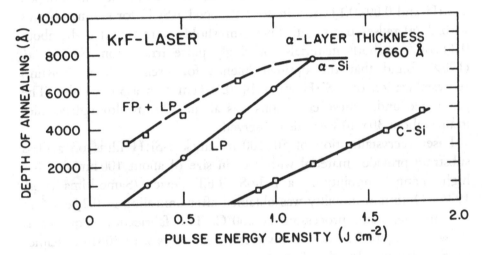

Figure 8.19. Depth of annealing as a function of pulse energy density for amorphous and crystalline samples. In the case of amorphous silicon, the thicknesses of (FP + LP) and LP regions are plotted (Narayan et al. 1985).

occurs after exposure to fluences near the melting threshold. At higher fluence, the distribution broadens significantly and dopant diffusion to both larger and smaller depths is observed (Young et al. 1983). The concentration of dopants in implanted laser-annealed samples can significantly exceed the solid solubility limit (White et al. 1980). Excess atoms above the solubility limit are found to occupy substitutional lattice sites. Some data on dopant concentrations in ion-implanted laser-annealed Si are given in Table 8.8. These data were obtained after annealing with a single pulse from a Q-switched ruby laser ($\lambda = 694\,nm$) (White et al. 1980). Similar concentrations, exceeding the solubility limit, can be maintained in As:Si implanted layers annealed with 308 nm excimer laser pulses.

Excimer laser annealing of amorphous a-Si:H has also been reported (Sameshima *et al.* 1989, Sera *et al.* 1989, Winer *et al.* 1990). Recrystallization of a 30 nm a-Si:H film occurred after exposure to a $0.21 \, \text{J cm}^{-2}$, 308 nm XeCl pulse with a 30 ns FWHM (Sameshima *et al.* 1989). Under these conditions it was found that the melted surface persisted for 80 ns and that the substrate was not exposed to a significant temperature rise. Winer *et al.* (1990) found that both the electrical conductivity and Hall mobilities are greatly increased in a-Si:H and P-doped a-Si:H irradiated at fluences above well defined thresholds. The threshold for melting was found to be 0.12–$0.14 \, \text{J cm}^{-2}$ in undoped a-Si:H and 0.09–$0.1 \, \text{J cm}^{-2}$ in 10^{-2} P-doped a-Si:H for scanned beams with 2560 pulses per point. These thresholds were found to be about $0.2 \, \text{J cm}^{-2}$ in both materials for single pulse irradiation. Sera *et al.* (1989) found that the optimal fluence for excimer laser (308 nm) recrystallization of a-Si:H was 0.14–$0.18 \, \text{J cm}^{-2}$ in a 35 ns pulse. The grain size under these conditions was about 200 nm although smaller grains (about 30 nm) were also observed.

Laser recrystallization of 50–100 nm thick a-Si:H films on a glass substrate provides material with a grain size of about 100 nm that has high channel mobility in a poly-Si TFT device (Sameshima *et al.* 1986). Optimum mobility was obtained after annealing at $0.2 \, \text{J cm}^{-2}$ in a 35 ns pulse and processing at 260°C. The fabrication sequence is shown in Figure 8.20 and involves two treatments with 308 nm excimer laser radiation. In the first treatment, laser radiation is used to recrystallize the a-Si:H deposited by plasma CVD. The second treatment occurs after p^+ implantation to form the source and drain regions and involves activation of implanted phosphorus using laser annealing. Laser-induced dehydrogenation of a-Si:H has been discussed by Mei *et al.* (1994).

Excimer laser radiation can also be used to amorphize crystalline Si (Sameshima *et al.* 1990a,b). The process occurs in the rapid melting followed by regrowth of a surface laser-heated layer and is reversible, i.e. the amorphous layer can be recrystallized by subsequent exposure to another laser pulse at slightly lower fluence. The interface between these two processing regimes occurred for a fluence near $0.24 \, \text{J cm}^{-2}$ with amorphization occurring at fluences $>0.24 \, \text{J cm}^{-2}$ and recrystallization of amorphized material occurring at fluences $<0.24 \, \text{J cm}^{-2}$ p-doped poly-Si films have also been crystallized in this way.

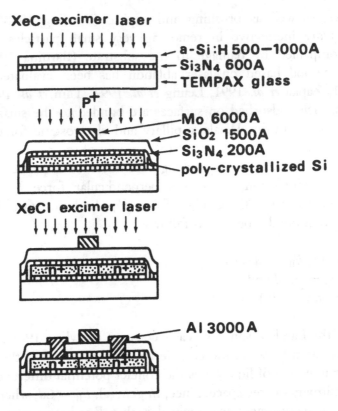

Figure 8.20. A schematic cross-section of the poly-Si TFT during the fabrication process (Sameshima *et al.* 1986).

8.6 PARTICLE REMOVAL

The presence of fine particulate matter on the surface of semiconductor devices can provide a severe limit in the form of device defects. This problem is particularly significant in the manufacture of ultra-large scale integrated circuits with sub-micrometer pattern dimensions. In such devices, particles with sizes as small as 0.1 µm can interfere with pattern generation and result in defective devices. Sub-micrometer particles are abundant in the normal atmosphere and are readily produced from liquids during device manufacture. Desorption of water vapor from Si using N_2 laser radiation has been reported by Tam and Schroeder (1988).

Conventional removal mechanisms for sub-micrometer particles involve such techniques as high pressure jet spraying, ultrasonics and megasonics, as well as brushing and wiping (Bowling 1988). These techniques are ineffective in removing very small particles and are often accompanied by damage to the substrate (Bardina 1988). An alternate method based on laser ablation has been evaluated (Imen et al. 1991, Zapka et al. 1991, Leung et al. 1992, Tam et al. 1992).

Small particles adsorbed on surfaces are bonded to the surface by a combination of van der Waals, capillary and electrostatic forces. Such forces are substantial and may exceed the weight of the particle by many orders of magnitude. If one envisages removal of a particle with a characteristic dimension, d, then a perpendicular force F must be applied, which exceeds the binding force, F_B. Analytic expressions for F_B for different bond types are as follows:

van der Waals, $F_B = hd/(16\pi z^2)$,
capillary, $F_B = 2\pi\gamma d$ and
ionic, $F_B = \pi\epsilon u^2 d/(2z)$,

where h is the Landau–van der Waals constant (Bowling 1988), z is the closest separation between the particle and the surface, γ is the surface energy per unit area of liquid, u is the contact potential difference and ϵ is the permittivity of free space. Since, in general, $F_B = Cd$, where C is a constant, the requirement for removal is that $F \simeq F_B$ or $ma = Cd$. As $m \simeq \rho d^3$, where ρ is particle density, the critical value for the acceleration, a, normal to the surface is $a \simeq (C/\rho)d^2$. Typically, $C/\rho \simeq 10^{-6}\,\mathrm{m^3\,s^{-2}}$ so that $a = 10^8\,\mathrm{m\,s^{-2}}$ for a particle with $d = 0.1\,\mu\mathrm{m}$. Such an acceleration is roughly $10^7\,g$, where g is the acceleration due to gravity.

Such accelerations can be produced during laser heating, either of the particle, or of the underlying surface. Direct absorption of incident laser radiation by the adsorbed particle (Kelly et al. 1991) can yield desorption via the rapid evolution of vapor from the particle. For this to occur, the particle must be strongly absorbing at the laser wavelength, whereas the substrate should be transparent. This limits the applicability of this technique. In addition, vaporization and particle breakup can result in redeposition of material on the substrate, reducing the overall efficiency of particulate removal.

Absorption of laser radiation by the substrate has been shown to result in efficient particle removal (Zapka et al. 1991, Tam et al. 1992). Removal has been postulated to occur as the surface expands in response to laser heating. For a temperature rise ΔT, the expansion normal to the surface, Δz is approximately

$$\Delta z = \frac{\beta \int I_{abs}(t)\, dt}{\rho C_p} \qquad (19)$$

where β is the expansion coefficient and C_p is the heat capacity. With $\beta = 10^{-6}\,K^{-1}$, $\rho = 3 \times 10^3\,kg\,m^{-3}$, $C_p = 300\,J\,kg^{-1}\,K^{-1}$ and $\int I_{abs}(t)\, dt = 10^4\,J\,m^{-2}$ one obtains $\Delta z \simeq 10^{-8}\,m$. For a pulse of 10 ns duration, the average velocity and acceleration of the surface would be $1\,m\,s^{-1}$ and $10^8\,m\,s^{-2}$, respectively. The latter is comparable to the normal acceleration required to exceed F_B. Tam et $al.$ (1992) reported that this expansion was sufficient to eject 0.3 µm Al_2O_3 particles from the surface or from grooves and trenches on Si.

Particle removal can be enhanced by the provision of a thin liquid layer (Imen et $al.$ 1991, Zapka et $al.$ 1991), which is rapidly heated directly by absorption of laser radiation, or by conduction from the substrate. A comprehensive study (Tam et $al.$ 1992) showed that optimized particle removal rates were obtained when strong absorption occurred in the Si substrate in the presence of a water layer. The water layer, which was typically a few micrometers thick, was deposited using a pulsed jet just prior to laser irradiation (Figure 8.21). Extensive particle removal was observed when irradiating with as little as $0.12\,J\,cm^{-2}$ at 248 nm under these conditions. Similar experiments have also been reported by Imen et $al.$ (1991) using pulses of 10.6 µm radiation from a TEA-type CO_2 laser to vaporize a water layer in contact with the adsorbed particles. Explosive vaporization of water adsorbed at the particle–substrate interface is probably the source of the normal force that liberates particles under these conditions. The fluence at 10.6 µm was about $30\,J\,cm^{-2}$ in a 1 µs pulse. A comprehensive study of the effect of rapid energy deposition into a thin liquid layer on an Si surface using KrF laser radiation has been reported by Leung et $al.$ (1992). They find that strong superheating of the liquid occurs at the liquid–substrate interface. The resulting explosion yields the high transient pressure required to eject particles. Liquid film thickness was found to be relatively unimportant as the primary interaction was found to occur with about 0.1 µm of the substrate–liquid surface.

Tsu et $al.$ (1991) have discussed the use of 248 nm excimer laser radiation to clean Si surfaces at the atomic level, specifically the removal of C and O impurities. It was found that irradiation at fluences

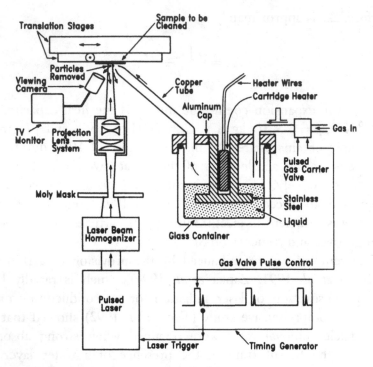

Figure 8.21. The experimental arrangement for liquid-film-enhanced pulsed laser cleaning to remove particulates on a surface (from A. C. Tam, W. P. Leung, W. Zapka and W. Ziemich, *Journal of Applied Physics* **71**, 3515, 1992).

$<0.5\,\mathrm{J\,cm^{-2}}$ did lead to a significant reduction in these species but some surface species were retained even after exposure to a substantial (about 200) number of overlapping pulses. However, at fluences sufficient to produce surface melting the concentration of C and O was reduced to very low levels. An analysis of the distribution of C and O impurities showed that both had been incorporated in the melted volume, so that the reduction in surface concentration of these species was due in part to diffusion into the melt and in part to desorption.

Overall, laser cleaning with or without the involvement of a liquid layer intermediary would appear to be a useful new technique in ultra-clean manufacturing of semiconductor and data storage devices.

8.7 REFERENCES

Alferov Z. I., Goryachev D. N., Gurevich S. A., Mizerov M. N., Portnoi E. L. and Ryvkin B. S., 1976. *Sov. Phys. Tech. Phys.* **21**, 857.

Antonenko A. K., Gerasimenko N. N., Drureihenski A. V., Sminov L. S. and Tseitlin G. M., 1976. *Sov. Phys. Semicond.* **10**, 81.

Arlinghaus H. F., Calaway W. F., Young C. E., Pellin M. J., Gruen D. M. and Chase L. L., 1989. *Appl. Phys. Lett.* **54**, 3176.

Arnone C., Rothschilde M. and Ehrlich D. J., 1986. *Appl. Phys. Lett.* **48**, 736.

Ashby C. I. H., 1984. *Appl. Phys. Lett.* **45**, 892.

Ashby C. I. H., 1991. 'Laser driven etching' in *Thin Film Processes* vol. II, Academic Press, New York.

Bäcklin L., 1987. *Electron. Lett.* **23**, 657.

Baller, T., Oostra D. J., de Vries A. E. and van Veen G. A. N., 1986. *J. Appl. Phys.* **60**, 2321.

Bardina J., 1988. in *Particles on Surfaces* ed. K. L. Mittal, Plenum, New York, Vol. 1, p. 327.

Bäuerle D., 1986. *Chemical Processing with Lasers*, Springer-Verlag, New York.

Bentini G. G., Bianconi M. and Summonte C., 1988. *Appl. Phys.* **A45**, 317.

Berman M. R., 1991. *Mater. Res. Soc. Symp. Proc.* **201**, 471.

Bialkowski M. M., Hurst G. S., Parks J. E., Lowndes D. H. and Jellison G. E., 1990. *J. Appl. Phys.* **68**, 4795.

Blum S. E., Brown K. H. and Srinivasan R., 1983. *Appl. Phys. Lett.* **43**, 1026.

Boulmer J., Budin J. P., Bourguignon B., Debarne D. and Desmur A., 1992 in *Laser Ablation of Electronic Materials* ed. E. Forarassy and S. Lazane, Elsevier, Amsterdam, p. 239.

Bowling R. A., 1988. in *Particles on Surfaces* ed. K. L. Mittal, Plenum, New York, vol. 1, p. 129.

Boyd I. W., 1989. *Mater. Res. Soc. Symp. Proc.* **129**, 421.

Boyd I. W. and Wilson J. I. B., 1982. *Appl. Phys. Lett.* **41**, 162.

Brewer P., Halle S. and Osgood R. M. Jr, 1984. *Appl. Phys. Lett.* **45**, 475.

Brewer P. D., McClure D. and Osgood R. M. Jr, 1985. *Appl. Phys. Lett.* **47**, 310.

Brewer P. D., McClure D. and Osgood R. M. Jr, 1986. *Appl. Phys. Lett.* **49**, 803.

Brewer P. D., Zinck J. J. and Olson G. L., 1990. *Mater. Res. Soc. Symp. Proc.* **191**, 67.

Brewer P. D., Zinck J. J. and Olson G. L., 1991. *Mater. Res. Soc. Symp. Proc.* **201**, 543.

Carey P. G., Bezjian K., Sigmon T. W., Gildea P. and Mager T. J., 1986. *IEEE Electron Device Lett.* **7**, 440.

Carey P. G., Sigmon T. W., Press R. L. and Fahlen T. S., 1985. *IEEE Electron Device Lett.* **6**, 291.

Carey P. G., Weiner K. H. and Sigmon T. W., 1988. *IEEE Electron Device Lett.* **9**, 542.

Chiang S. W., Liu Y. S. and Riehl R. F., 1981. *Appl. Phys. Lett.* **39**, 752.

Cros A., Salvan F. and Derrien J., 1982. *Appl. Phys.* **A28**, 241.

Cohen C., Siejka J., Berti M., Drigo A. V., Bentini G. G., Pribat D. and Jannitti E., 1984. *J. Appl. Phys.* **55**, 4081.

Davis G. M., Thomas D. W. and Gower M. C., 1988. *J. Phys. D: Appl. Phys.* **21**, 683.

Deutsch T. F., Ehrlich D. J., Osgood R. M. Jr and Liau Z. L., 1980. *Appl. Phys. Lett.* **36**, 847.

Deutsch T. F., Ehrlich D. J., Rathman D. D., Silversmith D. J. and Osgood R. M. Jr, 1981a. *Appl. Phys. Lett.* **39**, 825.

Deutsch T. F., Fan J. C. C., Turner G. W., Chapman R. L., Ehrlich D. J. and Osgood R. M. Jr, 1981b. *Appl. Phys. Lett.* **38**, 144.

Donnelly V. M. and Hayes T. R., 1990. *Appl. Phys. Lett.* **57**, 701.

Ehrlich D. J., Osgood R. M. Jr and Deutsch T. F., 1980a. *Appl. Phys. Lett.* **36**, 698.

Ehrlich D. J., Osgood R. M. Jr and Deutsch T. F., 1980b. *Appl. Phys. Lett.* **36**, 916.

Fairfield J. M. and Schwuttke G. H., 1968. *Solid State Electron.* **11**, 1175.

Fogarassy E. P., Lowndes D. H., Narayan J. and White C. W., 1985. *J. Appl. Phys.* **58**, 2167.

Fogarassy E., White C. W., Slaoui A., Fuchs C., Siffert P. and Pennycook S. J., 1988. *Appl. Phys. Lett.* **53**, 1720.

Foulon F., Fogarassy E., Slaoui A., Fuchs C., Unamuno S. and Siffert P., 1988. *Appl. Phys.* A**45**, 361.

Gat A. and Gibbons J. F., 1978. *Appl. Phys. Lett.* **32**, 143.

Georgiev M. and Singh J., 1992. *Appl. Phys.* A**55**, 175.

Gorodetsky G., Kanicki J., Kazyaka T. and Melcher R. L., 1985. *Appl. Phys. Lett.* **46**, 547.

Hanabusa M., 1993. *Mater. Res. Soc. Symp. Proc.* **285**, 447.

Harper F. I. and Cohen M. J., 1970. *Solid State Electron.* **13**, 1103.

Hattori K., Nakai Y. and Itoh N., 1990. *Surf. Sci. Lett.* **227**, L115.

Hattori K., Okano A., Nakai Y., Itoh N. and Haglund R. F., 1991. *J. Phys.: Condens. Matter* **3**, 7001.

Hattori K., Okano A., Nakai Y. and Itoh N., 1992. *Phys. Rev.* B**45**, 8424.

Hayasaka N., Okano H., Sekine M. and Horiihe Y., 1986. *Appl. Phys. Lett.* **48**, 1165.

Haynes R. W., Metze G. M., Kreismanis V. G. and Eastman L. F., 1980. *Appl. Phys. Lett.* **37**, 344.

Heaven M. C. and Clyne M. A. A., 1982. *J. Chem. Soc. Faraday Trans.* **78**, 1339.

Herman P. R., Chen B., Moore D. J. and Canaga-Retnam M., 1992. *Mater. Res. Soc. Symp. Proc.* **236**, 53.

Hoh K., Koyama H., Uda K. and Miura Y., 1980. *Jap. J. Appl. Phys.* **19**, L375.

Horike Y., Hayasaka N., Sekine M., Arikado T., Nakase M. and Okano H., 1987. *Appl. Phys.* A**44**, 313.

Ibbs K. G. and Lloyd M. L., 1983. *Mater. Res. Soc. Symp. Proc.* **17**, 243.

Ichige K., Matsumoto Y. and Nakiki A., 1988. *Nucl. Instrum. Methods* B**33**, 820.

Imen K., Lee S. J. and Allen S. D., 1991. *Appl. Phys. Lett.* **58**, 203.

Inui S., Nii T. and Matsumoto S., 1991. *IEEE Electron Device Lett.* **12**, 702.

Itoh N. and Nakayama T., 1982. *Phys. Lett.* A**92**, 471.

Jellison Jr G. E. and Lowndes D. J., 1985. *Appl. Phys. Lett.* **47**, 718.

Kasuya A. and Nishina Y., 1990. *Mater. Res. Soc. Symp. Proc.* **191**, 73.

Kato S., Nagahori T. and Matsumoto S., 1987a. *J. Appl. Phys.* **62**, 3656.

Kato S., Saeki H., Wada J. and Matsumoto S., 1987b. *J. Electrochem Soc.* **135**, 1030.

Kazor A. and Boyd I. W., 1994. *J. Appl. Phys.* **75**, 227.

Kelly J. D., Stuff M. I., Hovis F. E. and Linford G. J., 1991. *SPIE Proc.* **1415**, 211.

Kelly R., 1990. *J. Chem. Phys.* **92**, 5047.

Kelly R. and Dreyfuss R. W., 1988. *Surf. Sci.* **198**, 263.

Koren G. and Hurst J. E. Jr, 1988. *Appl. Phys.* A**45**, 301.

Kullmer R. and Bäuerle D., 1987. *Appl. Phys.* A**43**, 227.

Kullmer R. and Bäuerle D., 1988. *Appl. Phys.* A**47**, 377.

Leung P. T., Do N., Klees L., Leung W. P., Tong F., Lam L., Zapka W. and Tam A. C., 1992. *J. Appl. Phys.* **72**, 2256.

Liu Y. S., Chiang S. W. and Bacon F., 1981. *Appl. Phys. Lett.* **38**, 1006.

Lukes I., Sasik R. and Cerny R., 1992. *Appl. Phys.* A**54**, 327.

Maki P. A. and Erhlich D. J., 1989. *Appl. Phys. Lett.* **55**, 91.

Mei P., Boyce J. B., Hack M., Lujan R. A., Johnson R. I., Anderson G. B., Fork D. K. and Ready S. E., 1994. *Appl. Phys. Lett.* **64**, 1132.

Moison J. M. and Bensoussan M., 1982. *J. Vac. Sci. Technol.* **21**, 315.

Nakai Y., Hattori K., Okano A., Itoh N. and Haglund R., 1991. *Nucl. Instrum. Methods.* B**58**, 452.

Nakayama T., 1983. *Surf. Sci.* **133**, 101.

Nakayama T., Okigawa M. and Itoh N., 1982. *Radiat. Effects Lett.* **67**, 129.

Nakayama T., Okigawa M. and Itoh N., 1984. *Nucl. Instrum. Methods* B229, 301.

Namiki A., Cho S. and Ichige K., 1987. *Jap. J. Appl. Phys.* **26**, 315.

Namiki A., Katoh K., Yamashita Y., Matsumoto Y., Amano H. and Akasaki I., 1991. *J. Appl. Phys.* **70**, 3268.

Namiki A., Kawai T. and Ichige K., 1986. *Surf. Sci.* **166**, 129.

Narayan J., Holland O. W., White C. W. and Young R. T., 1984. *J. Appl. Phys.* **55**, 1125.

Narayan J., White C. W., Aziz M. J., Stritzker B. and Walthuis A., 1985. *J. Appl. Phys.* **57**, 564.

Narayan J., Young R. T., Wood R. F. and Christie W. H., 1978a. *Appl. Phys. Lett.* **33**, 338.

Narayan J., Young R. T. and White C. W., 1978b. *J. Appl. Phys.* **49**, 3912.

Okano A., Hattori K., Nakai Y. and Itoh N., 1991. *Surf. Sci.* **258**, L67.

Okano A., Matsuura A. Y., Hattori K., Itoh N. and Singh J., 1993. *J. Appl. Phys.* **73**, 3158.

Ong C. K., Sin E. H. and Tan H. S., 1986. *J. Opt. Soc. Am.* B3, 812.

Orlowski T. E. and Richter H., 1984. *Appl. Phys. Lett.* **45**, 241.

Osgood R. M. Jr, Sanchez-Rubio A., Ehrlich D. J. and Daneu V., 1982. *Appl. Phys. Lett.* **40**, 391.

Ostermayer F. W. Jr and Kohl P. A., 1981. *Appl. Phys. Lett.* **39**, 76.

Podlesnik D. V., Gilgen H. H. and Osgood R. M. Jr, 1983. *Appl. Phys. Lett.* **43**, 1083.

Podlesnik D. V., Gilgen H. H. and Osgood R. M. Jr, 1984. *Appl. Phys. Lett.* **45**, 564.

Podlesnik D. V., Gilgen H. H. and Osgood R. M. Jr, 1986a. *Appl. Phys. Lett.* **48**, 496.

Podlesnik D. V., Gilgen H. H., Willner A. E. and Osgood R. M., 1986b. *J. Opt. Soc. Am.* B3, 775.

Pospieszczyk A., Harith M. A. and Stritzer B., 1983. *J. Appl. Phys.* **54**, 3176.

Qin Q. Z., Li Y. L., Jin Z. K., Zhang Z. J., Yang Y. Y., Jia W. J. and Zheng Q. K., 1988. *Surf. Sci.* **207**, 142.

Raffel J. I. *et al.* 1985. *IEEE J. Solid State Circuits* **20**, 399.

Reksten G. M., Holber W. and Osgood R. M. Jr, 1986. *Appl. Phys. Lett.* **48**, 551.

Richter H., Orlowski T. E., Kelly M. and Margaritondo G., 1984. *J. Appl. Phys.* **56**, 2351.

Rothschild M., Arnone C. and Ehrlich D. J., 1987. *J. Mater. Res.* **2**, 244.

Sameshima T., Hara M. and Usui S., 1989. *Jap. J. Appl. Phys.* **28**, L2131.

Sameshima T., Hara M. and Usui S., 1990a. *Jap. J. Appl. Phys.* **29**, L548.

Sameshima T., Hara M. and Usui S., 1990b. *Jap. J. Appl. Phys.* **29**, L1363.

Sameshima T. and Usui S., 1988. *Mater. Res. Soc. Symp. Proc.* **101**, 491.

Sameshima T., Usui S. and Sekiya M., 1986. *IEEE Electron Device Lett.* **7**, 276.

Sameshima T., Usui S. and Sekiya M., 1987. *J. Appl. Phys.* **62**, 711.

Schafer S. A. and Lyon S. A., 1982. *J. Vac. Sci. Technol.* **21**, 422.

Seel M. and Bagus P. S., 1983. *Phys. Rev.* B28, 2023.

Sera K., Okumura F., Uchida H., Itoh S., Kaneko S. and Hotta K., 1989. *IEEE Trans. Electron Devices* **36**, 2868.

Sesselmann W., 1989. *Chemtronics* **4**, 135.

Shimizu T., Itoh N. and Matsunami N., 1988. *J. Appl. Phys.* **64**, 3663.

Shinn G. B., Steigerwald F., Stiegler R., Sauerbrey R., Tittel F. K. and Wilson W. L., 1986. *J. Vac. Sci. Technol.* B4, 1273.

Shtyrkov E. I., Khaibullin I. B., Zaripov M. M., Galyatudinov M. F. and Bayazitov R. M., 1976. *Sov. Phys. Semicond.* **9**, 1309.

Siejka J., Perriene J. and Srinivasan R., 1985. *Appl. Phys. Lett.* **46**, 773.

Singh J. and Itoh N., 1990. *Appl. Phys.* A**51**, 427.

Slaoui A., Fogarassy E., White C. W. and Siffert P., 1988. *Appl. Phys. Lett.* **53**, 1832.

Slaoui A., Foulon F., Stuck R. and Siffert P., 1990. *Appl. Phys.* A**50**, 479.

Smith R. T., Chlipala J. D., Bindels J. F. M., Nelson R. G., Fischer F. H. and Mantz T., 1981. *IEEE J. Solid State Circuits* **16**, 506.

Stritzker B., Pospieszczyk A. and Tagle J. A., 1981. *Phys. Rev. Lett.* **47**, 356.

Tam A. C., Leung W. P., Zapka W. and Ziemich W., 1992. *J. Appl. Phys.* **71**, 3515.

Tam A. C. and Schroeder H., 1988. *J. Appl. Phys.* **64**, 3667.

Tejedor P. and Briones F., 1991. *Mater. Res. Soc. Symp. Proc.* **201**, 141.

Tench R. J., Balooch M., Bernardez L., Allen M. J., Siekhaus W. J., Olander D. R. and Wang W., 1991. *J. Vac. Sci. Technol.* B**9**, 820.

Thompson M. O., Galvin G. J., Mayer J. W., Peercy P. S., Poate J. M., Jacobson D. C., Cullis A. G. and Chew N. G., 1984. *Phys. Rev. Lett.* **52**, 2360.

Tisone G. C. and Johnson A. W., 1983. *Appl. Phys. Lett.* **42**, 530.

Tomita H., Negishi M., Sameshima T. and Usui S., 1989. *IEEE Electron Device Lett.* **10**, 547.

Treyz G. V., Scarmozzino R., Burke R. and Osgood R. M. Jr, 1989. *Appl. Phys. Lett.* **54**, 561.

Tsu R., Lubben D., Bramblet T. R. and Greene J. E., 1991. *J. Vac. Sci. Technol.* A**9**, 223.

Uzan C., Legros R., Marfaing Y. and Triboulet R., 1984. *Appl. Phys. Lett.* **45**, 879.

Vega F., Afonso C. N., Ortega C. and Siejka J., 1993. *J. Appl. Phys.* **74**, 963.

Weiner K. H., Carey P. G., McCarthy A. M. and Sigmon T. W., 1992. *IEEE Electron Device Lett.* **13**, 369.

Weiner K. H. and Sigmon T. W., 1989. *IEEE Electron Device Lett.* **10**, 260.

White C. W., Wilson S. R., Appleton B. R. and Young F. W. Jr, 1980. *J. Appl. Phys.* **51**, 738.

Willner A. E., Ruberto M. N., Blumenthal D. J., Podlesnik D. V. and Osgood R. M. Jr, 1989. *Appl. Phys. Lett.* **54**, 1839.

Winer K., Anderson G. B., Ready S. E., Bachrach R. Z., Johnson R. I., Ponce F. A. and Boyce J. B., 1990. *Appl. Phys. Lett.* **57**, 2222.

Yaron G. and Hess L. D., 1980. *Appl. Phys. Lett.* **36**, 220.

Yoshioka S., Wada J., Saeki H. and Matsumoto S., 1989 *Mater. Res. Soc. Symp. Proc.* **129**, 597.

Young R. T., Narayan J., Christie W. H., van der Leeden G. A., Levatter J. I. and Cheng L. J., 1983. *Solid State Technol.* **26**, 183.

Zapka W., Ziemlich W. and Tam A. C., 1991. *Appl. Phys. Lett.* **58**, 2217.

Zhang H., Kusumoto N., Inushima T. and Yamazaki S., 1992. *IEEE Electron Device Lett.* **13**, 297.

CHAPTER 9

Laser deposition

9.1 INTRODUCTION

A wide variety of materials can be deposited from gaseous, solid and liquid precursors using laser techniques. Photothermal as well as photochemical routes are often available and range from the straightforward use of laser radiation as a vaporization source to photochemical decomposition of adsorbed layers. Laser deposition can be used for the creation of extended thin films or for selective deposition of specific features in localized regions with dimensions extending to less than 1 μm. The choice of deposition technique will depend on the required composition of the deposit together with the properties of the substrate. Laser wavelength may be of primary importance for photochemical deposition of sub-micrometer features. Some significant factors in the laser deposition of materials are:

1) chemical routes to the required deposit,
2) laser intensity and wavelength,
3) sensitivity of substrate to thermal/photochemical effects,
4) sensitivity of substrate to ambient atmosphere/chemical environment,
5) scale of features to be deposited, i.e. micro/macrofeatures,
6) required deposition rate,
7) sensitivity of deposited layer to contamination by secondary products and/or particulates, and
8) optical configuration required for deposition.

9.2 CHEMICAL ROUTES

An example is amorphous Si, which may be directly deposited by sputtering of an Si target using 248 nm KrF laser radiation at rates of

0.1–0.3 nm per pulse (Reddy 1986). An alternative route is via the photochemical decomposition of SiH_4

$$SiH_4 + h\nu \rightarrow Si\downarrow + 2H_2 \tag{1}$$

with KrF laser radiation (Gee *et al.* 1984). This proceeds at a rate of 1–2 nm s^{-1}. A similar reaction can be initiated photothermally in SiH_4 using CO_2 laser radiation to heat the substrate over the area in which deposition is to occur (Baranauskas *et al.* 1980). Other gaseous precursors such as $SiCl_4$ and SiH_2Cl_2 can be used instead of SiH_4.

Films of compounds with complex stoichiometry such as the non-linear material $KTa_{1-x}Nb_xO_3$ (KTN) can be produced by laser sputtering of the parent material; however, partial loss of volatile components such as K can occur. To overcome this difficulty, laser radiation can be used to ablate a segmented target, which provides additional quantities of the necessary component (Yilmaz *et al.* 1991). For KTN, Yilmaz *et al.* (1991) used a segmented target of KTN and KNO_3 with the latter material providing a higher level of K in the vapor condensing to form the KTN film.

Metallic alloys such as Fe–Nb, Fe–Ta, etc. can be produced by irradiating alloy targets or alternately by irradiating individual targets of the elements involved (Krebs and Bremert 1993). The latter irradiation geometry is shown schematically in Figure 9.1.

9.3 LASER INTENSITY AND WAVELENGTH

Al films may be deposited by photothermal decomposition of adsorbed molecular layers (Baum and Comita 1992). However, a study of the projection patterning of Al films grown from excimer-laser-induced decomposition of tri(isobutyl)aluminum (Higashi 1989) shows that feature definition is greatly reduced when irradiation occurs at 193 nm rather than at 248 nm (Figure 9.2). This effect is attributed to photochemical decomposition of gaseous molecules generating gas phase Al that is deposited less selectively on the substrate.

Direct writing of high conductivity Al interconnects on SiO_2 using a dimethylaluminum hydride precursor has been reported by Cacouris *et al.* (1988). UV radiation was obtained from a CW Ar$^+$ laser (275 nm) or a frequency-doubled Ar$^+$ laser (257 nm). At a scan speed of 42 μm s^{-1}, slow, photolytic, deposition was found to occur at laser intensities up to

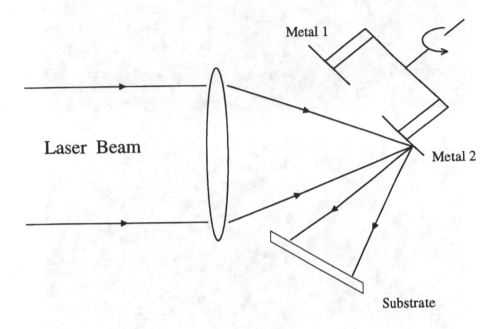

Figure 9.1. A schematic diagram showing a rotating target source for deposition
of metallic alloys by sequential ablation of elemental targets.

about $162\,\mathrm{kW\,cm^{-2}}$. At higher intensities, the deposition rate increased
exponentially with incident intensity. This behavior is consistent with
photothermal decomposition. Al lines deposited under photothermal
conditions were found to be much less conductive than those deposited
photolytically.

9.4 SUBSTRATE EFFECTS

Absorption of incident laser radiation by the substrate will act to
enhance thermally activated reactions and may obscure the more selective effects associated with photochemical deposition. Thermal heating
will also promote the migration of surface species, as well as the nucleation and growth of deposits (Yokoyama *et al.* 1985). Photochemical
effects in the substrate induced by UV laser radiation can act to create
defect centers that facilitate the initiation of deposition (Tsao and
Ehrlich 1984). For example, electron–hole pairs generated in Si by laser

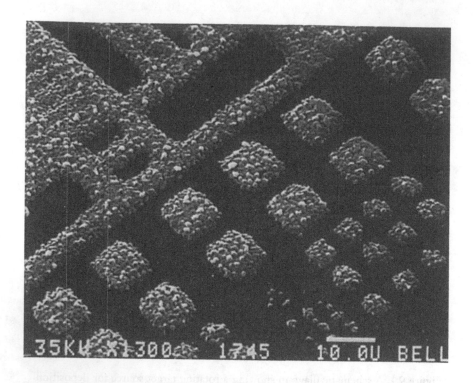

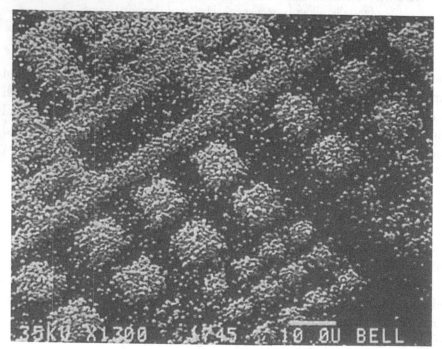

Figure 9.2. A comparison of Al films prepared by irradiation of adsorbed tri(isobutyl)-aluminum at 248 nm (upper picture) and at 193 nm (lower picture) (from Higashi 1989).

irradiation have been postulated to enhance the reaction between Si and radicals generated from the photochemical decomposition of NH_3 by ArF (193 nm) laser radiation (Sugii *et al.* 1988). This occurs as trapping of surface excitons weakens the bonds between Si atoms.

Under certain conditions, other parameters such as the thermal conductivity of the substrate may be an important factor in determining the onset of photothermal effects. Arnone *et al.* (1986) have found that the growth of chromium oxide films by radiation-induced decomposition of CrO_2Cl_2 is dominated by photochemical effects at low laser intensity for deposition either on SiO_2 or on Si substrates. However, the onset of the rapid photothermal deposition phase was greatly inhibited when deposition was carried out on Si (thermal conductivity, $K = 1.5 \, \mathrm{W \, cm^{-1} \, K^{-1}}$). Photothermal deposition occurred readily on SiO_2 ($K = 0.014 \, \mathrm{W \, cm^{-1} \, K^{-1}}$) at relatively low incident laser intensities.

9.5 SCALE OF FEATURES AND OPTICAL SYSTEMS

Tight focusing, together with scanning of the incident beam over the surface, can be used to create individual discrete depositions (lines, dots, simple geometrical features) (Bäuerle 1988). Line widths of $\lesssim 0.2 \, \mu m$ can be produced in this way (Ehrlich and Tsao 1984) although linear deposition rates are small (about $100 \, \mu m \, s^{-1}$).

Extended area films of approximately uniform thickness can be deposited by laser chemical vapor deposition (LCVD) from a precursor gas or by direct sputtering of a solid target. In the latter case the laser acts as a vaporization source and limitations are similar to those obtained with conventional sources. LCVD can be used to deposit films over areas as large as 10–$50 \, cm^2$, with the upper limit determined by the area of the laser beam. The beam may interact with the precursor gas only (parallel configuration), with the gas–surface interface (perpendicular configuration) or with both gas and surface (combined parallel and perpendicular incidence) (Figure 9.3). This flexibility offers a range of photothermal and photochemical processing regimes (Boyer *et al.* 1984).

For many applications, it is desirable to deposit films with predetermined structures. Such parallel patterned deposition schemes are necessary for full wafer fabricating applications and offer the advantage of high writing speeds and reproducibility. Patterning may be achieved by masking (contact or proximity), by image projection and by holographic

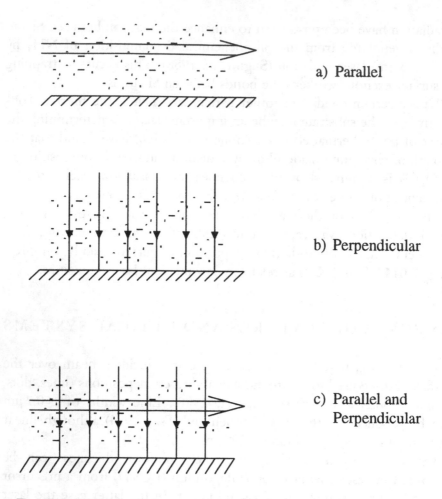

Figure 9.3. LCVD irradiation conditions: (a) parallel incidence, (b) perpendicular incidence and (c) combined parallel/perpendicular incidence.

image formation. The latter method is presently unsuitable for excimer laser deposition because of the low coherence of excimer laser sources. However, the large beam area of these devices, together with their poor temporal and spatial coherence, makes these sources suitable for image projection patterning and for the uniform irradiation of contact and proximity masks. Indeed, such large scale projection applications require that laser speckle by minimized. This is only possible with low coherence sources.

Although masking is a straightforward and readily applied technique, photodeposition of highly reactive species can lead to etching of the mask and to deposition on the mask. In addition, the mask may inhibit the mass flow rate of reaction products to or from the substrate. Reactant diffusion is also an important consideration both with masking and with image projection. This has the effect of reducing the sharpness of edges on the periphery of the deposition area. However, this effect can be minimized to some extent by increasing gas pressure when deposition occurs via a laser–surface reaction.

With image projection, interaction between the propagating laser beam and the gaseous precursor may be important under certain conditions (Higashi 1989) and can lead to image (i.e. pattern) degradation. A schematic diagram of a laser photodeposition system based on imaging of a mask is shown in Figure 9.4.

A novel laser deposition system, which involves mass transfer from a mask to the deposition surface, has been described by Bohandy *et al.* (1986, 1988). The technique utilizes thermal ablation at the interior surface of the mask to mechanically transfer material from the mask to a nearby substrate (Figure 9.5). This technique can yield clean deposits but edge definition would appear to be poor.

9.6 DEPOSITION RATES

Deposition rates using the laser sputtering technique are limited by the ablation rate, Δ, of the target material. This is typically 0.1–0.2 μm per pulse for many materials. If A is the ablated area on the target and R is the target–substrate distance then

$$\delta \simeq \frac{\Delta A}{2\pi R^2} \qquad (2)$$

is the deposition rate per pulse on the substrate. With $\Delta = 1\,\mu m$ per pulse, $A = 1\,mm \times 1\,mm = 10^{-6}\,m^2$, and $R = 0.1\,m$, $\delta \simeq 1.6 \times 10^{-11}\,m = 0.016\,nm$ per pulse. Deposition rates of 10^{-3}–$1\,nm$ per pulse are typically obtained using laser sputtering. Under these conditions each pulse deposits up to a monolayer. The kinetics of film growth under these conditions have been discussed by Metev and Meteva (1992).

Much greater deposition rates can be obtained using LCVD, where under certain conditions, with CW laser sources, rates of about $1\,mm\,s^{-1}$ may be possible (Baum and Comita 1992). Mass transport surface adsorption/desorption phenomena and surface reaction rates are important factors in determining the overall rate of deposition. These

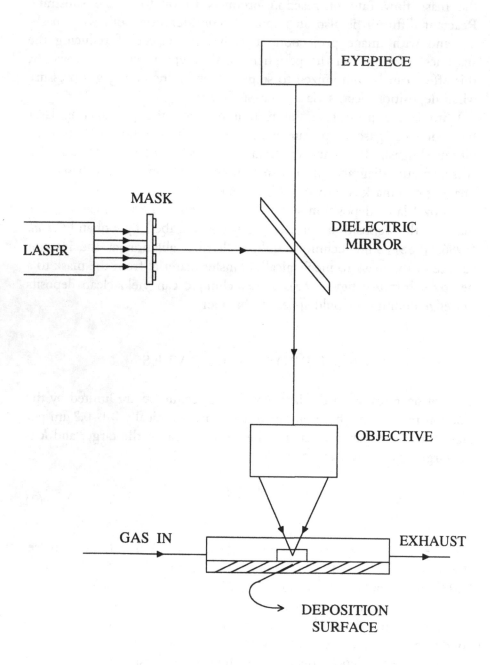

Figure 9.4. An LCVD system based on image projection.

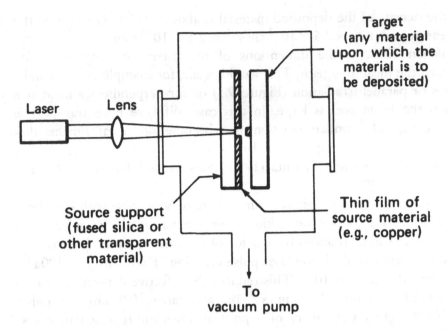

Figure 9.5. A schematic description of the apparatus for metal deposition from a solid-phased precursor. The source material and target are in contact during an actual experiment (from Bohandy *et al.* 1986).

are determined by the gaseous mean free path, characteristic dimensions of the system, and the focal area of the beam. The Knudsen number, Kn, is a useful parameter in describing various deposition regimes. Kn is the ratio of the gas mean free path to the characteristic dimension of the system.

LCVD with a tightly focused laser beam irradiating a substrate at low reactant pressure ($\lesssim 10^{-3}$ atm) corresponds to Kn $\gg$ 1. Under these conditions, there will be no concentration gradients in the gas and no interference between reaction products leaving the surface and reactants. The reaction rate may be limited by the rate of collision, R_c, of the reactants with the surface. This is

$$R_c \simeq \tfrac{1}{4} n v \tag{3}$$

where n is the gas density and v is the average molecular speed. With $n = 10^{16}\,\mathrm{cm}^{-3}$ and $v = 3 \times 10^4\,\mathrm{cm\,s}^{-1}$ as typical values under these conditions,

$$R_c \simeq 7.5 \times 10^{19}\,\mathrm{cm}^{-2}\,\mathrm{s}^{-1} \tag{4}$$

If the density of the deposited material is about $3 \times 10^{22} \, cm^{-3}$, then the linear growth rate is $2.5 \times 10^{-3} \, cm \, s^{-1} = 2.5 \times 10^{-2} \, mm \, s^{-1}$.

When $Kn \ll 1$, the dimensions of the interaction region greatly exceed the mean free path. This would occur, for example, under conditions of parallel irradiation (Figure 9.3) or for perpendicular incidence when the beam area is large. In this case, diffusive mass transport is dominant, and secondary reactions may become significant. Further discussions of mass transport in relation to LCVD can be found in the reports of Kodas and Comita (1990), Skouby and Jensen (1988) and Kodas *et al.* (1987).

Deposition rates as high as 1 mm s^{-1} are only possible with CW laser irradiation. For pulsed deposition sources such as the excimer laser, these rates will be reduced by a factor of $f\tau$, where f is the laser repetition frequency and τ is the laser pulse duration. Then with $f = 100 \, Hz$, and $\tau = 10^{-8} \, s$, $f\tau = 10^{-6}$. This reduces the effective deposition rate to about $10^{-6} \times 1 \, mm \, s^{-1} = 1 \, nm \, s^{-1}$ or, alternately, 0.01 nm per pulse. Table 9.1 gives a summary of deposition rates for representative systems. As expected, deposition rates using LCVD are larger than those obtained with laser sputtering. However, laser sputtering is a more generally applicable technique and has been successfully utilized to deposit complex materials such as YBCO and some polymers. LCVD routes to these materials are generally not available.

9.7 CONTAMINATION

The primary source of contamination in thin films deposited using laser sputtering involves small particles. These are liberated from the target during sputtering and are ejected at speeds of up to 300 m s^{-1} (Dupendant *et al.* 1989). Particle sizes can range up to several micrometers and particles may be spherical or irregular in shape. Metallic particles are usually spherical but particles ejected from non-metallic materials often exhibit random morphologies. Irregular shaped particles tend to be located in regions where the overall particle density is high, whereas spherical particles are more thinly spread over the deposition area. It is thought that the irregular particles are ejected explosively as the target decomposes whereas spherical particles may be the result of a hydrodynamical interaction between the laser plume and liquid on the target. An extensive discussion of particulate formation has been given by Chen (1994).

In LCVD, film contamination with photochemical or photothermal decomposition products of the precursor materials is common and may compromise film properties. Carbon is a common contaminant during LCVD of metals (Higashi 1988) and can affect the conductivity of deposits. Many metallic deposits such as Cr and Ti act as efficient getters for oxygen and other species that may be present at trace levels in the LCVD system. Figure 9.6 shows how the composition of Cr films deposited by KrF LCVD from $Cr(CO)_6$ depends on laser intensity (Yokoyama *et al.* 1985). It is evident that, whereas the Cr content of these films is approximately constant, the concentration of C and O is highly intensity-dependent. Yokoyama *et al.* (1985) suggested that these changes may accompany a shift from a photochemical to a photothermal processing regime at higher laser intensities.

One would expect that organic polymer films deposited by laser sputtering of a parent target would be heavily contaminated by a variety of reaction and decomposition products. Deposition of organic polymeric films using this technique has been discussed by Hansen and Robitaille (1988), Blanchet and Shah (1993) and Ogale (1994). However, this does not appear to be the case, because deposited films of PTFE, PMMA and PET, as well as other polymers, appear to have similar compositions and structures to those of the parent material. It appears that, although laser sputtering leads to depolymerization and the ejection of low molecular weight fragments, these recombine to reconstitute the parent material at the deposition surface. This is similar to the behavior of other complex solids such as $YBa_2Cu_3O_7$, which also exhibit congruent deposition in response to excimer laser sputtering of a parent material.

9.8 DEPOSITION BY SPUTTERING AND ABLATION

Deposition of solid films by laser evaporation of a solid precursor has been investigated since 1965 (Smith and Turner 1965). Ruby, Nd:YAG and CO_2 lasers have all been shown to be effective in laser deposition by vaporization (see Duley 1983 for a review). The CO_2 laser is particularly useful in vaporizing insulators and some semiconductors (Duley 1976) whereas ruby and Nd:YAG laser radiation are more efficient in vaporizing metals. CW laser radiation with power as low as about 25 W can be used for controlled deposition of highly absorbing materials

Table 9.1. *Representative deposition rates for different materials: X, excimer laser; A, argon ion laser; HC, He–Cd laser; L, LCVD, perpendicular ($\perp$) and parallel ($\parallel$) configurations; SO, solution; and A, ablation source.*

Solid	Precursor	Laser	Deposition rate ($\mathrm{nm\,s^{-1}}$)	Type	Reference
Al	$Al_2(CH_3)_6$	X	3.3	$L\parallel$	Solanki et al. (1983)
	Me_2AlH	A275	10^5	$L\perp$	Cacouris et al. (1988)
Cr	$Cr(CO)_6$	X	10	$L\perp,\parallel$	Flynn et al. (1986)
Cu	CuO	X	0.4	A	Ogale et al. (1993)
Mo	$Mo(CO)_6$	X	4	$L\perp,\parallel$	Boyer et al. (1983)
Pd	$MeCn:MeOH:Pd_2(CNMe)_6(PF_6)_2$	HC325	0.1	$SO\perp$	Montgomery and Mantei (1986)
Si	Si	X	0.2	SO	Reddy (1986)
	SiH_4	X	2	$L\perp,\parallel$	Gee et al. (1984)
Ti	$TiCl_4$	X	5	$L\perp$	Lavoie et al. (1991)
W	$W(CO)_6$	X	2.8	$L\perp$	Solanki et al. (1982)
AlN	AlN	X	0.025	A	Norton et al. (1991)
BN	BN	X	7×10^{-3}	A	Friedmann et al. (1993)
SiC	SiC	X	0.2	A	Tench et al. (1990)
Si_3N_4	Si_3N_4	X	0.02	A	Xu et al. (1993)
	Si_2H_6/NH_3	X	0.1	$L\perp$	Sugii et al. (1988)
SiO_2	SiH_4/N_2O	X	1.5	$L\perp,\parallel$	Boyer et al. (1984)
	SiO_2	X	5×10^{-3}	A	Slaoui et al. (1992)

BaTiO$_3$	BaTiO$_3$	X	0.03	A	Davis and Gower (1989)
KTiOPO$_4$ (KTP)	KTP	X	0.2	A	Xiong et al. (1994)
LiNbO$_3$	LiNbO$_3$	X	0.1	A	Shibata et al. (1993)
Pb$_5$Ge$_3$O$_{11}$	Pb$_5$Ge$_3$O$_{11}$	X	2.2	A	Peng and Krupanidhi (1992)
SrTiO$_3$	SrTiO$_3$	X	3	A	Hiratani et al. (1993)
YBa$_2$Cu$_3$O$_7$ (YBCO)	YBCO	X			(see Table 7.1)
Ge	GeH$_4$	X	0.2	⊥	Andreatta et al. (1982b)
Si	SiH$_4$	X	0.2	⊥	Suzuki et al. (1985)
Hg$_{1-x}$Cd$_x$Te	Me$_2$Cd, Hg, Te	X	1	L∥	Morris (1986)
C	Graphite	X	0.2	A	Richter and Klose (1992)
	C$_2$H$_2$	X	0.06	L⊥	Dischler and Bayer (1990)
PMMA	PMMA	X	≃0.05	A	Hansen and Robitaille (1988)
Nylon	Nylon	X	≃0.05	A	Hansen and Robitaille (1988)
PET	PET	X	≃0.05	A	Hansen and Robitaille (1988)
Polyimide (PI)	PI	X	≃0.05	A	Hansen and Robitaille (1988)
Polyamide/Sn	Polyamide/Sn	X	≃0.05	A	Gitay et al. (1991)

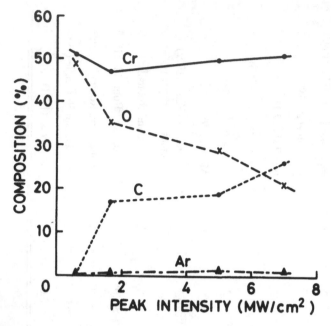

Figure 9.6. The dependence of the composition of Cr film deposited by 248 nm LCVD of $Cr(CO)_6$ on laser intensity. The fluence was $140\,mJ\,cm^{-2}$ with a projected slit width of $400\,\mu m$. The $Cr(CO)_6$ pressure was $0.7\,Torr$ (from Yokoyama *et al.* 1985).

(Groh 1968, Hess and Milkosky 1972), but optimized results are usually obtained using pulsed laser sources. However, these sources can lead to the sputtering of large particles, which can contaminate the deposited film. In addition, the structure and composition of thin films deposited by laser vaporization may differ significantly from those of the parent material.

The high intensity, short wavelength and narrow pulse width of excimer laser radiation have made these devices ideal as vaporization sources for metals, insulators, semiconductors and superconductors (Paine and Bravman 1990, Ashby *et al.* 1992, Braren *et al.* 1993). A comparison of typical parameters for deposition with free-running Nd:YAG (1.064 μm) and KrF (248 nm) laser sources is given in Table 9.2 (Lynds and Weinberger 1990). The useful characteristics of excimer laser deposition sources are then:

1)   high ablation rates,
2)   stoichiometric deposition,
3)   monolayer deposition and
4)   ability to vaporize all materials.

Table 9.2. *A comparison of Nd:YAG and KrF ablation sources for the deposition of YBa$_2$Cu$_3$O$_{7-x}$ films by laser ablation (Lynds and Weinberger 1990).*

	Nd:YAG (1.064 μm)	KrF (248 nm)
Ablation rate (atoms s^{-1})	10^{20}	10^{22}
Gas phase density (atoms cm^{-3})	10^{15}–10^{16}	10^{15}–10^{16}
Ablation rate per pulse (atoms/pulse)	10^{17}	10^{15}
Kinetic energy of ejecta (eV)	2–20	40–850
Composition of ejecta	Neutral clusters	Single atoms or ions

A comparative study of the effect of wavelength and pulse duration on the properties of laser deposited BaTiO$_3$ films (Gibson *et al.* 1990) has shown that films deposited with picosecond pulses contain many spattered particles. These particles were observed when ablation occurred either at 532 or at 266 nm, although the morphology of the deposited film was roughest at 532 nm. Deposition using nanosecond pulses at 532 nm resulted in films with a number of 0.5–3 μm inclusions. Clean, clear films were deposited with 266 nm nanosecond pulses. Shorter wavelength excitation for ablative deposition sources then optimizes film quality and minimizes particulate contamination when the pulse length lies in the nanosecond range. A summary of materials that have been deposited in thin film form by direct laser ablation of a solid precursor is given in Table 9.3. In many instances, optimum results are obtained using either 193 or 248 nm radiation. This is particularly true when ablating materials with a wide bandgap (e.g. SiO$_2$).

Despite the large number of reports on film deposition by laser ablation, limited quantitative data have been published on the physical properties of such deposits, other than their structures and stoichiometries. Where comparisons have been made, laser ablation seems to yield deposits that have properties similar to or somewhat less favorable than those of samples prepared by conventional plasma or photodeposition (Ishikawa *et al.* 1994, Rao and Krupanidhi 1994). Under most conditions, near stoichiometric deposition is possible, even for deposition of compounds such as BaTiO$_3$ and PbZrTiO$_3$ (PZT). Substrate-oriented growth has been demonstrated in many systems.

Deposition of carbon films by UV laser sputtering yields diamond-like carbon (DLC) with a composition intermediate between pure diamond and crystalline graphite. This variability can be expressed in terms of the ratio of diamond-like bonding (sp^3 hybridized bonding) to

Table 9.3. *Representative data on the deposition of thin films by direct laser ablation.*

Material	Substrate	Wavelength (nm)	Fluence ($J\,cm^{-2}$)	Rate (nm/pulse)	Reference
AlN	Sapphire (500–670°C)	248	2	0.025	Norton et al. (1991)
B$_4$C	(100) Si (300°C)	248	≃2		Donley et al. (1992)
BN	(100) Si	248	3.9	0.018	Doll et al. (1990)
	(100) Si	266		6×10^{-3}	Knapp (1992)
	(100) Si (400°C)	248	2.5	7×10^{-3}	Friedmann et al. (1993)
BaTiO$_3$	MgO (700°C)	266	3	4×10^{-3}	Gibson et al. (1990)
	SiO$_2$ (620°C)	308		0.03	Davis and Gower (1989)
BaFe$_{12}$O$_{19}$	Sapphire (900°C)	248	2		Horwitz et al. (1993)
Bi$_4$Ti$_3$O$_{12}$	(100) SrTiO$_3$ (700–800°C)	248			Ramesh et al. (1990)
	(100) MgO (675°C)	248	2	0.5	Buhay et al. (1991)
	Pt/Si	248	≃3		Maffei and Krupanidhi (1992)
CaAl$_2$Si$_2$O$_8$	Al$_2$O$_3$ (530–750°C)	248		0.01	Mallamaci et al. (1993)
CdS	Al$_2$O$_3$	532	1.1		Shi et al. (1991)
a-CN	Si	193	15	0.1–0.2	Xiong et al. (1993)
CoSi$_2$	(100) Si (200–600°C)	248			Tiwari et al. (1993)
CuO/Cu$_2$O	(100) Si	193	≃10		Ortiz et al. (1992)
GeO	(100) Si	193	≃1		Wolf et al. (1993)
InSnO (ITO)	Glass	193	≃1	0.01	Zheng and Kwok (1993)

Material	Substrate (temperature)	λ (nm)			Reference
KTaNbO₃ (KTN)	SrTiO₃ (700–750°C) (100) MgO (300–700°C)	248	≃1	0.02	Yilmaz et al. (1991)
		248	1.5	0.08	Cotell and Leuchtner (1993)
LiNbO₃	GaAs (680°C) Al₂O₃ (500–800°C)	308 193	0.8–1.3	0.1	Fork and Anderson (1993) Shibata et al. (1993)
LiTaO₃	Sapphire	248	1–2	0.1	Agostinelli et al. (1993)
MgO	(100) Si	248			Kanetkar et al. (1991)
MoS₂	440°C stainless steel	248, 193	0.5–1		John et al. (1991)
NbSe₂	440°C stainless steel	248	1.3–2.8		Day et al. (1993)
Pb₅Ge₃O₁₁	Sapphire	248	5	2.2	Peng and Krupanidhi (1992)
PbNbMgO_x (PMN)	Pt/SiO₂	248	1.5	0.05	Saenger et al. (1993)
PbO	440°C stainless steel	248	0.53		Zabinski et al. (1992)
PbTiO₃/Pb(MgW)O₃	Pt/Si (300–650°C)	193	16	0.3–0.4	Lee et al. (1993)
PbZrTiO₃ (PZT)	Pt/Si	248	1–3		Roy et al. (1991)
	MgO (550°C)	248	0.5–1.4		Leuchtner et al. (1992)
	SrTiO₃ (200–550°C)	248	2		Chrisey et al. (1990)
	Sapphire (400–750°C)	193	3–8.5		Morimoto et al. (1990)
SiC	Si (800°C)	308	8.7	0.2	Balooch et al. (1990)
	Si (800°C)	308	8.7	0.5	Tench et al. (1990)
	SiO₂ Sapphire (800°C)	351	1.5		Rimai et al. (1993)

Table 9.3 (*cont*).

Material	Substrate	Wavelength (nm)	Fluence (J cm^{-2})	Rate (nm/pulse)	Reference
Si$_3$N$_4$	(100) Si	193	0.5–10	0.01	Fogarassy et al. (1993)
	(100) Si (210°C)	248	1–6	0.02	Xu et al. (1993)
SiO	(100) Si	193	0.5–10	0.04–0.05	Fogarassy et al. (1993)
SiO$_2$	(100) Si	193	1–10	$(5-10) \times 10^{-3}$	Slaoui et al. (1992)
SrFeO$_{2.5}$	Quartz	248	0.7–3.9	0.25–0.4	Sanders and Post (1993)
SrTiO$_3$	(001) MgO (600°C)	248		3	Hiratani et al. (1993)
TiC	440°C stainless steel	248	≃2		Donley et al. (1992)
TiN	(100) Si (25–550°C)	308	4–5	0.05	Biunno et al. (1989)
	(100) GaAs (350°C)	248	10		Zheleva et al. 1993
TiO$_x$N$_y$	(100) Si (400–800°C)	248, 532	4–5		Craciun et al. (1993)
TiO$_2$	(100) Si (500°C)	532			Chen and Murray (1990)
ZnO	(100) Si (250–600°C)	248	2.5–3		Amirhaghi et al. (1993)
ZnS/ZnSe	(001) GaAs (300–325°C)	248	0.35	0.2	McCamy et al. (1993)
	GaAs (420°C)	308	2–8	≃0.3	Rajakarunanayake et al. (1993)
ZrO$_2$	Al$_2$O$_3$ (20–500°C)	248	1–4		Smith et al. (1992)

Table 9.4. *Optical and electrical properties of DLC films prepared by laser ablation of graphite (data from Duley 1984, Ogmen and Duley 1988, Pappas et al. 1992); n and k at 632 nm.*

Laser	Intensity ($W\,cm^{-2}$)	n	k	$\%sp^2$	E_g (eV)	Resistivity ($\Omega\,cm$)
Nd : YAG (1.064 µm)	5×10^{11}	2.35	0.32	25	1.0	$>3.3 \times 10^7$
XeCl (308 nm)	1.25×10^8	2.40	0.13		1.27	
	8×10^8	2.40	0.60		0.30	
	3×10^8	2.2	0.042		1.4	
KrF	1.4×10^8	2.55	0.035	15–27	1.70	$>10^6$
(248 nm)	1.4×10^8	2.53	0.14	32	1.50	$>10^6$
C_2H_2 plasma		1.74	0.0065		2.2	10^{10}

graphitic bonding (sp^2 hybridized bonding). This variability is reflected in such parameters as the value of the optical bandgap energy E_g, hardness, resistivity and IR absorption. These parameters depend on laser fluence, the composition of the ambient atmosphere and the temperature of the deposition substrate (Duley 1984, Ogmen and Duley 1988, Collins *et al.* 1989, Krishnaswamy *et al.* 1989, Davanloo *et al.* 1990, Richter and Klose 1992, Pappas *et al.* 1992, Charyshkin and Sakipov 1992, Collins and Davanloo 1994). Some properties of DLC films deposited from graphite using excimer laser ablation are summarized in Table 9.4. It is evident that the imaginary component, k, of the complex refractive index is particularly sensitive to deposition conditions. In the visible part of the spectrum, k, which is proportional to the optical absorption coefficient, is a strong function of the average size of aromatic clusters (Robertson 1986). An increase in k then accompanies an increase in the sp^2 content of the film and a decrease in E_g.

Although direct laser sputtering is effective in the production of DLC films, there have been several studies of the effect of an electrical bias on the properties of the deposited film (Krishnaswamy *et al.* 1989, Collins *et al.* 1989, Davanloo *et al.* 1990, Collins *et al.* 1993). Such systems use an external voltage to extract ions from the laser plume over the sputter target. This voltage can be either continuous or in the form of a pulse. There has been some discussion on the effects of this treatment on film hardness and uniformity (Davanloo *et al.* 1990, Pappas *et al.* 1992).

Table 9.5. *Properties of a-C : N films prepared by ArF laser
sputtering of pyrolytic graphite in N_2 gas (from Xiong et al. 1993).*

Deposition rate ($nm\,s^{-1}$)	0.5–2
N/C ratio	0.3–0.66
Microhardness (GPa)	8–18
Density ($N/C = 0.45$) ($g\,cm^{-3}$)	2.5 ± 0.2
Young's modulus ($N/C = 0.45$) (GPa)	153
Bandgap (eV)	0.25
Refractive index (600 nm)	$n = 2.7$
	$k = 0.6$

Laser sputtering of graphite in an atmosphere of N_2 can yield carbon films with up to 66% N content (Xiong *et al.* 1993). The a-C:N films have high mechanical hardness, good wear resistance and chemical inertness. Nitrogen content increases linearly with N_2 pressure over the range 30–100 mTorr. These films are essentially hydrogen-free and contain less sp^3 bonded carbon than pure DLC films prepared under the same conditions. Some properties of these deposited films are summarized in Table 9.5. Surprisingly, a study of the preparation of DLC films by KrF laser ablation of graphite in the presence of H atoms (Thebert-Peeler *et al.* 1992) has shown that the characteristics of the deposited film are independent of the presence of H atoms.

9.9 LASER CHEMICAL VAPOR DEPOSITION

LCVD is readily suited to the deposition of many metals (Deutsch *et al.* 1979), elemental semiconductors (Andreatta *et al.* 1982a) and other simple solids such as hydrogenated amorphous silicon, SiO_2, Si_3N_4 and oxides such as SnO_2, TiO_2 and ZnO. Compound semiconductors such as InP, GaAs, HgTe, HgCdTe and ZnSe have also been deposited in this way. LCVD can be performed with all available laser wavelengths between 10.6 μm (CO_2) and 157 nm (F_2). Pyrolysis dominates at long wavelengths but is not entirely absent even at the shortest available laser wavelength. This is particularly true when the perpendicular LCVD configuration is used (Figure 9.3) since some heating of the deposition surface is invariably present in this configuration.

Both CW and pulsed laser sources may be utilized but CW laser excitation is most appropriate for the deposition of discreet lines and points.

Extended structures have also been generated (Lehmann and Stuke 1991). Ar and Kr ion lasers are well suited to these applications whereas doubling of the Ar^+ 514 nm line to 257 nm permits deposition from a wider variety of molecular precursors. A comprehensive review of early LCVD studies, which surveys results obtained at a variety of laser wavelengths, can be found in the book by Bäuerle (1986). A more recent review of metal deposition (Baum and Comita 1992) provides a thorough outline of LCVD techniques utilized to deposit particular metals.

A common feature of LCVD at UV wavelengths involves the liberation of the desired element or molecule from a gaseous precursor thermally or by photodissociation. The mechanisms by which this occurs are not well understood even for simple systems. For example, possible photochemical routes to Al from trimethylaluminum, $Al(CH_3)_3$, include the following sequence (Motooka et $al.$ 1985):

$$2Al(CH_3)_3 \rightarrow Al_2(CH_3)_6 \tag{5}$$

$$Al_2(CH_3)_6 + h\nu \rightarrow 2Al(CH_3)_3 \tag{6}$$

$$Al(CH_3)_3 + h\nu \rightarrow Al(CH_3)_2 + CH_3 \tag{7}$$

$$Al(CH_3)_2 + h\nu \rightarrow Al(CH_3) + CH_3 \tag{8}$$

$$Al(CH_3) + h\nu \rightarrow Al + CH_3 \tag{9}$$

which would involve the sequential absorption of four laser photons during the laser pulse. Under the conditions of these experiments the laser photo flux would be $F \simeq 10^{18}$–10^{19} cm^{-2} per pulse and the cross-section for absorption at the KrF wavelength used is $\sigma = 10^{-18}$–10^{-19} cm^2. The product $\sigma F \simeq 0.1$–10, hence such multiple absorption may be possible. However, gas phase thermal decomposition of $Al_2(CH_3)_3$ following the laser pulse must also be considered to be likely since such decomposition is frequently seen when strongly absorbing molecular gases are irradiated with intense short duration laser pulses (Isenor and Richardson 1971, Duley 1983).

Direct evidence for the dissociation of adsorbed $Al_2(CH_3)_6$ after irradiation with 248 and 193 nm pulses has been obtained from changes in the infrared absorption spectrum (Higashi and Rothberg 1985). They found that the 2960 cm^{-1} absorption band due to CH_3 is weakened significantly after irradiation. In addition, the concentration of surface

OH groups is reduced, probably due to the gettering effect of the Al liberated in the dissociation of $Al(CH_3)_3$. This suggests that secondary surface reactions can be significant, even in relatively simple chemical systems.

A careful study of the photodissociation of dimethylaluminum hydride (DMAlH) (Ohashi *et al.* 1993) has shown that wavelength selective effects can be important even for adsorbed species. DMAlH adsorbed on (100)Si at 150 K was found to decompose in response to irradiation at 193 nm. Cleavage occurred at an Al–C bond due to direct photo-excitation. Irradiation at 351 nm, a wavelength well outside the absorption band of the Al–C bond, had little effect and led to no photodecomposition of DMAlH. Selective photodissociation of $In(CH_3)_3$ on Si by 193 nm excimer laser radiation has also been demonstrated and assigned to photocleavage of the In–C bond (Shogen *et al.* 1992). Calloway *et al.* (1983) have shown that metal deposits can be obtained from trimethylaluminum even under irradiation at very low intensities with incoherent radiation when the photon energy lies within the photodissociation band of this molecule. The primary effect of laser irradiation is then to accelerate this process, at least at laser intensities that preclude significant thermal effects either in the gas or on the deposition surface.

Cacouris *et al.* (1988) have reported on the writing of Al lines using a DMAlH precursor and frequency-doubled Ar^+ laser radiation ($\lambda = 275$ nm) as the writing source. They found that the vertical deposition rate of lines with a width of about $2\,\mu m$ was typically $0.09\,\mu m\,s^{-1}$ under photolytic deposition conditions and as high as $7\,\mu m\,s^{-1}$ under photothermal, i.e. pyrolytic, conditions. A change from photolytic to pyrolytic deposition conditions was found to occur at a laser intensity of about $1.6 \times 10^5\,W\,cm^{-2}$. The activation energy for pyrolytic growth from DMAlH was estimated to be 67–$92\,kJ\,mol^{-1}$. The thickness of the deposited layer was found to decrease inversely with scan speed both in photolytic and in photothermal deposition regimes. Typical writing speeds were 1–$10\,\mu m\,s^{-1}$ (photolytic) and 20–$100\,\mu m\,s^{-1}$ (pyrolytic).

Deposition of Ti lines from $TiCl_4$ using 248 nm laser radiation in the form of a series of overlapping pulses has been reported by Izquierdo *et al.* (1990). Deposition occurred onto $LiNbO_3$ at writing speeds of $0.2\,\mu m\,s^{-1}$. Under these conditions, with low pulse energy ($<0.10\,J\,cm^{-2}$) and low repetition rate (2 Hz), thermal heating is not significant and deposition should occur photolytically. It is likely that prenucleation effects would be significant under these conditions.

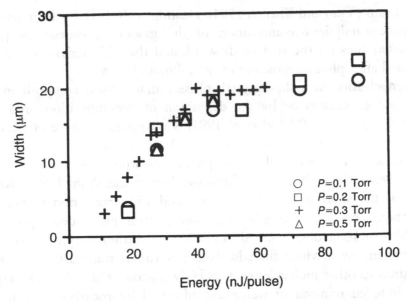

Figure 9.7. Titanium linewidth variation as a function of the pulse energy and TiCl$_4$ pressure for LCVD at 248 nm onto a LiNbO$_3$ substrate. The scanning speed was 0.2 μm s^{-1} at a laser frequency of 2 Hz (from Izquierdo *et al.* 1990).

Figure 9.7 shows the dependence of the line width of the deposit on laser pulse energy and the pressure of TiCl$_4$ obtained in this experiment.

Growth of deposits is probably initiated by the formation of prenucleation centers that act as preferred sites for subsequent growth (Ehrlich *et al.* 1981, Tsao and Ehrlich 1984). These prenucleation sites are clusters containing perhaps as many as 10^7 atoms with dimensions as small as 10^{-6} cm. These clusters are formed by the photo-dissociation of molecules in an adsorbed layer. Growth of clusters occurs when atoms adsorbed from the gas move to within a distance of $(D\tau)^{1/2}$ of a cluster, where D is a diffusion coefficient and τ is the residence time of the atom on the surface. The existence of a prenucleation stage eliminates the need for spontaneous nucleation at the surface from the atomic vapor. Ehrlich *et al.* (1981) conclude that the rate of formation of critical nuclei on the surface under LCVD conditions is typically 10^{15}–10^{17} atoms cm^{-2} s^{-1}. These nuclei may also act as catalytic sites for the activation of the surface reaction of precursor molecules. This effect has been confirmed in an experiment in which enhanced deposition of Al from tri-isobutylaluminum was observed along lines photo-deposited using UV laser radiation and then irradiated with a low intensity CO$_2$

laser beam (Tsao and Ehrlich 1984). Heating with the CO_2 laser acts to promote catalytic decomposition of the gaseous precursor at pre-nucleated sites on the surface. It was found that this activity could be initiated after photo-deposition of several monolayers.

Prenucleation in a slow growth phase initiated by decomposition of adsorbed precursors or by the deposition of decomposition products from the gas phase (Wood et al. 1983) is a common feature in LCVD (Rytz-Froidevaux et al. 1982, Lavoie et al. 1991). Prenucleation can occur in a variety of ways including two-photon absorption (Rytz-Froidevaux et al. 1982) or as a photothermal process (Esrom and Wahl 1989). How-ever, at the shortest laser wavelengths and with irradiation within an absorption band of the adsorbate, one-photon photo-decomposition would seem to be dominant. Absorption cross-sections may be enhanced at wavelengths at which the adsorbate is normally transparent by com-plexation to other molecules such as H_2O (Lavoie et al. 1991), although laser-induced removal of water and adsorbed hydrocarbons has been shown to yield enhanced film deposition under certain conditions (Coombe and Wodarczyk 1980).

Film deposition by prenucleation is favored in the perpendicular LCVD mode (Figure 9.3). This is probably one of the reasons why the adherence of deposited films is enhanced using perpendicular inci-dence geometry (Flynn et al. 1986). By irradiating through a mask, the prenucleation technique has been successfully used to prepare a variety of substrates for subsequent metallization using electroless plating solutions (Esrom and Wahl 1989). Prenucleation of Pd metal was obtained by irradiation of coated palladium II acetate films with several low fluence $(0.05\,J\,cm^{-2})$ pulses from an ArF laser (193 nm). This resulted in the deposition of about 6 nm thick Pd layers over areas of up to a few cm^2. With a suitable choice of mask, or by projection imaging, these layers could be utilized for the growth of well adhered Cu layers on top of the Pd surface. Pd layers deposited using this exci-mer laser prenucleation technique were highly active in electroless copper solutions. Copper layers of up to 6 μm thickness could be grown at rates of about $0.1\,\mu m\,min^{-1}$ from a Shipley copper bath. Copper has also been deposited on polyimide using the prenucleation method (Cole et al. 1988).

Prenucleation for electroless metal deposition has also been per-formed using excimer UV lamp radiation (Esrom and Kogelschatz 1992). These are incoherent sources based on UV emission from dimers excited in a silent discharge. They are characterized by high efficiency, and relatively narrow bandwidth. They emit over a wide area and can be run at high repetition rates. Esrom and Kogelschatz (1992) used these lamps to prepare surfaces prenucleated with Pd by decomposition of palladium acetate. They concluded that the decomposition process with excimer lamps is purely photolytic in nature and exhibits no threshold fluence effects.

Techniques developed for prenucleation can also be applied in the generation of finely divided metallic catalyst particles. Ehrlich and Tsao (1985) irradiated mixed $TiCl_4$ and $Al_2(CH_3)_6$ adsorbates with 257.2 nm radiation obtained from the doubled output of the Ar^+ laser.

The overall reaction is

$$TiCl_4(\text{adsorbed}) + h\nu \rightarrow TiCl_n(\text{adsorbed}) + (4 - n)Cl \qquad (10)$$

$$TiCl_n(\text{adsorbed}) + Al_2(CH_3)_6(\text{adsorbed}) \rightarrow Ti{:}Cl{:}Al{:}CH_3 \qquad (11)$$

yielding the catalytically active Ti complex $Ti{:}Cl{:}Al{:}CH_3$. This catalyst was demonstrated to be active in the polymerization of ethylene and acetylene at room temperature from the gas phase. Patterned depositions of polymer could be obtained from patterned prenucleated catalytic deposits. Deposits containing variable amounts of Ni and Fe have been obtained in a similar way by ArF laser irradiation of mixed vapors of iron and nickel ferrecene (Armstrong et al. 1987).

The electrical resistivity of metals deposited by LCVD has been reviewed by Baum and Comita (1992). Generally resistivities are at least 2–3 times larger than that of the bulk material. Lowest resistivities occur on prenucleated surfaces that have been thermally annealed following deposition. For example, the resistivity of 3.5 μm wide gold lines deposited on alumina by photolytic decomposition of Au metallopolymer at 257 nm decreases from about 10 μΩ cm after deposition to 4.5 μΩ cm after annealing to 275°C for 150 min (Beeson and Clements 1988). Such low resistivity was not obtained on a consistent basis on samples that had not been prenucleated. Under these conditions, deposited layers show a high degree of cracking and have resistivities that rarely drop below 100 μΩ cm.

A correlation between film resistivity and microstructure has also been observed in W films deposited by LCVD from a WF_6/H_2 mixture irradiated at 193 nm (Deutsch and Rathman 1984). They found that a sharp drop in resistivity at a deposition temperature above 325°C accompanies the creation of the low resistivity α-W phase in which crystallites with diameters in the 150–250 nm range are dominant. This suggests that a photothermal component in the photo-deposition process may be useful in increasing grain size with a corresponding reduction in resistivity. Indeed, a study of pyrolytic deposition of Ni (Kräuter *et al.* 1983) from $Ni(CO)_6$ using CW Kr^+ laser radiation has shown that grain sizes increase with increasing laser power. Pyrolysis also results in a reduction of the concentration of hydrocarbons and other carbon contaminants within the deposited layer. Such contaminants have a serious deleterious effect on resistivity and are readily produced in purely photolytic reactions.

The thermal conductivity of the deposition substrate can also have a profound effect on the resistivity of LCVD metal films. For example, Gilgen *et al.* (1987) find that W stripes deposited by LCVD from $W(CO)_6$ using the UV lines from the Ar^+ laser ($\lambda = 351$–364 nm) are much more highly conducting when irradiation is carried out at high power on a glass substrate. Under similar conditions the resistivity of W stripes deposited on a GaAs substrate was found to be about a factor of 10^2 higher. This effect can be attributed to the high thermal conductivity of GaAs compared with glass so that photothermal effects will be enhanced on the glass substrate. In the case of W, a reduction in resistivity can arise from the phase change to α-W material when the deposition temperature rises above 325°C. W can also be deposited using LCVD from precursors such as WF_6 (Deutsch and Rathman 1984) and WCl_6 (Kullmer *et al.* 1992), which yield deposits that are less likely to be contaminated by C and O and perhaps can provide a source of lower resistivity deposits.

Deposits of insulating materials such as Al_2O_3, SiO_2, AlN, etc. can be obtained by LCVD of the metallic precursor in the presence of an oxidant or nitriding agent. Al_2O_3 films have been deposited using LCVD with a gaseous mixture of $Al_2(CH_3)_6$ and N_2O (Solanki *et al.* 1983) whereas SiO_2 has been prepared by irradiation of SiH_4 plus N_2O (Boyer *et al.* 1982, 1984) and SiH_4 plus O_2 plus N_2/H_2 (Nishino *et al.* 1986). Al_2O_3 films could be deposited at rates of up to 3.3 nm s^{-1} using this technique in the parallel irradiation configuration. Deposition occurred over a 2.5 cm × 10 cm area. Film properties were excellent (Table 9.6) compared with those obtained using RF sputtering. It was found that application of a negative bias between the substrate and the LCVD region resulted in films with a refractive index $n > 1.6$ whereas a positive bias gave $n \simeq 1.52$. Electrical biasing has also been shown to increase hardness and uniformity in diamond-like

Table 9.6. *A comparison between properties of Al_2O_3 LCVD films deposited at 350°C and planar RF magnetron sputtered films deposited at 300°C (from Solanki et al. 1983).*

	LCVD	RF sputtered
Deposition rate ($nm\,s^{-1}$)	8.3	0.6
Adhesion ($N\,m^{-2}$)	$>6.5 \times 10^7$	Strongly adherent
Pinhole defects	<1 in $5\,cm^2$	$31/cm^2$
	(110 nm film)	(250 nm film)
Stress ($N\,m^{-2}$) (compressive)	$<6 \times 10^8$ (tensile)	2.8×10^8
Refractive index	1.63	1.66
Etching rate in 10% HF ($nm\,s^{-1}$)	0.16	
Stoichiometry	Al_2O_3	$Al_{2.1}O_3$
Resistivity ($\Omega\,cm$)	10^{11}	10^{12}
Dielectric constant at 1 MHz	9.74	9.96

carbon films deposited using laser sputtering (Krishnaswamy *et al.* 1989).

With a suitable precursor, oxide films can be deposited directly without the requirement for several reaction steps. Arnone *et al.* (1986) have reported on the direct writing of chromium oxide lines on fused silica by Ar^+ LCVD of CrO_2Cl_2. Deposition under these conditions involves a combination of photolytic and photothermal effects. A mixture of chromium oxides rather than Cr_2O_3 was deposited. 193 nm LCVD under similar conditions, but in the presence of HCl gas, was found to yield films whose Cr:O ratio could be varied between $2:3$ and $6:5$.

Li and Tansley (1990) has reported on the production of polycrystalline AlN films using LCVD from a mixture of ammonia and trimethylaluminum with 193 nm laser radiation. By careful elimination of oxygen, pure AlN films could be deposited with no evidence for the presence of either Al_2O_3 or Al-oxynitride. Some properties of these AlN films are given in Table 9.7. They are characterized by high dielectric strength, high resistivity and low dielectric loss.

The composition of AlN films prepared by LCVD is highly sensitive to the presence of traces of oxygen. Incorporation of small amounts of O_2 during deposition results in a mixed $(Al_2O_3)_{1-c}(AlN)_c$ composite (Demiryont *et al.* 1986a,b) with c varying over the range $0 \lesssim c \lesssim 1$. Parameters such as the Tauc bandgap were found to depend strongly on c with Al_2O_3 properties dominating for $c \lesssim 0.3$ and AlN properties

Table 9.7. *Properties of AlN films grown on GaAs and Si*
substrates by LCVD of ammonia and trimethylaluminum
(from Li and Tansley 1990).

Film thickness (μm)	0.3–1.2
Bandgap energy (eV)	6.00 ± 0.03
Resistivity (300 K) (Ω cm)	$>2 \times 10^{13}$
Breakdown field (V cm^{-1})	3×10^6
Relative permittivity (1 MHz)	8
Loss tangent (1 MHz)	$\simeq 0.01$

dominating above this value. Electrical characteristics of $(Al_2O_3)_{1-c}(AlN)_c$ composite films show both 'original state' i.e. virgin sample and 'programmed state' properties with both capacitance–voltage and current–voltage characteristics sensitive to the parameter c.

By alternating precursor gases, LCVD can also be used to prepare multilayer depositions. Deposition of a-Si:H/a-Al$_x$O$_{1-x}$ superlattices by LCVD using this method has been reported by Uwasawa *et al.* (1992).

In another application, refractory Si$_3$N$_4$ films have been created on Si by LCVD from Si$_2$H$_6$ and NH$_3$ gases using 193 nm radiation (Sugii *et al.* 1988). The absorption cross-section of NH$_3$ at 193 nm is $\sigma \simeq 2 \times 10^{-18}$ cm^2, which ensures that strong photoexcitation of NH$_3$ is present. The primary result of this photoexcitation is probably the NH and NH$_2$ radicals

$$NH_3(X\,^1A_1) \rightarrow NH_3(A\,^1A_2) \qquad (12)$$

$$NH_3(A\,^1A_2) \rightarrow NH_2(X\,^2B_1) + H \qquad (13)$$

$$\rightarrow NH(a^1) + H_2 \qquad (14)$$

With SiH$_4$, the primary photodecomposition also involves loss of hydrogen:

$$SiH_4 + h\nu \rightarrow SiH_3 + H \qquad (15)$$

$$\rightarrow SiH_2 + H_2 \qquad (16)$$

H atoms will continue to attack Si–H bonds, abstracting hydrogen. The overall result is the liberation of Si atoms or SiH molecules, which then react with NH or NH_2 to form the nitride. For example,

$$SiH + NH \rightarrow SiN + H_2 \qquad (17)$$

$$SiH_2 + H \rightarrow SiH + H_2 \qquad (18)$$

$$SiH + H \rightarrow Si + H_2 \qquad (19)$$

$$Si + NH_2 \rightarrow SiN + H_2 \qquad (20)$$

Infrared spectra of LCVD films of Si_3N_4 prepared in this way show absorption bands attributable to N–H bonds, but not to Si–H bonds.

An alternate method, based on direct nitriding of Si by photo-generated NH and NH_2, was also described by Sugii et al. (1988). Radicals were generated from NH_3 in the perpendicular LCVD mode by irradiating the surface of Si with 193 nm excimer laser pulses.

Titanium silicide films can be prepared by ArF LCVD of $TiCl:SiH_4$ mixtures (Gupta et al. 1985). Stoichiometry depends on the partial pressure of the gaseous reactants. $TiSi_2$ films could be deposited using a 2:1 $SiH_4/TiCl_4$ mixture. Deposited films were found to be a mixture of amorphous material and a metastable $TiSi_2$ phase. This mixture converts to polycrystalline $TiSi_2$ on annealing. Deposition at 400–500°C was found to yield adherent, smooth films.

Si and Ge in pure and doped form can be deposited using LCVD from a suitable gaseous precursor (typically SiH_4, Si_2H_6 or GeH_4) (Andreatta et al. 1982b,c, Gee et al. 1984, Yoshikawa and Yamaga 1984, Fowler et al. 1991). The growth rate has been found to be a strong function of laser wavelength as well as intensity. In the experiments of Andreatta et al. (1982b) both Si and Ge were found to be polycrystalline with a random orientation and a grain size of up to 0.5 μm. The sheet resistivity of as-deposited films was typically $>10^6 \, \Omega/\square$. Simultaneous LCVD of GeH_4 and $Al(CH_3)$ was shown to yield Ge samples doped with up to 1 at% Al.

Yoshikawa and Yamaga (1984) measured the optical bandgap energy, E_g, of LCVD Si films deposited on a glass substrate kept at a temperature between 100 and 350°C. They found that E_g ranges from about 2.7 eV for films deposited at a substrate temperature of 100°C to about 1.8 eV for deposition at 350°C. The photoconductivity was found to be largest in films deposited at 250°C. The dark conductivity was also largest at this temperature. Yoshikawa and Yamaga (1984) also demonstrated that these films could be doped with P by dissociation of gaseous PH_3.

Detailed studies of the physical processes that lead to LCVD of Si and Ge films have been reported by Osmundsen *et al.* (1985) and by Motooka and Greene (1986). For Ge photo-deposition using 248 nm KrF laser radiation exciting GeH_4, Osmundsen *et al.* (1985) observe a two-photon dependence on laser intensity. Since the absorption cross-section of GeH_4 at 248 nm is small ($\sigma \simeq 6 \times 10^{-23}$ cm^2), they suggest that the initial decomposition occurs from a two-photon excitation of GeH_4. The reaction

$$GeH_4 + 2h\nu \rightarrow GeH_2 + 2H \qquad (21)$$

forms GeH_2, which can react with GeH_4 to form Ge_2H_6, which is deposited on the substrate and is then photo-dissociated. A complementary reaction route involves H atom abstraction from GeH_4:

$$GeH_4 + H \rightarrow GeH_3 + H_2 \qquad (22)$$

Subsequently GeH_3 is photo-dissociated:

$$GeH_3 + 2h\nu \rightarrow GeH + H_2 \qquad (23)$$

$$GeH + h\nu \rightarrow Ge\downarrow + H \qquad (24)$$

Since LCVD was carried out in the perpendicular configuration, these photochemical reactions must compete with thermal dissociation such as the direct pyrolysis of GeH_4. Motooka and Greene (1986) suggested that film growth involves GeH_2 insertion into the Ge surface followed by H_2 desorption.

Irradiation of Si films with 248 nm KrF laser radiation *during* conventional CV deposition leads to melting and resolidification of thin Si layers (Suzuki *et al.* 1985). This has the effect of facilitating the deposition of polycrystalline layers with preferred (111) orientation when the laser pulse energy, repetition rate and film deposition rate are adjusted so that the melt depth slightly exceeds the film thickness deposited between laser pulses. The 300 K electrical conductivity of In-doped films prepared in this way was found to be about 10^5 times higher than that of unirradiated films prepared under identical conditions.

John *et al.* (1992) have shown that GaN can be successfully deposited on basal plane sapphire and GaAs(100) substrates using ArF LCVD from a Ga(CH$_3$) plus NH$_3$ plus Ar mixture in the parallel configuration. Trimethylaluminum was also found to be a useful gaseous precursor for Al to form an AlN buffer layer. Films could be grown with thicknesses of up to 3.5 μm using substrate temperatures of 600–700°C. Growth rates of up to 2 μm h^{-1} were reported. GaN films deposited on sapphire were observed to be fine-grained polycrystalline in nature with a preferential (0001) orientation. Those deposited on GaAs(100) were found to be preferentially oriented (00$\bar{1}$0).

Photodecomposition of P(CH$_3$) and (CH$_3$)$_3$InP(CH$_3$)$_3$ with 193 nm ArF laser radiation has been found to yield a stoichiometric InP deposit when irradiation is carried out in the perpendicular configuration on GaAs and other substrates (Donnelly *et al.* 1984, 1985). The deposition rate was 0.02 nm per pulse. Under certain conditions, the surface layer of the deposit was found to be contaminated with carbon, which was probably the result of a high CH and CH$_2$ concentration accompanying the photo-decomposition of the gaseous precursor molecules.

A comprehensive study of the crystallographic properties of InP films deposited in this way (Donnelly *et al.* 1985) has shown that InP can be deposited and annealed to form epitaxial films on (100) InP substrates. The degree of crystallinity was found to be sensitive to laser fluence over the range 0.01–0.25 J cm^{-2}. However, the optimum fluence for epitaxial growth was reported to be in the range 0.1–0.2 J cm^{-2} at a substrate temperature of about 320°C. The upper fluence limit was provided by the tendency of the InP substrate surface to exhibit damage. In addition, the central deposition area in this case was found to be deficient in P. Such depletion in a volatile component is a common effect near the threshold for macroscopic surface damage during laser irradiation of binary compounds.

LCVD of Hg$_{1-x}$Cd$_x$Te has been reported by Morris (1986) using dimethylmercury, dimethylcadmium and dimethyltellurium as gaseous precursors. 193 nm ArF laser radiation is effective in the decomposition of these compounds with the decomposition postulated to proceed in two stages

$$M(CH_3)_2 + h\nu \rightarrow M(CH_3)^* + CH_3 \tag{25}$$

$$M(CH_3)^* \rightarrow M + CH_3 \tag{26}$$

Table 9.8. *Deposition parameters for LCVD growth of*
$Hg_{1-x}Cd_xTe$ *thin films using 193 nm laser radiation and*
dimethyl (Hg, Cd, and Te) precursors (Morris 1986).

Substrate	CdTe
Substrate temperature (°C)	150
Total pressure (Torr)	666.5
Flow rate (std cm^3 min^{-1})	5.0 Te(CH$_3$)$_2$
	10.0 Hg(CH$_3$)$_2$
	1.0 Cd(CH$_3$)$_2$
	64.0 He
Growth rate (μm h^{-1})	4

where M is Hg, Cd or Te. Growth parameters using the parallel irradiation configuration are summarized in Table 9.8. Growth rates of about
$4\,\mu$m h^{-1} were reported. Typically the stoichiometry was $Hg_{0.2}Cd_{0.8}Te$
in a crystalline film.

Amorphous superlattices consisting of a large number of layers with
alternating composition have also been prepared by LCVD (Lowndes
et al. 1988, Uwasawa *et al.* 1991). Lowndes *et al.* (1988) identified the
following advantages of LCVD in this regard:

1)   high depth resolution due to single pulse deposition,
2)   high deposition rates at low temperature due to high pulse
      repetition rates, and
3)   substrate temperature can be chosen to minimize deleterious
      effects such as impurity or dopant diffusion.

The 'digital' nature of the LCVD deposition process using excimer
laser pulses provides close control over film thickness. Lowndes *et al.*
(1988) have prepared 32-layer Si_3N_4/a-Si amorphous structures by
LCVD from the parent gases using 193 nm radiation and Uwasawa *et al.*
(1991) have deposited a ten-layer a-Si:H/a-Al$_{1-x}$O$_x$ superlattice. Both
structures show sharp transitions between adjacent layers. However,
thin films (>10 layers) showed some instability with respect to lateral
rippling. As expected, the optical bandgap of superlattice LCVD a-
Si:H/Al$_{1-x}$O$_x$ films shows a strong dependence on well layer thickness
(Uwasawa *et al.* 1991) (Figure 9.8).

LCVD of carbon from hydrocarbon feedstocks was first reported by
Nelson and Richardson (1972) and by Lydtin (1972). Deposition was
carried out on surfaces heated with 10.6 μm radiation from a CW CO$_2$

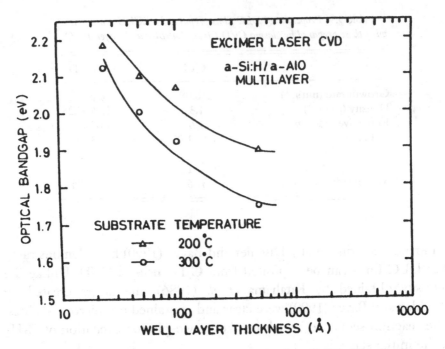

Figure 9.8. The optical bandgap from a Tauc plot for a-Si:H/a-Al$_{1-x}$O$_x$ films
for various well layer thicknesses (from Uwasawa *et al.* 1991).

laser. Subsequent experiments at visible wavelengths using Ar$^+$ laser
radiation (Leyendecker *et al.* 1981) demonstrated that carbon films can
be deposited from C$_2$H$_2$ and CH$_4$ onto tungsten and Al$_2$O$_3$ substrates.
This deposition proceeds as the hydrocarbon decomposes pyrolytically
at the point on the substrate heated by laser radiation.

Since C$_2$H$_2$ does not absorb at 488 nm, the wavelength of the Ar$^+$
laser, photochemical LCVD is not possible at this wavelength. However, C$_2$H$_2$ does absorb weakly at 193 nm and this led Kitahama *et al.*
(1986) to suggest that excimer laser LCVD of diamond-like carbon
might be feasible. They found that LCVD of C$_2$H$_2$ resulted in the
generation of a carbon deposit of (111) Si for substrate temperatures of
40–800°C. The deposit took the form of particles with sizes <1 µm,
which grew preferentially along scratch lines on the Si substrate. Optimum results were obtained in the perpendicular LCVD configuration
with a tightly focused laser beam and fluences of up to 0.05 J cm^{-2}. This
material is probably a diamond-like carbon (DLC) which consists of a
disordered mixture of sp^2 and sp^3 bonded material (Kitahama 1988). It
is likely that pyrolytic decomposition of C$_2$H$_2$ was the source of this
material in the experiments of Kitahama *et al.* (1986).

Table 9.9. *Properties of DLC films prepared by LCVD compared with those prepared by plasma CVD (from Dischler and Bayer 1990).*

	LCVD	PCVD
Growth rate ($nm\,s^{-1}$)	$\lesssim 0.08$	Up to 5
Density ($g\,cm^{-3}$)	1.3	1.75–2.25
Refractive index, n	1.6	1.65–2.2
H/C ratio	0.8–1.0	0.3–0.5
sp^3/sp^2 ratio	6.7–0.1	1.1–2.5
Resistivity ($\Omega\,cm$)	10^{12}–10^{13}	$\simeq 10^{12}$
Microhardness (GPa)	3–6	12–44

Further experiments by Dischler and Bayer (1990) have demonstrated that DLC films can be deposited from C_2H_2 using LCVD. Unlike the deposits obtained by Kitahama *et al.* (1986), the layers created by Dischler and Bayer (1990) were clear and contained no microcrystallites. The reaction sequence, which invokes a two-photon excitation of C_2H_2 at the initial stage, is

$$C_2H_2 + 2h\nu \rightarrow CH^* + CH \qquad (27)$$

$$Si + (CH^*, CH) \rightarrow Si + a\text{-}C\text{:}H\downarrow \qquad (28)$$

with the reverse reaction

$$(CH^*, CH) + a\text{-}C\text{:}H \rightarrow C_2H_2\uparrow \qquad (29)$$

causing an etching of the resulting a-C:H deposit. Since the chemistry involved in plasma deposition of a-C:H films is complex, the above reactions are only indicative of the overall processes involved.

The hydrogen content, as well as the ratio of sp^3 (diamond-like) to sp^2 (graphite-like) bonding in these films was found to be sensitive to substrate temperature during deposition. Hydrogen content and the sp^3/sp^2 ratio were found to be largest for substrate temperatures in the range 200–300°C. The maximum value of the H/C ratio was about unity at 200–250°C. The hardest films were deposited with substrate temperatures near 300°C. A summary of some of the physical properties of LCVD DLC films is given in Table 9.9. In general, LCVD was found to yield films with a higher H/C ratio than those deposited using plasma CV. In addition, the sp^3/sp^2 ratio was very large compared with that of plasma-deposited films. Films were of relatively low density and were highly transparent with an optical bandgap of about 2.5 eV.

9.10 LASER PLANARIZATION

Planarization is used to smooth and flatten the surface of large area integrated circuits at various stages of fabrication. For multilayer systems with vias, planarization must include a mechanism for filling of these vias. Although technologies such as metal vapor deposition or metal deposition by sputtering are available, the filling of vias using these techniques is often incomplete due to shadowing. On the other hand, the small diameter of vias (about 1 μm) ensures that surface tension will cause liquid metal to flow into these structures. This has led to experimental studies in which pulsed laser radiation has been used to melt deposited metal films (Tuckerman and Weisberg 1986a,b, Mukai *et al*. 1987, Spiess and Strack 1989, Bernhardt *et al*. 1989). The resulting liquid flow, together with the rapid quenching of the liquid phase, ensures a planar surface and the filling in of vias. Optimized conditions require that the laser pulse length be small enough to inhibit deleterious metallurgical reactions, while being long enough to fully melt the metal layer and allow high frequency spatial modes to be damped (Tuckerman and Weisberg 1986a,b).

Calculations indicate that pulse lengths $\lesssim 1\,\mu s$ are acceptable (Marella *et al*. 1989a). Excellent results have been obtained with both excimer (Mukai *et al*. 1987) and dye laser (Tuckerman and Weisberg 1986a,b) radiation. A comprehensive theoretical study of heating and liquid flow for 1–2 μm thick Au films on SiO_2 layers (Marella *et al*. 1989a,b) has shown that, although both types of laser radiation are effective in yielding planarization, recoil effects associated with the impact of excimer laser pulses may enhance surface rippling under certain conditions. Rippling in excimer-laser-treated metal surfaces has been discussed in Chapter 4. A simple analysis of long wavelength topographic variations in a molten metal surface (Tuckerman and Weisberg 1986b) predicts that the decay time, τ, for a structure with a spatial period L in a layer of thickness h is

$$\tau = \frac{3\mu L^4}{16\pi^4 \gamma h^3} \qquad (30)$$

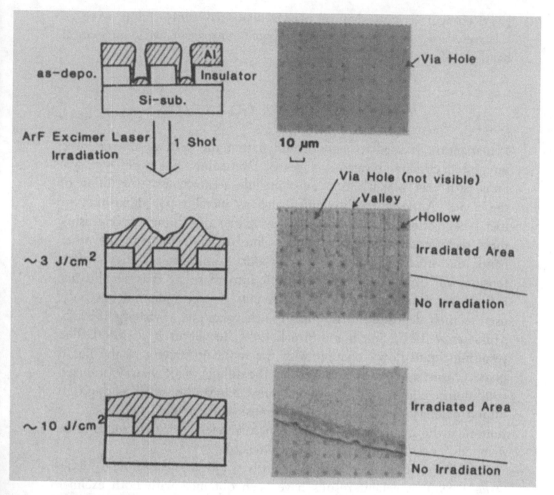

Figure 9.9. Schematic cross sections and optical micrographs of a 1 μm thick aluminum film, sputter-deposited over vertical-walled 2 μm diameter via holes etched in 1 μm of insulator. The vias connect to a silicon substrate (not visible). The effect of increasing optical fluence on the planarization of aluminum film is evident (Mukai *et al.* 1987).

where μ is the dynamic viscosity and γ is surface tension. For a 1 μm thick Al layer equation (30) predicts that structures with $L < 20$ μm will be damped over a timescale $\tau \simeq 1$ μs.

The rôle of laser fluence in planarization and via filling can be seen in Figure 9.9 (Mukai *et al.* 1987). The vias in this case were 2 μm in diameter and were formed in a 1 μm thick insulating layer on Si. A 1 μm thick Al layer was sputter-deposited on top of this structure and melted with pulses from an ArF laser. At low fluence (about $3\,\mathrm{J\,cm^{-2}}$) the vias were filled but the surface around the vias was ridged. This

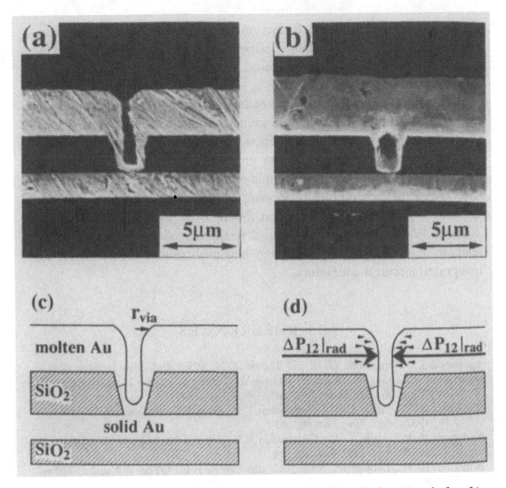

Figure 9.10. The typical roughly 1.25 aspect ratio Au-to-Au via before (a) and after (b) irradiation with a $3.8 \, J \, cm^{-2}$ flashlamp-pumped dye laser pulse. This via contains a void beneath its fully planarized surface after irradiation. All such vias yielded this characteristic void at all planarizing fluences using either a flashlamp-pumped dye or an excimer laser. These voids result from the geometry of this via (c), which gives rise to a flow (d) that closes off the top of the via without filling it (Marella *et al.* 1990).

irregularity disappeared at higher fluences (about $10 \, J \, cm^{-2}$) and the surface was planarized. At higher fluences, the Al layer was vaporized.

Under certain irradiation conditions and particularly in vias with higher aspect ratios, void formation may occur (Figure 9.10). These are created when surface tension causes liquid to close the entrance to the via (Bernhardt *et al.* 1989, Marella *et al.* 1990, Baseman *et al.* 1992). Voids are observed after planarization either with dye laser (600 ns) or with excimer laser (35 ns) pulses; however, the fluence threshold for void formation is much higher with dye laser pulses than with excimer pulses. An analysis of void formation by Bernhardt *et al.* (1989) concluded that voids occur when vaporization at the front surface of the metal film produces hydrodynamical instabilities in the liquid at the entrance to the via.

Baseman (1990) has shown that collective effects involving liquid flow coupled to groups of vias may dominate under certain fluence conditions. In all cases, optimized fluences for planarization depend on the size of individual features and the proximity of these features to each other. Planarization at higher sample temperature was found to increase the range of acceptable processing fluences. Planarization of Cu deposited over polyimide has been found to be feasible (Baseman and Turene 1990), although the thermal sensitivity of the polyimide often leads to defect formation, which may make this process unacceptable for integrated circuit manufacture.

9.11 REFERENCES

Agostinelli J. A., Braunstein G. H. and Blanton T. N., 1993. *Appl. Phys. Lett.* **63**, 123.

Amirhaghi S., Craciun V., Beech F., Vichers M., Tarling S., Barnes P. and Boyd I. W., 1993. *Mater. Res. Soc. Symp. Proc.* **285**, 489.

Andreatta R. W., Abele C. C., Osmundsen J. F., Eden J. G., Lubben D. and Greene J. E., 1982a. *Appl. Phys. Lett.* **40**, 183.

Andreatta R. W., Abele C. C., Osmundsen J. F., Eden J. G., Lubben D. and Greene J. E., 1982b. *Appl. Phys. Lett.* **40**, 184.

Andreatta R. W., Lubben D., Eden J. G. and Greene J. E., 1982c. *J. Vac. Sci. Technol.* **20**, 740.

Armstrong J. V., Burk A. A., Coly J. M. D. and Moorjani K., 1987. *Appl. Phys. Lett.* **50**, 1231.

Arnone C., Rothschild M., Black J. G. and Ehrlich D. J., 1986. *Appl. Phys. Lett.* **48**, 1018.

Ashby C. I. H., Brannon J. H. and Pang S. W. (eds) 1992. *Photons and Low Energy Particles in Surface Processing*, Materials Research Society, Orlando.

Balooch M., Tench R. J., Siekhaus W. J., Allen M. J., Connor A. L. and Olander D. R., 1990. *Appl. Phys. Lett.* **57**, 1540.

Baranauskas V., Mammana C. I. Z., Klinger R. E. and Greene J. E., 1980. *Appl. Phys. Lett.* **36**, 930.

Baseman R. J., 1990. *J. Vac. Sci. Technol.* B8, 84.

Baseman R. J., Kuan T. S., Aboelfotoh M. O., Andreshak J. C., Turene F. E., Previti-Kelly R. A. and Ryan J. G., 1992. *Mater. Res. Soc. Symp. Proc.* **236**, 361.

Baseman R. J. and Turene F. E., 1990. *J. Vac. Sci. Technol.* B8, 1097.

Bäuerle D., 1986. *Chemical Processing with Lasers*, Springer-Verlag, New York.

Bäuerle D., 1988. *Appl. Phys.* B46, 261.

Baum T. H. and Comita P. B., 1992. *Thin Solid. Films* **218**, 80.

Beeson K. W. and Clements N. S., 1988. *Appl. Phys. Lett.* **53**, 547.

Bernhardt A. F., Contolini R. J., Tuckerman D. B. and Weisberg A. H., 1989. *Mater. Res. Soc. Symp. Proc.* **129**, 559.

Biunno N., Narayan J., Hofmeister S. K., Srivatsa A. R. and Singh R. K., 1989. *Appl. Phys. Lett.* **54**, 1519.

Blanchet G. B. and Shah S. I., 1993. *Appl. Phys. Lett.* **62**, 1026.

Bohandy J., Kim B. F. and Adrian F. J., 1986. *J. Appl. Phys.* **60**, 1538.

Bohandy J., Kim B. F., Adrian F. J. and Jette A. N., 1988. *J. Appl. Phys.* **63**, 1158.

Boyer P. K., Emery K. A., Zarnani H. and Collins G. J., 1984. *Appl. Phys. Lett.* **45**, 979.

Boyer P. K., Moore C. A., Solanki R., Ritchie W. K., Roche G. A. and Collins G. J., 1983. In *Laser Diagnostics and Photochemical Processing for Semiconductor Devices* ed. R. M. Osgood, S. R. J. Brueck and H. R. Schlossberg, North Holland, New York, p. 119.

Boyer P. K., Roche G. A., Ritchie W. H. and Collins G. J., 1982. *Appl. Phys. Lett.* **40**, 716.

Braren B., Dubowski J. J. and Norton D. P., 1993. *Mater. Res. Soc. Symp. Proc.* **285**, 1.

Buhay, H., Sinharoy S., Kasner W. H., Francombe M. H., Lampe D. R. and Stephe E., 1991. *Appl. Phys. Lett.* **58**, 1470.

Cacouris T., Scelsi G., Shaw P., Scarmozzino R., Osgood R. M. and Krchnavek R. R., 1988. *Appl. Phys. Lett.* **52**, 1865.

Calloway A. R., Galantowicz T. A. and Fenner W. R., 1983. *J. Vac. Sci. Technol.* **A1**, 534.

Charyshkin Y. V. and Sakipov N. Z., 1992. *J. Appl. Phys.* **72**, 2508.

Chen L. Y., 1994. In *Pulsed Laser Deposition of Thin Films* ed. D. B. Chrisey and G. K. Hubler, J. Wiley, New York, p. 167.

Chen M. Y. and Murray P. T., 1990. *Mater. Res. Soc. Symp. Proc.* **191**, 43.

Chrisey D. B., Horwitz J. S. and Grabowski K. S., 1990. *Mater. Res. Soc. Symp. Proc.* **191**, 25.

Coombe R. D. and Wodarczyk F. J., 1980. *Appl. Phys. Lett.* **37**, 846.

Collins C. B. and Davanloo F., 1994. In *Pulsed Laser Deposition of Thin Films* ed. D. B. Chrisey and G. K. Hubler, J. Wiley, New York, p. 417.

Collins C. B., Davanloo F., Juengerman E. M., Osborn W. R. and Jander D. R., 1989. *Appl. Phys. Lett.* **54**, 216.

Collins C. B., Davanloo F., Lee T. J., You J. H. and Park H., 1993. *Mater. Res. Soc. Symp. Proc.* **285**, 547.

Cole H. S., Liu Y. S., Rose J. W. and Guida R., 1988. *Appl. Phys. Lett.* **53**, 2111.

Cotell C. M. and Leuchtner R. E., 1993. *Mater. Res. Soc. Symp. Proc.* **285**, 367.

Craciun V., Craciun D., Amirhaghi S., Vickers M., Tarling S., Barnes P. and Boyd I. W., 1993. *Mater. Res. Soc. Symp. Proc.* **285**, 337.

Davanloo F., Juengermann E. M., Jander D. R., Lee T. J. and Collins C. B., 1990. *J. Appl. Phys.* **67**, 2081.

Davis G. M. and Gower M. C., 1989. *Appl. Phys. Lett.* **55**, 112.

Day A. E., Laube S. J. P., Donley M. S. and Zabinski J. S., 1993. *Mater. Res. Soc. Symp. Proc.* **285**, 539.

Demiryont H., Thompson L. R. and Collins G. J., 1986a. *J. Appl. Phys.* **59**, 3235.

Demiryont H., Thompson L. R. and Collins G. J., 1986b. *Appl. Opt.* **25**, 1311.

Deutsch T. F., Ehrlich D. J. and Osgood R. M., 1979. *Appl. Phys. Lett.* **35**, 175.

Deutsch T. F. and Rathman D. D., 1984. *Appl. Phys. Lett.* **45**, 623.

Dischler B. and Bayer E., 1990. *J. Appl. Phys.* **68**, 1237.

Doll G. L., Sell J. A., Salamanca-Riba L. and Ballal A. K., 1990. *Mater. Res. Soc. Symp. Proc.* **191**, 55.

Donnelly V. M., Braren D., Appelbaum A. and Geva M., 1985. *J. Appl. Phys.* **58**, 2022.

Donnelly V. M., Geva M., Long J. and Karlicek R. F., 1984. *Appl. Phys. Lett.* **44**, 951.

Donley M. S., Zabinski J. S., Sessler W. J., Dyhouse V. J., Walck S. D. and McDevitt N. T., 1992. *Mater. Res. Soc. Symp. Proc.* **236**, 461.

Duley W. W., 1976. *CO₂ Lasers: Effects and Applications*, Academic Press, New York.

Duley W. W., 1983. *Laser Processing and Analysis of Materials*, Plenum Press, New York.

Duley W. W., 1984. *Astrophys. J.* **287**, 694.

Dupendant H., Gavigan J. P., Givard D., Lienard A., Rebouillat J. and Sonche Y., 1989. *Appl. Surf. Sci.* **43**, 369.

Ehrlich D. J., Osgood R. M. and Deutsch T. F., 1981. *Appl. Phys. Lett.* **38**, 946.

Ehrlich D. J. and Tsao J. Y., 1984. *Appl. Phys. Lett.* **44**, 267.

Ehrlich D. J. and Tsao J. Y., 1985. *Appl. Phys. Lett.* **46**, 198.

Esrom H. and Kogelschatz U., 1992. *Thin Solid Films* **218**, 231.

Esrom H. and Wahl G., 1989. *Chemtronics* **4**, 216.

Flynn D. K., Steinfeld J. I. and Sethi D. S., 1986. *J. Appl. Phys.* **59**, 3914.

Fogarassy E., Fuchs C., Slaoui A., de Unamuno S., Stoquert J. P. and Marine W., 1993. *Mater. Res. Soc. Symp. Proc.* **285**, 319.

Fork D. K. and Anderson G. B., 1993. *Mater. Res. Soc. Symp. Proc.* **285**, 355.

Fowler B., Lian T., Bulloch D. and Banerjee S., 1991. *Mater. Res. Soc. Symp. Proc.* **201**, 153.

Friedmann T. A., McCarty K. F., Klaus E. J., Johnsen H. A., Medlin D. L., Mills M. J., Ottesen D. K. and Stulen R. H., 1993. *Mater. Res. Soc. Symp. Proc.* **285**, 513.

Gee J. M., Hargis P. J., Carr M. J., Tallant D. R. and Light R. W., 1984. In *Laser Controlled Chemical Processing of Surfaces* ed. A. W. Johnson, D. J. Ehrlich and H. R. Schlossberg, North Holland, New York, p. 15.

Gibson U. J., Ruffner J. A., McNally J. J. and Peterson G., 1990. *Mater. Res. Soc. Symp. Proc.* **191**, 19.

Gilgen H. H., Cacouris T., Shaw P. S., Karlicek R. R. and Osgood R. M., 1987. *Appl. Phys.* B**42**, 55.

Gitay M., Joglekar B. and Ogale S. B., 1991. *Appl. Phys. Lett.* **58**, 197.

Groh G., 1968. *J. Appl. Phys.* **39**, 5807.

Gupta A., West G. A. and Beeson K. W., 1985. *J. Appl. Phys.* **58**, 3573.

Hansen S. G. and Robitaille T. E., 1988. *Appl. Phys. Lett.* **52**, 81.

Hess M. S. and Milkosky J. F., 1972. *J. Appl. Phys.* **43**, 4680.

Higashi G. S., 1988. *J. Chem. Phys.* **88**, 422.

Higashi G. S., 1989. *Chemtronics* **4**, 123.

Higashi G. S. and Rothberg L. J., 1985. *Appl. Phys. Lett.* **47**, 1288.

Hiratani M., Tarutani M., Fukazawa T., Okamoto M. and Takagi K., 1993. *Thin Solid Films* **227**, 100.

Horwitz J. S., Chrisey D. B., Grabowski K. S., Carosella C. A., Lubitz P. and Edmondson C., 1993. *Mater. Res. Soc. Symp. Proc.* **285**, 391.

Isenor N. R. and Richardson M. C., 1971. *Appl. Phys. Lett.* **18**, 224.

Ishikawa A., Tanahashi K., Yahisa Y., Hosoe Y. and Shiroishi Y., 1994. *J. Appl. Phys.* **75**, 5978.

Izquierdo R., Lavoie C. and Meunier M., 1990. *Appl. Phys. Lett.* **57**, 647.

John P. C., Alwan J. J. and Eden J. G., 1992. *Thin Solid Films* **218**, 75.

John P. J., Dyhouse V. J., McDevitt N. T., Safriet A., Zabinski J. S. and Donley M. S., 1991. *Mater. Res. Soc. Symp. Proc.* **201**, 117.

Kanetkar S. M., Sharan S., Tiwari P., Matera J. and Narayan J., 1991. *Mater. Res. Soc. Symp. Proc.* **201**, 189.

Kitahama K., 1988. *Appl. Phys. Lett.* **53**, 1814.

Kitahama K., Hirata K., Nakamatsu H., Kawai S., Fujimore N., Imai T., Yoshino H. and Doi A., 1986. *Appl. Phys. Lett.* **49**, 634.

Knapp J. A., 1992. *Mater. Res. Soc. Symp. Proc.* **236**, 473.

Kodas T. T., Baum T. H. and Comita P. B., 1987. *J. Appl. Phys.* **62**, 281.

Kodas T. T. and Comita P. B., 1990. *Accounts Chem. Res.* **23**, 188.

Kräuter W., Bäuerle D. and Fimberger F., 1983. *Appl. Phys.* A**31**, 13.

Krebs H. U. and Bremert O., 1993. *Mater. Res. Soc. Symp. Proc.* **285**, 527.

Krishnaswamy J., Rengan A., Narayan J., Vedam K. and McHargue C. J., 1989. *Appl. Phys. Lett.* **54**, 2455.

Kullmer R., Kargl P. and Bäuerle D., 1992. *Thin Solid Films* **218**, 122.

Lavoie C., Meunier M., Izquierdo R., Boivin S. and Desjardins P., 1991. *Appl. Phys.* A**53**, 339.

Lee B. W., Lee H. M., Cook L. P., Schenck P. K., Paul A., Wong-Ng W., Chiang C. K., Brody P. S., Rod B. J. and Bennett K. W., 1993. *Mater. Res. Soc. Symp. Proc.* **285**, 403.

Lehmann O. and Stuke M., 1991. *Appl. Phys.* A**53**, 343.

Leuchtner R. E., Grabowski K. S., Chrisey D. B. and Horwitz J. S., 1992. *Appl. Phys. Lett.* **60**, 1193.

Leyendecker G., Bäuerle D., Geittner P. and Lydtin H., 1981. *Appl. Phys. Lett.* **39**, 922.

Li X. and Tansley T. L., 1990. *J. Appl. Phys.* **68**, 5369.

Lowndes D. H., Geohagen D. B., Eres D., Pennycook S. J., Mashburn D. N. and Jellison G. E., 1988. *Appl. Phys. Lett.* **52**, 1868.

Lydtin H., 1972. *Int. Conf. Chem. Vapor. Deposition, 3rd Salt Lake City Conf.* ed. F. A. Glaski, American Nuclear Society, Hinsdale, p. 121.

Lynds L. and Weinberger B. R., 1990. *Mater. Res. Soc. Symp. Proc.* **191**, 3.

Maffei N. and Krupanidhi S. B., 1992. *Appl. Phys. Lett.* **60**, 781.

Mallamaci M. A., Bentley J. and Carter C. B., 1993. *Mater. Res. Soc. Symp. Proc.* **285**, 433.

Marella P. F., Tuckerman D. B. and Pease R. F., 1989a. *Mater. Res. Soc. Symp. Proc.* **129**, 569.

Marella P. F., Tuckerman D. B. and Pease R. F., 1989b. *Appl. Phys. Lett.* **54**, 1109.

Marella P. F., Tuckerman D. B. and Pease R. F., 1990. *Appl. Phys. Lett.* **56**, 2626.

McCamy J. W., Lowndes D. H., Budai J. D., Jellison G. E., Herman I. P. and Kim S., 1993. *Mater. Res. Soc. Symp. Proc.* **285**, 471.

Metev S. and Meteva K., 1992. *Mater. Res. Soc. Symp. Proc.* **236**, 417.

Montgomery R. K. and Mantei T. D., 1986. *Appl. Phys. Lett.* **48**, 493.

Morimoto A., Otsubo S., Shimizu T., Miniamikawa T., Yonezawa Y., Kidoh H. and Ogawa T., 1990. *Mater. Res. Soc. Symp. Proc.* **191**, 31.

Morris B. J., 1986. *Appl. Phys. Lett.* **48**, 867.

Motooka T. and Greene J. E., 1986. *J. Appl. Phys.* **59**, 2015.

Motooka T., Gorbatkin S., Lubben D. and Greene J. E., 1985. *J. Appl. Phys.* **58**, 4397.

Mukai R., Sasaki N. and Nakano M., 1987. *IEEE Electron Device Lett.* **8**, 76.

Nelson L. S. and Richardson N. L., 1972. *Mater. Res. Bull.* **7**, 971.

Nishino S., Honda H. and Matsunami J., 1986. *Jap. J. App. Phys.* **25**, L87.

Norton M. G., Kotula P. G. and Carter C. B., 1991. *J. Appl. Phys.* **70**, 2871.

Ogale S. B., 1994. In *Pulsed Laser Deposition of Thin Films* ed. D. B. Chrisey and G. K. Hubler, J. Wiley, New York, p. 567.

Ogale S. B., Joshi S., Bilurkav P. G. and Mate N., 1993. *J. Appl. Phys.* **74**, 6418.

Ogmen M. and Duley W. W., 1988. *J. Phys. Chem. Solids* **50**, 1221.

Ohashi M., Shogen S., Kawasaki M. and Hanabusa M., 1993. *J. Appl. Phys.* **73**, 3549.

Ortiz C., Vega F. and Solis J., 1992. *Thin Solid Films* **218**, 182.

Osmundsen J. F., Abele C. C. and Eden J. G., 1985. *J. Appl. Phys.* **57**, 2921.

Paine D. C. and Bravman J. C., 1990. *Mater. Res. Soc. Symp. Proc.* **191**.

Pappas D. L., Saenger K. L., Bruley J., Krakow W., Cuomo J. J., Gu T. and Collins
    R. W., 1992. *J. Appl. Phys.* **71**, 5675.
Peng C. J. and Krupanidhi S. B., 1992. *Thin Solid Films* **219**, 162.
Rajakarunanayake Y., Luo Y., Adkins B. T. and Compaan A., 1993. *Mater. Res. Soc.
    Symp. Proc.* **285**, 477.
Ramesh R., Luther K., Wilkens B., Hart D. L., Wang E., Tarascon J. M., Inam A., Wu
    X. D. and Venkatesan T., 1990. *Appl. Phys. Lett.* **57**, 1505.
Rao G. M. and Krupanidhi S. B., 1994. *J. Appl. Phys.* **75**, 2604.
Reddy K. V., 1986. *J. Opt. Soc. Am.* B3, 801.
Richter A. and Klose M., 1992. *Opt. Laser Technol.* **24**, 215.
Rimai L. R., Ager R., Hangas J., Logothetis E. M., Abu-Agell N. and Aslam M., 1993.
    Mater. Res. Soc. Symp. Proc. **285**, 695.
Robertson J., 1986. *Adv. Phys.* **35**, 317.
Roy D., Krupanidhi S. B. and Dougherty J. P., 1991. *J. Appl. Phys.* **69**, 7930.
Rytz-Froidevaux Y., Salathe R. P., Gilgen H. H. and Wever H. P., 1982. *Appl. Phys.*
    A27, 133.
Saenger K. L., Roy R. A., Beach D. B. and Etzold K. F., 1993. *Mater. Res. Soc. Symp.
    Proc.* **285**, 421.
Sanders B. W. and Post M. L., 1993. *Mater. Res. Soc. Symp. Proc.* **285**, 427.
Shi L., Hashishin Y., Dong S. Y. and Kwok H. S., 1991. *Mater. Res. Soc. Symp. Proc.*
    201, 171.
Shibata Y., Kaya K., Akashi K., Kanai M., Kawai T. and Kawai S., 1993. *Mater. Res. Soc.
    Symp. Proc.* **285**, 361.
Shogen S., Matsumi Y. and Kawasaki M., 1992. *Thin Solid Films* **218**, 58.
Skouby D. C. and Jensen K. F., 1988. *J. Appl. Phys.* **63**, 198.
Slaoui A., Fogarassy E., Fuchs C. and Siffert P., 1992. *J. Appl. Phys.* **71**, 590.
Smith G. A., Chen Li-C. and Chuang M. C., 1992. *Mater. Res. Soc. Symp. Proc.* **236**, 429.
Smith H. M. and Turner A. F., 1965. *Appl. Opt.* **4**, 147.
Solanki R., Boyer P. K. and Collins G. J., 1982. *Appl. Phys. Lett.* **41**, 1048.
Solanki R., Ritchie W. H. and Collins G. J., 1983. *Appl. Phys. Lett.* **43**, 454.
Spiess W. and Strack H., 1989. *J. Vac. Sci. Technol.* B7, 127.
Sugii T., Ito T. and Ishikawa H., 1988. *Appl. Phys.* A46, 249.
Suzuki K., Lubben D. and Greene J. E., 1985. *J. Appl. Phys.* **58**, 979.
Tench R. J., Balooch M., Connor A. L., Bernardez L., Olson B., Allen M. J., Siekhaus
    W. J. and Olander D. R., 1990. *Mater. Res. Soc. Symp. Proc.* **191**, 61.
Thebert-Peeler D., Murray P. T., Petry L. and Haas T. W., 1992. *Mater. Res. Soc.
    Symp. Proc.* **236**, 467.
Tiwari P., Chowdhury R. and Narayan J., 1993. *Mater. Res. Soc. Symp. Proc.* **285**, 533.
Tsao J. Y. and Ehrlich D. J., 1984. *Appl. Phys. Lett.* **45**, 617.
Tuckerman D. B. and Weisberg A. H., 1986a. *Solid State Technol.* **29**, 129.
Tuckerman D. B. and Weisberg A. H., 1986b. *IEEE Electron Device Lett.* **7**, 1.
Uwasawa K., Ishihara F. and Matsumoto S., 1992. *Thin Solid Films* **218**, 62.
Uwasawa K., Ishihara F., Wada J. and Matsumoto S., 1991. *Mater. Res. Soc. Symp. Proc.*
    201, 147.
Wolf P. J., Christensen T. M., Coit N. G. and Swinford R. W., 1993. *Mater. Res. Soc.
    Symp. Proc.* **285**, 439.
Wood T. H., White J. C. and Thacker B. A., 1983. *Appl. Phys. Lett.* **42**, 408.
Xiong F., Chang R. P. H. and White C. W., 1993. *Mater. Res. Soc. Symp. Proc.* **285**, 587.
Xiong F., Chang R. P. H., Hagerman M. E., Kozhevnikov U. L., Poeppeimeier K. R.,
    Zhoy H., Wong G. K., Ketterson J. B. and White C. W., 1994. *Appl. Phys. Lett.*
    64, 161.

Xu X., Seki K., Chen N., Okabe H., Frye J. M. and Halpern J. B., 1993. *Mater. Res. Soc. Symp. Proc.* **285**, 331.

Yilmaz S., Venkatesan T. and Gerhard-Multhaupt R., 1991. *Appl. Phys. Lett.* **58**, 2479.

Yokoyama H., Uesugi F., Kishida S. and Washio K., 1985. *Appl. Phys.* A**37**, 25.

Yoshikawa A. and Yamaga S., 1984. *Jap. J. Appl. Phys.* **23**, L91.

Zabinski J. S., Donley M. S., Dyhouse V. J., Moore R. and McDevin N. T., 1992. *Mater. Res. Soc. Symp. Proc.* **236**, 437.

Zheleva T., Jagannadham K., Kumer A. and Narayan J., 1993. *Mater. Res. Soc. Symp. Proc.* **285**, 343.

Zheng J. P. and Kwok H. S., 1993. *Appl. Phys. Lett.* **63**, 1.

Index